MEMOIRS
of the
American Mathematical Society

ISSN 0065-9266 (print)
ISSN 1947-6221 (online)

Number 1238

Variations on a Theorem of Tate

Stefan Patrikis

March 2019 • Volume 258 • Number 1238 (second of 7 numbers)

AMS AMERICAN MATHEMATICAL SOCIETY

Library of Congress Cataloging-in-Publication Data

Cataloging-in-Publication Data has been applied for by the AMS. See http://www.loc.gov/publish/cip/.

DOI: https://doi.org/10.1090/memo/1238

Memoirs of the American Mathematical Society

This journal is devoted entirely to research in pure and applied mathematics.

Subscription information. Beginning with the January 2010 issue, *Memoirs* is accessible from `www.ams.org/journals`. The 2019 subscription begins with volume 257 and consists of six mailings, each containing one or more numbers. Subscription prices for 2019 are as follows: for paper delivery, US$1041 list, US$832.80 institutional member; for electronic delivery, US$916 list, US$732.80 institutional member. Upon request, subscribers to paper delivery of this journal are also entitled to receive electronic delivery. If ordering the paper version, add US$20 for delivery within the United States; US$80 for outside the United States. Subscription renewals are subject to late fees. See `www.ams.org/help-faq` for more journal subscription information. Each number may be ordered separately; *please specify number* when ordering an individual number.

Back number information. For back issues see `www.ams.org/backvols`.

Subscriptions and orders should be addressed to the American Mathematical Society, P. O. Box 845904, Boston, MA 02284-5904 USA. *All orders must be accompanied by payment.* Other correspondence should be addressed to 201 Charles Street, Providence, RI 02904-2213 USA.

Memoirs of the American Mathematical Society (ISSN 0065-9266 (print); 1947-6221 (online)) is published bimonthly (each volume consisting usually of more than one number) by the American Mathematical Society at 201 Charles Street, Providence, RI 02904-2213 USA. Periodicals postage paid at Providence, RI. Postmaster: Send address changes to Memoirs, American Mathematical Society, 201 Charles Street, Providence, RI 02904-2213 USA.

This publication is indexed in *Mathematical Reviews®*, *Zentralblatt MATH*, *Science Citation Index®*, *Science Citation Index™-Expanded*, *ISI Alerting Services℠*, *SciSearch®*, *Research Alert®*, *CompuMath Citation Index®*, *Current Contents®/Physical, Chemical & Earth Sciences*.

This publication is archived in *Portico* and *CLOCKSS*.

Printed in the United States of America.

∞ The paper used in this book is acid-free and falls within the guidelines established to ensure permanence and durability.

Visit the AMS home page at `https://www.ams.org/`

10 9 8 7 6 5 4 3 2 1   24 23 22 21 20 19

To my parents

Contents

Abstract

Let F be a number field. These notes explore Galois-theoretic, automorphic, and motivic analogues and refinements of Tate's basic result that continuous projective representations $\mathrm{Gal}(\overline{F}/F) \to \mathrm{PGL}_n(\mathbb{C})$ lift to $\mathrm{GL}_n(\mathbb{C})$. We take special interest in the interaction of this result with algebraicity (for automorphic representations) and geometricity (in the sense of Fontaine-Mazur). On the motivic side, we study refinements and generalizations of the classical Kuga-Satake construction. Some auxiliary results touch on: possible infinity-types of algebraic automorphic representations; comparison of the automorphic and Galois "Tannakian formalisms"; monodromy (independence-of-ℓ) questions for abstract Galois representations.

Received by the editor April 22, 2013 and, in revised form, December 6, 2013, December 2, 2015, and July 31, 2016.

Article electronically published on February 7, 2019.

DOI: `https://doi.org/10.1090/memo/1238`

2010 *Mathematics Subject Classification.* Primary 11R39, 11F80, 14C15.

Key words and phrases. Galois representations, algebraic automorphic representations, motives for motivated cycles, monodromy, Kuga-Satake construction, hyperkähler varieties.

Partially supported by an NDSEG fellowship and NSF grant DMS-1303928.

CHAPTER 1

Introduction

1.1. Introduction

Let F be a number field, with $\Gamma_F = \mathrm{Gal}(\overline{F}/F)$ its Galois group relative to a choice of algebraic closure. Tate's theorem that $H^2(\Gamma_F, \mathbb{Q}/\mathbb{Z})$ vanishes (see [**Ser77**, Theorem 4] or Theorem 2.1.1 below) encodes one of the basic features of the representation theory of Γ_F: since $\mathrm{coker}(\mu_\infty(\mathbb{C}) \to \mathbb{C}^\times)$ is uniquely divisible, $H^2(\Gamma_F, \mathbb{C}^\times)$ vanishes as well, and therefore all obstructions to lifting projective representations $\Gamma_F \to \mathrm{PGL}_n(\mathbb{C})$ vanish. More generally, as explained in [**Con11**] (Lemma 2.1.4 below), for any surjection $\widetilde{H} \twoheadrightarrow H$ of linear algebraic groups over $\overline{\mathbb{Q}}_\ell$ with central torus kernel, any homomorphism $\Gamma_F \to H(\overline{\mathbb{Q}}_\ell)$ lifts (continuously) to $\widetilde{H}(\overline{\mathbb{Q}}_\ell)$. The simplicity of Tate's theorem is striking: replacing $\mathbb{Q}/\mathbb{Z}$ by some finite group $\mathbb{Z}/m$ of coefficients, the vanishing result breaks down.

Two other groups, more or less fanciful, extend Γ_F and conjecturally encode the key structural features of, respectively, automorphic representations and (pure) motives. The more remote automorphic Langlands group, $\mathcal{L}_F$, at present exists primarily as heuristic, whereas the motivic Galois group $\mathcal{G}_F$ would take precise form granted Grothendieck's Standard Conjectures, and in the meantime can be approximated by certain unconditional substitutes, for instance via Deligne's theory of absolute Hodge cycles, or André's theory of motivated cycles (see §4.1.3). In either case, however, we can ask for an analogue of Tate's lifting theorem and attempt to prove unconditional results. The basic project of this book is to pursue these analogies wherever they may lead; the fundamental importance of Tate's theorem is affirmed by the fruitfulness of these investigations, which lead us to reconsider and reinvigorate classic works of Weil ([**Wei56**]) and Kuga and Satake ([**KS67**]).

We begin by translating the heuristic lifting problem for the (conjectural) automorphic Langlands group into a concrete and well-defined question about automorphic representations. Let $G \subset \widetilde{G}$ be an inclusion of connected reductive F-groups having the same derived group; basic cases to keep in mind are $\mathrm{SL}_n \subset \mathrm{GL}_n$ or $\mathrm{Sp}_{2n} \subset \mathrm{GSp}_{2n}$. For simplicity, we will throughout assume these groups are split, although many of the results described continue to hold in greater generality. We let $G^\vee$ and $\widetilde{G}^\vee$ denote the respective Langlands dual groups, which are related by a surjective homomorphism $\widetilde{G}^\vee \to G^\vee$ whose kernel is a central torus in $\widetilde{G}^\vee$ (in the $\mathrm{SL}_n \subset \mathrm{GL}_n$ case, this dual homomorphism is the usual surjection $\mathrm{GL}_n(\mathbb{C}) \to \mathrm{PGL}_n(\mathbb{C})$); thus on the dual side we have a setup analogous to the one in Tate's lifting theorem, and indeed the whole point of the Langlands group $\mathcal{L}_F$ is that, loosely, automorphic representations of $G(\mathbf{A}_F)$ should correspond to maps $\mathcal{L}_F \to G^\vee$. The automorphic analogue of Tate's theorem is then the assertion that automorphic representations of $G(\mathbf{A}_F)$ extend to automorphic representations of $\widetilde{G}(\mathbf{A}_F)$, i.e. the fibers of the functorial transfer associated to $\widetilde{G}^\vee \to G^\vee$ are non-empty. We show (Proposition 3.1.4) that cuspidal representations of $G(\mathbf{A}_F)$ do indeed lift to $\widetilde{G}(\mathbf{A}_F)$. This result is elementary, but our analogies insist that we investigate it further. For instance, the analogous lifting result is false for $\mathcal{G}_F$, and *a priori* there is no reason to believe that 'lifting problems' for the three groups $\Gamma_F, \mathcal{L}_F$, and $\mathcal{G}_F$ should qualitatively admit the same answer: a general automorphic representation has seemingly no connection with either ℓ-adic representations or motives; and a general ℓ-adic representation has no apparent connection with classical automorphic representations or motives. Two quite distinct kinds of transcendence–one complex, one ℓ-adic–prevent these overlapping theories from being reduced to one another. A key problem, therefore, is to identify the overlap and ask how the lifting problem behaves when restricted to this (at least conjectural) common ground; it is only in posing this refined form of the lifting problem that the relevant structures emerge.

What automorphic representations, then, do we expect to correspond to ℓ-adic Galois representations or motives? Putting off for the time being precise definitions, we now informally state the fundamental conjectures, due to Fontaine-Mazur, Langlands, Tate, Serre, and Grothendieck, that lay out the basic expectations.

CONJECTURE 1.1.1 (Preliminary form: see Conjecture 1.2.1 for the precise version). *There is a bijection, characterized by local compatibilities, between*

- *"Algebraic" cuspidal automorphic representations of* $\mathrm{GL}_n(\mathbf{A}_F)$.
- *Irreducible "geometric" Galois representations* $\Gamma_F \to \mathrm{GL}_n(\overline{\mathbb{Q}}_\ell)$.
- *Irreducible pure motives over F (with $\overline{\mathbb{Q}}$-coefficients).*

(The only direction of this correspondence whose construction is known is from motives to Galois representations: simply take ℓ-adic cohomology.) This book will focus on studying the analogues of Tate's lifting theorem at this conjectural intersection of the three subjects of automorphic forms, motives, and Galois representations. The automorphic and Galois-theoretic aspects will be more or less fully treated, and we will find that the motivic aspect is a deep and largely unconsidered problem; but our Galois-theoretic and automorphic investigations will allow us to frame a precise conjectural response to the motivic lifting question.

To continue, then, we must discuss notions of algebraicity for automorphic representations; then we can see whether the automorphic lifting analogous to Tate's theorem can be made to respect this 'algebraicity.' Weil laid the foundation for this discussion in his paper [**Wei56**], by showing that for Hecke characters (automorphic representations of $\mathrm{GL}_1(\mathbf{A}_F)$), an integrality condition on the archimedean component suffices to imply algebraicity of the coefficients of the corresponding L-series (i.e. algebraicity of its Satake parameters). Waldschmidt ([**Wal82**]) later proved necessity of this condition. Weil, Serre, and others also showed that a subset, the 'type A_0' Hecke characters (Definition 2.3.2), moreover give rise to compatible systems of Galois characters $\Gamma_F \to \overline{\mathbb{Q}}_\ell^\times$ (and, via the theory of CM abelian varieties, motives underlying these compatible systems). The general feeling since has been that the most obvious analogue of Weil's type A_0 condition should govern the existence of associated ℓ-adic representations. To be precise, let π be an automorphic representation of our F-group G, and let T be a maximal torus of G. We will use the terminology '*L-algebraic*' of [**BG11**] as the general analogue of type A_0 characters. That is, fixing at each $v|\infty$ an isomorphism $\iota_v \colon \overline{F}_v \xrightarrow{\sim} \mathbb{C}$, we can write (in

Langlands' normalization of [**Lan89**]) the restriction to $W_{\overline{F}_v}$ of its L-parameter as

$$\mathrm{rec}_v(\pi_v)\colon z \mapsto \iota_v(z)^{\mu_{\iota_v}} \bar{\iota}_v(z)^{\nu_{\iota_v}} \in T^\vee(\mathbb{C}).$$

with $\mu_{\iota_v}, \nu_{\iota_v} \in X^\bullet(T)_{\mathbb{C}}$ and $\mu_{\iota_v} - \nu_{\iota_v} \in X^\bullet(T)$ (here and throughout, $\bar{\iota}_v$ denotes the complex conjugate of ι_v). Unless there is risk of confusion, we will omit reference to the embedding ι_v, writing $\mu_v = \mu_{\iota_v}$, etc.

DEFINITION 1.1.2. The automorphic representation π is L-algebraic if for all $v|\infty$, μ_v and ν_v lie in $X^\bullet(T)$.

The naïve reason for focusing on this condition is that, for $G = \mathrm{GL}_n$, it lets us see the Hodge numbers of the (hoped-for) corresponding motive. Clozel ([**Clo90**]), noticing that cohomological representations need not satisfy this condition, but have parameters $\mu_v, \nu_v \in \rho + X^\bullet(T)$, where ρ is the half-sum of the positive roots,[1] studied this alternative integrality condition for $G = \mathrm{GL}_n$. Buzzard and Gee ([**BG11**]) have recently elaborated on this latter condition ('*C-algebraic*') and its relation with L-algebraicity. For GL_1, the two notions C and L-algebraic coincide, and in general they are both useful and distinct generalizations of Weil's type A_0 condition.

But Weil's 'integrality condition,' which he calls 'type A' (Definition 2.3.2), is more general than the type A_0 condition, and its analogues in higher rank seem largely to have been neglected.[2] It is of independent interest to resuscitate this condition, but we moreover find it essential for understanding the interaction of descent problems and algebraicity, including our original lifting question. By this we mean the following problem: given connected reductive (say quasi-split) F-groups H and G with a morphism of L-groups ${}^L H \to {}^L G$, and given a cuspidal L-algebraic representation π of $G(\mathbf{A}_F)$ which is known to be in the image of the associated functorial transfer, is π the transfer of an L-algebraic representation? The answer is certainly 'no,' even for GL_1, but it fails to be 'yes' in a tightly constrained way. I believe the most useful general notion is the following:

DEFINITION 1.1.3. We say that π is *W-algebraic* if for all $v|\infty$, μ_v and ν_v lie in $\frac{1}{2}X^\bullet(T)$.

This is more restrictive than Weil's type A condition (which allows twists $|\cdot|^r$ for $r \in \mathbb{Q}$ as well), and it is easy to concoct examples of automorphic representations that have algebraic Satake parameters but are not even twists of W-algebraic representations. It is one of our guiding principles, suggested by the (archimedean) Ramanujan conjecture, that all such examples are degenerate, and up to unwinding these degeneracies, all representations with algebraic Satake parameters should be built up from W-algebraic pieces.

We can now describe the algebraic refinement for automorphic representations of the aforementioned (Proposition 3.1.4) lifting result. The answer depends on whether the field F has real embeddings, so we state it in two parts:

PROPOSITION 1.1.4 (See Proposition 3.1.12). *Let F be a CM field, and for simplicity let G be a split semi-simple F-group, and let $\widetilde{G}$ be a split F-group containing*

[1] For some choice of Borel containing T. Such a choice was implicit in defining a dual group, with its Borel $B^\vee$ containing a maximal torus $T^\vee$, and the above archimedean L-parameters.

[2] One substantial use of type A but not A_0 Hecke characters since the early work of Weil and Shimura occurs in the paper [**BR93**], in which Blasius-Rogawski associate motives to certain tensor products of Hilbert modular representations.

G as its derived group. Let π be a cuspidal representation of $G(\mathbf{A}_F)$. Assume π_∞ is tempered.

(1) *If π is L-algebraic, then there exists an L-algebraic lift to a cuspidal automorphic representation $\tilde{\pi}$ of $\widetilde{G}(\mathbf{A}_F)$.*
(2) *If π is W-algebraic, then there exists a W-algebraic lift $\tilde{\pi}$.*

In §2.4 we develop a conjectural framework that allows this result to be extended to all totally imaginary F. See page 10 of this introduction. In contrast, over totally real fields, we have the following:

PROPOSITION 1.1.5 (See Proposition 3.1.14). *Now suppose F is totally real, with π as before. Continue to assume π_∞ is tempered. Write $\widetilde{Z}$ for the center of $\widetilde{G}$, and $\widetilde{Z}^0$ for its connected component.*[3]

(1) *If π is L-algebraic, then it admits a W-algebraic lift $\tilde{\pi}$.*
(2) *For the 'only if' direction of the following statement assume Hypothesis 3.1.8. Then the images of μ_v and ν_v under $X^\bullet(T) \to X^\bullet(Z_G \cap \widetilde{Z}^0)$ lie in $X^\bullet(Z_G \cap \widetilde{Z}^0)[2]$, and π admits an L-algebraic lift if and only if these images are independent of $v|\infty$.*

The basic example in which an L-algebraic representation does not lift to an L-algebraic representation is given by a mixed-parity Hilbert modular form π on $\mathrm{GL}_2(\mathbf{A}_F)$ (F a totally real field not equal to $\mathbb{Q}$). That is, π is the cuspidal automorphic representation associated to a classical Hilbert modular cusp form whose weights (in the classical sense) at two different infinite places of F have different parities: such a π restricts to an L-algebraic representation of $\mathrm{SL}_2(\mathbf{A}_F)$, but no twist of π itself is L-algebraic. In contrast Proposition 1.1.5 tells us that, when F is totally real, there is no obstruction to finding L-algebraic lifts for G a simple split group of type A_{2n}, E_6, E_8, F_4 or G_2. For other groups the obstructions are in fact realizable: Corollary 3.1.15 applies limit multiplicity formulas (as in [**Clo86**]) to produce many discrete-series examples generalizing the basic case of mixed-parity Hilbert modular forms. These examples imply that Arthur's conjectural construction ([**Art02**]) of a morphism $\mathcal{L}_F \to \mathcal{G}_F$ requires modification (see Remark 3.1.16).

Propositions 1.1.4 and 1.1.5 give the basic answer to our refined lifting question for automorphic representations. We next turn to the corresponding question for geometric Galois representations; Conjecture 1.1.1 leads us to expect conclusions analogous to those of these two automorphic results. First, we recall the definition (due to Fontaine-Mazur) of a geometric Galois representation:

DEFINITION 1.1.6. A semi-simple representation $\rho\colon \Gamma_F \to \mathrm{GL}_n(\overline{\mathbb{Q}}_\ell)$ is geometric if it is unramified at all but a finite set of primes and is de Rham at all places $v|\ell$. (See §2.2.1 for further discussion of the de Rham condition.) More generally, if H is a linear algebraic group over $\overline{\mathbb{Q}}_\ell$, then a representation $\rho\colon \Gamma_F \to H(\overline{\mathbb{Q}}_\ell)$ is geometric if it is almost everywhere unramified, de Rham at all places above ℓ, and if the Zariski closure of $\rho(\Gamma_F)$ in H is reductive.

The Galois question parallel to our refined automorphic lifting results has been raised by Brian Conrad in [**Con11**]. In that paper, Conrad, building on previous

[3]Note that we use somewhat different notation in §3.1; see the introduction to that section.

work of Wintenberger ([**Win95**]) addresses lifting problems of the form

$$\begin{array}{ccc} & & \widetilde{H}(\overline{\mathbb{Q}}_\ell)\,, \\ & \overset{\tilde{\rho}}{\nearrow} & \downarrow \\ \Gamma_F & \xrightarrow{\rho} & H(\overline{\mathbb{Q}}_\ell) \end{array}$$

where $\widetilde{H} \twoheadrightarrow H$ is a surjection of linear algebraic groups with central kernel. He discusses existence (a local-global principle), ramification control, and ℓ-adic Hodge theory properties, and the results are comprehensive, except for one question (see Remark 1.6 and Example 6.8 of [**Con11**]):

QUESTION 1.1.7. Suppose that the kernel of $\widetilde{H} \twoheadrightarrow H$ is a torus. If ρ is geometric, when does there exist a geometric lift $\tilde{\rho}$?

This, in the case $H = G^\vee$, $\widetilde{H} = \widetilde{G}^\vee$, is the natural Galois analogue of the question addressed by Propositions 1.1.4 and 1.1.5; indeed, it provided much of the motivation to understand those automorphic questions.

Conrad's Example 6.8 shows the answer is at least 'not always;' he produces a character $\hat{\psi}\colon \Gamma_L \to \overline{\mathbb{Q}}_\ell^\times$ over certain CM fields L (with F the totally real subfield) such that $\mathrm{Ind}_L^F(\hat{\psi})$ reduces to a geometric projective representation that has no geometric lift. We give a general solution to Question 1.1.7, assuming the representations in question satisfy a certain 'generalized Hodge-symmetry' property, which is automatic in automorphic or motivic contexts. We formalize this generalized Hodge-Tate weight symmetry in Hypothesis 3.2.4 and then prove (Theorem 3.2.10 and remarks following):

THEOREM 1.1.8. *Let F be totally imaginary, and let $\rho\colon \Gamma_F \to H(\overline{\mathbb{Q}}_\ell)$ be a geometric representation satisfying Hypothesis* 3.2.4. *Then ρ admits a geometric lift $\tilde{\rho}\colon \Gamma_F \to \widetilde{H}(\overline{\mathbb{Q}}_\ell)$.*

There is an analogue in the totally real case (Corollary 3.2.8 and Remark 3.2.11), which is parallel to Proposition 1.1.5. Again, the reader may want to keep in mind the key example of mixed parity Hilbert modular representations: to these one can associate geometric PGL_2-valued representations that do not lift to geometric GL_2-valued representations. In the preliminary Chapter 2, we work this particular example (and, more generally, similar examples for automorphic representations of symplectic groups) through in detail, since it is instructive to see precisely how the automorphic and Galois theories match one another. Then in Chapter 3 we proceed to the general arguments, which of course cannot rely on any such automorphic-Galois correspondence, as this remains entirely conjectural in general. Finally, we remark that it is somewhat surprising that the Galois theory is not more complicated than the automorphic theory: roughly speaking, the automorphic input of temperedness is missing, and the obstruction to finding a geometric lift, say for the simple yet decisive case $\mathrm{GL}_n \to \mathrm{PGL}_n$, seems to be a question of $\frac{1}{n}$-integrality rather than $\frac{1}{2}$-integrality; this first appearance is, however, a red herring–it is precisely Hodge symmetry that provides the Galois-theoretic substitute for temperedness–and W-algebraicity remains the condition of basic importance on the Galois side as well.

We also include a local version of Theorem 1.1.8. Namely, for $K/\mathbb{Q}_\ell$ finite, we can ask the same sorts of lifting questions with Γ_K in place of Γ_F. A theorem of

Wintenberger ([**Win95**]), in the case of $\widetilde{H} \twoheadrightarrow H$ a central isogeny, generalized by Conrad to $\widetilde{H} \twoheadrightarrow H$ with kernel of multiplicative type, asserts that for any ℓ-adic Hodge theory property **P** (i.e. Hodge-Tate, crystalline, semi-stable, or de Rham) a ρ of type **P** admits a type **P** lift if and only if ρ restricted to the inertia group I_K admits a Hodge-Tate lift. This Hodge-Tate lift need not exist in the isogeny case, but we can complete this story by showing it exists unconditionally for central torus quotients: see Corollary 3.2.12. The argument is a simpler version of that of Theorem 3.2.10, requiring no additional assumption on 'Hodge-Tate symmetry.' This question too was suggested by Conrad.

Having addressed the refined ("algebraic" and "geometric," respectively) lifting problems in the automorphic and Galois-theoretic settings–and, reassuringly, having found the answers match up under the correspondence of Conjecture 1.1.1 (or, more precisely, Conjecture 1.2.1)–it remains to consider the analogous motivic lifting problem. Before formulating this precisely, however, let us explain how quite concrete Galois-theoretic considerations already point to a fundamental gap in the Galois-theoretic story, whose implications will demand that we embrace the formalism of the motivic Galois group. If $\ell \neq \ell'$ are distinct primes, recall that, in spite of the incompatibility of the ℓ- and ℓ'-adic topologies, there is sometimes a meaningful way to compare a pair of Galois representations $\rho_\ell \colon \Gamma_F \to \mathrm{GL}_n(\overline{\mathbb{Q}}_\ell)$ and $\rho_{\ell'} \colon \Gamma_F \to \mathrm{GL}_n(\overline{\mathbb{Q}}_{\ell'})$. Namely, having fixed embeddings $\overline{\mathbb{Q}} \hookrightarrow \overline{\mathbb{Q}}_\ell$ and $\overline{\mathbb{Q}} \hookrightarrow \overline{\mathbb{Q}}_{\ell'}$, ρ_ℓ and $\rho_{\ell'}$ are said to be *weakly compatible* if for all but a finite set of primes S of F, both representations are unramified outside S, and for $v \notin S$ the characteristic polynomials of $\rho_\ell(\mathrm{fr}_v)$ and $\rho_{\ell'}(\mathrm{fr}_v)$ have coefficients in $\overline{\mathbb{Q}}$ (with respect to the fixed embeddings) and are in fact equal. We then extend this to a notion of weak compatibility for any collection of ℓ-adic representations as the prime ℓ varies. Deligne's work on the last of the Weil conjectures ([**Del74**]) implies that this compatibility holds for the Galois representations arising from ℓ-adic cohomology (for all ℓ) of smooth projective varieties over F.

Similarly, we can formulate a notion of compatibility for a pair of representations $\rho_\ell \colon \Gamma_F \to H(\overline{\mathbb{Q}}_\ell)$ and $\rho_{\ell'} \colon \Gamma_F \to H(\overline{\mathbb{Q}}_{\ell'})$, where now H is any linear algebraic group over $\overline{\mathbb{Q}}$. In fact, there are a few possible notions, which may not be equivalent for groups other than GL_n; so as not to burden the discussion, we put off precise formulations until §1.2.4 and §3.3.1. In light of Theorem 1.1.8, it is then natural to ask the following question:

QUESTION 1.1.9. Let $\widetilde{H} \to H$ be a surjection of linear algebraic groups over $\overline{\mathbb{Q}}$ with kernel equal to a central torus. When does a weakly compatible system of geometric representations $\rho_\ell \colon \Gamma_F \to H(\overline{\mathbb{Q}}_\ell)$ lift to a weakly compatible system of geometric representations $\tilde{\rho} \colon \Gamma_F \to \widetilde{H}(\overline{\mathbb{Q}}_\ell)$?

Even when, as for instance ensured by Theorem 1.1.8, all ρ_ℓ lift to geometric $\widetilde{H}$-valued representations, compatible lifts may not exist! See Example 2.1.10 for the prototypical counterexample; this phenomenon is quite familiar from the automorphic theory, where it is closely related to questions of automorphic multiplicities and endoscopy. How, then, can we address Question 1.1.9? The only way I know involves both assuming and proving much more: granting that the ρ_ℓ are ℓ-adic avatars of a (pro-algebraic) representation of the motivic Galois group $\mathcal{G}_{F,E}$ of motives over F with coefficients in a number field E (implicit here are specified E-forms of H and $\widetilde{H}$), and showing that this motivic Galois representation lifts to $\widetilde{H}$, possibly after enlarging E. We have therefore come full-circle to the motivic lifting-problem raised at the beginning of this introduction. This is far more difficult than the corresponding automorphic and Galois questions, but we can unconditionally treat certain examples.

In order to pose the question precisely, we need a category of 'motives' in which the Galois formalism of $\mathcal{G}_F$ is unconditional. There are two common constructions, one based on Deligne's theory ([**DMOS82**]) of absolute Hodge cycles, the other based on André's theory ([**And96b**]), much inspired by Deligne's work, of motivated cycles. We work with motivated cycles, since the inclusions

$$\text{algebraic cycles} \subseteq \text{motivated cycles} \subseteq \text{absolute Hodge cycles} \subseteq \text{Hodge cycles}$$

more or less imply that results about motivated cycles (eg, 'Hodge cycles are motivated') follow *a fortiori* for absolute Hodge cycles. Note that the Hodge conjecture asserts that each $\subseteq$ is an equality; Deligne's 'espoir' ([**Del79**, 0.10]) asserts that the last $\subseteq$ is an equality. We let $\mathcal{M}_{F,E}$ denote the category of motives for motivated cycles over F with coefficients in E (see §4.1.3 and 4.1.4). This is a semi-simple E-linear Tannakian category, and, choosing an embedding $F \hookrightarrow \mathbb{C}$, we can associate (via the E-valued Betti fiber functor) a (pro-reductive) motivic Galois group $\mathcal{G}_{F,E}$. Note that Serre has asked ([**Ser94**, 8.3]) whether homomorphisms $\mathcal{G}_{\overline{F}} \to \mathrm{PGL}_2$ (note the algebraically closed ground field) lift to GL_2. This is closely related to the question, also raised by Serre, of whether the derived group of $\mathcal{G}_{\overline{F}}$ is simply-connected, but Serre notes that questions of this sort do not have obvious conjectural answers, even if we assume we are in *le paradis motivique*. Our geometric Galois and algebraic automorphic lifting results (in combination with Conjecture 1.1.1 or Conjecture 1.2.1) suggest the following sharpening and generalization of Serre's question, that this motivic lifting phenomenon is utterly general:

CONJECTURE 1.1.10. *Let F and E be number fields, and let $\widetilde{H} \twoheadrightarrow H$ be a surjection, with central torus kernel, of linear algebraic groups over E. Suppose we are given a homomorphism $\rho\colon \mathcal{G}_{F,E} \to H$. Then if F is imaginary, there is a finite extension E'/E and a homomorphism $\tilde{\rho}\colon \mathcal{G}_{F,E'} \to \widetilde{H}_{E'}$ lifting $\rho \otimes_E E'$. If F is totally real, then such a lift exists if and only if the Hodge number parity obstruction of Corollary* 3.2.8 *vanishes.*

But what does this conjecture mean? There is in fact one classical construction in complex algebraic geometry, of a very special Hodge-theoretic nature, that fits into this framework. Namely, Kuga and Satake ([**KS67**]) showed in the late 1960's how, given a complex K3 surface X, one could construct a complex abelian variety $KS(X)$ related to X by an inclusion of rational Hodge structures

$$H^2(X, \mathbb{Q}) \subset H^1(KS(X), \mathbb{Q})^{\otimes 2}.$$

Implicit in Deligne's work ([**Del72**]) on the Weil conjectures for K3 surfaces, in which he used the Kuga-Satake construction to reduce the case of K3 surfaces to the (known) case of abelian varieties, is the perspective that the Kuga-Satake construction arises from lifting an SO-valued representation (of the Deligne torus of Hodge theory, for instance) through the surjection GSpin $\to$ SO. André ([**And96a**]), building on work of Deligne, showed that the above inclusion of Hodge structures is motivated, i.e. derives from a morphism in $\mathcal{M}_{\mathbb{C}}$, making the motivic Galois group appear more explicitly in the discussion of the Kuga-Satake construction. He also proved a weak arithmetic descent result, showing that if the K3 X is defined over some subfield $F \subset \mathbb{C}$, then $KS(X)$ can be descended to some finite extension of F. Our Conjecture 1.1.10 suggests a sharpening of this arithmetic descent (down to the field F itself), and we will indeed take up this question in Chapter 4.

Before describing these results, however, let us pause to consider the scope of Conjecture 1.1.10. The classical Kuga-Satake construction depends on a lucky numerical coincidence: it is straightforward to lift the Hodge structure associated to the K3 to a GSpin-valued Hodge structure, and it turns out we get exactly the right weights (in the natural representation of GSpin on the Clifford algebra) to have a Hodge structure that looks like it comes from H^1 of an abelian variety. And then Riemann's theorem tells us that all such Hodge structures *do* come from abelian varieties! Riemann's remarkable theorem, however, is a miraculous exception; in general, we understand almost nothing about the essential image of the Hodge-Betti realization functor from $\mathcal{M}_{\mathbb{C}}$ to rational Hodge structures. For example, the reader could write down almost any smooth projective variety X/F, say of even dimension d, and the analogous motivic lifting problem (or even its base-change to $\mathbb{C}$, let alone the arithmetic refinement)

$$\begin{array}{ccc} & & \mathrm{GSpin}(H^d(X)(\frac{d}{2})) \\ & \overset{\tilde{\rho}}{\nearrow} & \downarrow \\ \mathcal{G}_F & \xrightarrow[\rho]{} & \mathrm{SO}(H^d(X)(\frac{d}{2})) \end{array}$$

would be totally out of reach. (Here we write $H^d(X)(\frac{d}{2})$ for the object of $\mathcal{M}_F$ corresponding to $\frac{d}{2}$-fold Tate twist of the degree d cohomology of X; and we have possibly made a quadratic base-change on F to obtain an SO-valued rather than O-valued ρ.) In essence, each new case of the "generalized Kuga-Satake" Conjecture 1.1.10 requires us, in light of our Galois-theoretic lifting results (Theorem 1.1.8) to prove a new case of the Fontaine-Mazur conjecture.

That said, let us turn to some positive results toward Conjecture 1.1.10. The crucial limitation shared by all results to be discussed here is that they take place entirely in the Tannakian category of motives generated by abelian varieties (and Artin motives), the reliance on Riemann's theorem being the essential crutch. We will return at the end of the introduction to mention some later work that begins to push the boundaries of the conjecture. In Chapter 4 we prove a lifting result of the form

$$\begin{array}{ccc} & & \mathrm{GSpin}(V_E) \\ & \overset{\tilde{\rho}}{\nearrow} & \downarrow \\ \mathcal{G}_{F,E} & \xrightarrow{\rho} & \mathrm{SO}(V_E) \end{array}$$

for certain ρ arising from degree 2 primitive cohomology $V_{\mathbb{Q}}$ of a hyperkähler variety (see Definition 4.2.1) over F (or, rather, for ρ having analogous formal properties); the extension of scalars $V_E = V_{\mathbb{Q}} \otimes_{\mathbb{Q}} E$ is essential for the lifting result to hold. The starting-point for this refined motivic lifting result is, again, the work of André ([**And96a**]) on the 'motivated' theory of hyperkählers; as mentioned above, his results imply a version of this lifting statement with F replaced by a large but quantifiable (with considerable work: see Theorem 8.4.3 of [**And96a**]) finite extension F'/F. Our contribution is the arithmetic descent from F' to F, and for this we generalize a technique introduced by Ribet ([**Rib92**]) to study so-called '$\mathbb{Q}$-curves,' elliptic curves over $\overline{\mathbb{Q}}$ that are isogenous to all of their $\Gamma_{\mathbb{Q}}$-conjugates. Passing from a softer geometric statement (compare: 'a variety over F has a point over some finite extension') to a precise arithmetic version (compare: 'when does a variety over F have a point over F?') typically requires some deep input; in this case, Faltings's isogeny theorem ([**Fal83**]) does the hard work. The method requires a case-by-case analysis (depending on the motivic group of the transcendental lattice), and I have decided not to pursue all the cases here, but here is a partial result (see Theorem 4.2.13, Theorem 4.2.31, and Proposition 4.2.33 and following for more cases and more precise versions):

THEOREM 1.1.11. *Let (X, η) be a polarized variety over F satisfying André's conditions A_k, B_k, B_k^+ (see §4.2.1). For instance, with $k = 1$, X can be a hyperkähler variety with second Betti number greater than 3. Suppose that the transcendental lattice $T_{\mathbb{Q}} \subset V_{\mathbb{Q}} = Prim^{2k}(X_{\mathbb{C}}, \mathbb{Q})(k)$ satisfies either*

- $\mathrm{End}_{\mathbb{Q}\text{-HS}}(T_{\mathbb{Q}}) = \mathbb{Q}$*; or*
- *$T_{\mathbb{Q}}$ is odd-dimensional.*

For simplicity, moreover assume that $\det V_\ell = \det T_\ell = 1$ *as Γ_F-representations.*[4] *Then after some finite scalar extension to $V_E = V_{\mathbb{Q}} \otimes_{\mathbb{Q}} E$ there is a lifting of the motivic Galois representation $\rho^V : \mathcal{G}_F \to \mathrm{SO}(V_{\mathbb{Q}})$:*

$$\begin{array}{ccc} & & \mathrm{GSpin}(V_E) \\ & \overset{\tilde{\rho}}{\nearrow} & \downarrow \\ \mathcal{G}_{F,E} & \xrightarrow{\rho^V} & \mathrm{SO}(V_E). \end{array}$$

Moreover, this lifting gives rise on λ-adic realizations (for all finite places λ of E) to lifts $\tilde{\rho}_\lambda : \Gamma_F \to \mathrm{GSpin}(V_\lambda)$ whose compositions with the spin (or sum of half-spin) representations form a weakly compatible system.

For more cases of Theorem 1.1.11 in which $T_{\mathbb{Q}}$ is even-dimensional, see §4.2.7; the omitted cases should yield to similar methods. We have not pursued in depth whether Question 1.1.9 can be answered in this setting, i.e. whether the $\tilde{\rho}_\lambda$ can actually be taken to be weakly compatible as GSpin-valued representations (in the sense of Definition 1.2.3 below). For a partial result, see Corollary 4.2.11. We stress that this compatibility of λ-adic realizations is *not* automatic for André's motives (nor for absolute Hodge motives), so we need to exploit an explicit description of the lift. Moreover, the excluded case $b_2(X) = 3$ of Theorem 1.1.11 is related to a more general result for potentially abelian motives: we prove such a lifting result across an arbitrary surjection $\widetilde{H} \to H$ with central torus kernel: see Proposition 4.1.30 and Lemma 4.2.5.

Our next example seems to be a novel result even in complex algebraic geometry, where many authors have studied variants of the Kuga-Satake construction for Hodge structures of '$K3$-type' (see, for example, [**Mor85**], [**Voi05**], [**Gal00**]). Generalizing the well-known case ([**Mor85**]) of H^2 of an abelian surface–which has a Hodge structure of $K3$ type, so falls within the ken of classical Kuga-Satake

[4]This can be achieved in each case after a quadratic, independent-of-ℓ extension on F. It is only so we can work with $\mathrm{SO}(V_\ell)$ rather than $\mathrm{O}(V_\ell)$, but since our abstract Galois-lifting results apply to non-connected groups (see Theorem 1.1.8, for instance), this hypothesis is not essential.

theory[5]–we prove an analogous GSpin $\to$ SO motivic lifting result for H^2 of any abelian variety (see Corollary 4.3.3). Moreover, we can describe the corresponding motive (in the spin representation) as a Grothendieck motive, without assuming the Standard Conjectures.

In all, the Galois-theoretic, automorphic, and limited motivic evidence (the cases of potentially abelian motives, hyperkähler motives, and abelian varieties) should encourage optimism about Conjecture 1.1.10. The crucial next step would be to verify interesting "non-classical" examples of Conjecture 1.1.10. After the initial writing of this book, I found two ways to produce such examples, both relying on the theory of rigid local systems: the interested reader can consult [**Pat**] and [**Pat14a**] for these developments. There is clearly vast terrain waiting to be discovered here, some of which is not entirely hostile to exploration.

I have also made progress, in [**Pat16**], on a weaker version of Question 1.1.9. In the present book, we do not emphasize questions of ramification control. The geometric representations $\rho\colon \Gamma_F \to H(\overline{\mathbb{Q}}_\ell)$ of Theorem 1.1.8 are unramified outside some finite set S of primes of F, and the lifts $\tilde{\rho}\colon \Gamma_F \to \widetilde{H}(\overline{\mathbb{Q}}_\ell)$ produced by that theorem are unramified outside some, possibly larger, finite set $\widetilde{S} \supseteq S$. Achieving sharp (quantitative) control of the ramification set $\widetilde{S}$ is a deep problem, but a softer version still provides a more demanding Galois-theoretic test of the generalized Kuga-Satake conjecture. Namely, if $\{\rho_\iota\colon \Gamma_F \to H(\overline{\mathbb{Q}}_\ell)\}_{\iota\colon \overline{\mathbb{Q}} \hookrightarrow \overline{\mathbb{Q}}_\ell}$ is the collection of ℓ-adic realizations of a motivic Galois representation $\rho\colon \mathcal{G}_{F,\overline{\mathbb{Q}}} \to H$ valued in a $\overline{\mathbb{Q}}$-group H, then there is a finite set S such that each ρ_ι is unramified outside the union of S and the places S_ℓ above ℓ (where $\iota\colon \overline{\mathbb{Q}} \hookrightarrow \overline{\mathbb{Q}}_\ell$). Certainly a necessary condition for Conjecture 1.1.10 to hold is that the representations ρ_ι should all admit geometric lifts $\tilde{\rho}_\iota$ unramified outside $\widetilde{S} \cup S_\ell$ for a *finite, independent-of-ℓ*, set $\widetilde{S}$. A strengthening of this necessary condition, including the fact that the lifts $\tilde{\rho}_\iota$ can be taken to be crystalline outside of the bad set $\widetilde{S}$, is proven in [**Pat16**, Corollary 1.1]; the basic difficulty of such a result is that one has to kill cohomological obstructions lying in infinitely many different cohomology groups, one for each ι, but in an independent-of-ι manner. This result provides the natural generalization of the theorem of Wintenberger (Theorem 2.1.8 below) to the case of central torus quotients; and indeed it implies (with a different proof) Theorem 2.1.8: see [**Pat16**, Corollary 3.20, Corollary 3.22]. We also remark that the method of proof yields a new demonstration ([**Pat16**, Corollary 3.12]) of the basic theorem of Tate, whose classical proof is recalled in Theorem 2.1.1 below.

In the remainder of this introduction, we describe a few other general themes that recur throughout these notes, in automorphic, Galois-theoretic, and motivic variants. The first is the systematic exploitation of 'coefficients,' and the general principle that our arithmetic objects will naturally have coefficients in CM fields. In §2.4, we use this principle to formulate (and prove, for regular representations; see Proposition 2.4.8) a conjectural generalization of Weil's result that a type A Hecke character of a number field F descends, up to finite order twist, to the maximal CM subfield F_{cm}; this generalization asserts that the *infinity-type* of an L-algebraic cuspidal automorphic representation of $\mathrm{GL}_n(\mathbf{A}_F)$, over a totally imaginary field F, necessarily descends to F_{cm}. In §4.1.5, we use this principle–descent

[5]The paper [**Gal00**] treats the case of a particular family of abelian four-folds, where a constraint on the Mumford-Tate group allows one to extract a Hodge structure of $K3$-type; as we show, though, the lifting phenomenon is completely general.

of infinity-types to the CM subfield, which in the motivic context follows from the Hodge index theorem–to show that André's motives satisfy the Hodge-Tate weight symmetries needed for the Galois lifting Theorem 1.1.8. Finally, Corollary 3.4.14 gives an example of how having CM coefficients can be exploited even in the study of abstract compatible systems of ℓ-adic representations. Since the writing of this paper, the same principles have been applied in [**PT15**] to establish new results on the potential automorphy of regular motives, and the irreducibility of automorphic Galois representations.

Corollary 3.4.14 depends on the following abstract independence-of-ℓ result, which may be of independent interest (see Proposition 3.4.9 for a more general statement):

PROPOSITION 1.1.12. *Let $\rho_\lambda \colon \Gamma_F \to \mathrm{GL}_n(\overline{E_\lambda})$ be a compatible system of irreducible, Lie-multiplicity-free representations of Γ_F with coefficients in a number field E, so that $\rho_\lambda \cong \mathrm{Ind}_{L^\lambda}^F(r_\lambda)$ for some Lie-irreducible representation r_λ of Γ_{L^λ}, where the number field L^λ a priori depends on λ. Then the set of places of F having a split factor in L^λ is independent of λ. If we further assume that the L^λ/F are Galois, then L^λ is independent of λ.*

By Lie-multiplicity-free, we mean multiplicity-free after all finite restrictions. One application is a converse to a theorem of Rajan (Theorem 4 of [**Raj98**]), also generalizing a result of Serre (Corollaire 2 to Proposition 15 of [**Ser81**]). See §3.4 for further discussion, as well as the aformentioned application (Corollary 3.4.14), which is a weak Galois-theoretic shadow of the automorphic Proposition 2.4.8. In §3.4, we also record a variant for number fields of a result of Katz (for affine curves over a finite field; see [**Kat87**]), which in particular clarifies the place in the general theory occupied by Lie-multiplicity-free (or more specifically, Lie-irreducible) representations:

PROPOSITION 1.1.13. *Let $\rho \colon \Gamma_F \to \mathrm{GL}_n(\overline{\mathbb{Q}}_\ell)$ be an irreducible representation. Then either ρ is induced, or there exists d dividing n, a Lie irreducible representation τ of dimension $\frac{n}{d}$, and an Artin representation ω of dimension d such that $\rho \cong \tau \otimes \omega$. This decomposition is unique up to twisting by a finite-order character. Consequently, any (irreducible) ρ can be written in the form*

$$\rho \cong \mathrm{Ind}_L^F(\tau \otimes \omega)$$

for some finite L/F and irreducible representations τ and ω of Γ_L, with τ Lie-irreducible and ω Artin.

This is a very handy result, which despite its basic nature does not seem to be widely-known. Some of our descent arguments in §4.2 rely on it.

Finally, time and again our arguments are buttressed by some explicit knowledge of the constraints on infinity-types of automorphic representations, and on Hodge-Tate weights of Galois representations. More generally, we find that asking finer structural questions about the interaction of functoriality and algebraicity naturally leads to existence (and non-existence) problems that are often dissociated from functoriality; as we will see in §2.1, this theme is present even in the proof of Tate's original vanishing theorem $H^2(\Gamma_F, \mathbb{Q}/\mathbb{Z}) = 0$. We certainly have many more questions than answers (scattered through §'s 2.3-2.7; see eg Corollary 2.7.8 for a curious positive result), although fortunately our main results depend largely on detailed study of GL_1 (§2.3), which provides both the technical ingredients for later arguments and the motivation for higher-rank results (namely, after a close reading of Weil's [**Wei56**], the definition of W-algebraicity and the key descent principle of Proposition 2.4.8). In this latter respect, I do not believe that GL_1 has yielded all of its fruit–see the discussion surrounding Corollary 2.3.9–but the extreme difficulty of establishing the (non-)existence of automorphic representations with given infinity-types, even in qualitative form, necessarily tempers further conjecture. We also use the GL_1-theory, in combination with Conjecture 1.2.1, to show how the conjectural automorphic-Galois correspondence for the group GL_n implies an analogous correspondence for SL_n: see §3.3.2. For an unconditional construction, building on the arguments of §3.3.2, of geometric projective Galois representations associated to certain 'mixed-parity' automorphic representations on higher-rank Spin-groups, see [**Pat14b**].

For a reader interested only in certain aspects of these notes, I hope that the table of contents is a clear reference to the points of interest. Roughly, Chapter 2 contains preliminary material and many examples arising from the groups GL_1 and GL_2. Chapter 3 contains the general automorphic and Galois-theoretic lifting theorems, as well as 'abstract' results about monodromy groups of Galois representations. Chapter 4 contains the discussion of motivic lifting problems, culminating in the statement of the generalized Kuga-Satake conjecture, Conjecture 4.3.1.

1.2. What is assumed of the reader: Background references

Because this book studies all three vertices–and, to a far lesser extent, the edges–of the mysterious Galois-automorphic-motivic triangle, the reader will need some limited familiarity with all three subjects. In this subsection, we indicate what background will be assumed, give some useful references, and also explain that while the background required may be broad, it is not terribly deep; indeed, the arguments are largely elementary, once some basic definitions are assimilated. Let us deal individually with each of these three topics.

1.2.1. Automorphic representations. Preliminary to the study of automorphic representations, one must have some familiarity with the theory of reductive algebraic groups. For simplicity, we always restrict to split groups, so it is sufficient to know the theory over algebraically closed fields. Many arguments in this paper rest on the manipulation of root data; Springer's survey [**Spr79**] and his book [**Spr09**] (especially chapters 7-10) are very clear, and provide more than enough background. We also assume familiarity with the algebraic representation theory of (connected) reductive groups (i.e., elementary highest-weight theory).

We use a little of the representation theory of reductive groups over local fields. Familiarity with unramified representations of split p-adic groups and the Satake correspondence ([**Gro98**] is a beautiful guide; [**Car79**, III] is a thorough treatment of unramified representations; [**Cas**] is the best general introduction to the theory of admissible smooth representations), and with some aspects of the formalism of the archimedean local Langlands correspondence will suffice. As for the global theory of automorphic representations themselves, and the theory of the L-group, the standard and more than adequate reference is [**Bor79**]; [**Bor79**, §9-11] would be particularly useful background reading, sketching the archimedean theory and, crucially for our purposes, explaining the desired, in some cases proven, connection between central characters and L-parameters.

1.2.2. ℓ-adic Galois representations. Apart from some very elementary notions, we only require some familiarity with the formal, non-geometric, aspects of ℓ-adic Hodge theory, which is to say the study of ℓ-adic representations of Γ_K for a finite extension $K/\mathbb{Q}_\ell$ (when studied in a purely local context, ℓ is traditionally replaced by p, as in 'p-adic Hodge theory'). An excellent overview, with references to detailed proofs, is [**Ber04**, §I-II]. A 'text-book-style' reference (very useful but not quite in final form) for everything we do is [**BC**].[6] Finally, we should remark to a reader new to p-adic Hodge theory that the subject has been much in flux in the last 10 years, and that one should take heed of recent conceptual advances (eg [**Bei12**], [**Bha12**], [**FF12**], [**Sch13**]) before reading *too* deeply into the 'classical' theory.

1.2.3. Motives. Here we recommend some familiarity with the notion of motive, as conceived by Grothendieck, and with Grothendieck's Standard Conjectures on algebraic cycles. Kleiman's articles [**Kle68**] and [**Kle94**] provide a clear, careful, and concise introduction. Certainly familiarity with the various cohomological realizations (Betti, de Rham, ℓ-adic), and the relations between them, of a smooth projective variety is necessary; the first section, 'Review of Cohomology,' of Deligne's article [**DMOS82**, §I.1] will provide quick and easy orientation for someone new to the subject. We will also require (some of) the theory of Tannakian categories; we will give a brief introduction, which should be enough for a reader to follow all of our arguments, but a thorough treatment is [**DM11**] (originally [**DMOS82**]). Finally, although not necessary for the present work, the excellent book [**And04**] provides a broad survey of the theory of motives, from its inception to more recent developments.

1.2.4. Connecting the dots. These notes rely on the systematic transfer of intuition back and forth between these three areas. We now make precise the towering conjectures that dominate this conceptual landscape. The following collection of conjectures (crudely stated before as Conjecture 1.1.1), and the consequences of combining the various parts, summarize the principal problems in the field (in contrast to Conjecture 1.1.1, here we give an asymmetric formulation in order to distinguish the original Fontaine-Mazur conjecture from stronger versions).

CONJECTURE 1.2.1 (Fontaine-Mazur, Langlands, Tate, Grothendieck-Serre). *Let $\rho\colon \Gamma_F \to \mathrm{GL}_n(\overline{\mathbb{Q}}_\ell)$ be an irreducible geometric Galois representation. Recall from the introduction that this means ρ is almost everywhere unramified, and is de Rham at all places above ℓ. Then:*

(1) *(Fontaine-Mazur) There exists a smooth projective variety X/F and an integer r such that ρ is isomorphic to some sub-quotient of $H^j(X_{\overline{F}}, \overline{\mathbb{Q}}_\ell)(r)$.*
(2) *(Grothendieck-Serre) For any smooth projective variety X/F, each $H^j(X_{\overline{F}}, \overline{\mathbb{Q}}_\ell)$ is a semi-simple Γ_F-representation.*
(3) *(Above plus Tate) Such a ρ is motivic in the sense that it is cut out by $\overline{\mathbb{Q}}_\ell$-linear combinations of (homological) algebraic cycles. More precisely, for any embedding $\iota_\ell\colon \overline{\mathbb{Q}} \hookrightarrow \overline{\mathbb{Q}}_\ell$, there is an idempotent project e in the algebra $C^0_{\mathrm{hom}}(X, X)_{\overline{\mathbb{Q}}}$ (see* [**Kle68**, §1.3.8]*, where this is denoted $\mathcal{A}(X)$) of*

[6] It is also highly recommended to visit Laurent Berger's website to check on the status of the course-notes from his course at IHP in the Galois Trimestre of 2010.

self-correspondences of X with $\overline{\mathbb{Q}}$-coefficients, such that

$$\rho \cong e(H^*(X)(r)_{\overline{\mathbb{Q}}}) \otimes_{\overline{\mathbb{Q}},\iota_\ell} \overline{\mathbb{Q}}_\ell$$

as Γ_F-representations.

(4) *(Above plus Langlands) Fix an embedding $\iota_\infty \colon \overline{\mathbb{Q}} \hookrightarrow \mathbb{C}$. There exists a cuspidal automorphic representation Π of $\mathrm{GL}_n(\mathbf{A}_F)$ such that for almost all v unramified for ρ, the eigenvalues of $\rho(\mathrm{fr}_v)$ agree with the Satake parameters of Π_v, viewed in $\overline{\mathbb{Q}}_\ell$ via the composition $\iota_\ell \circ \iota_\infty^{-1}(\mathrm{rec}_v(\Pi_v))$.*[7] *Moreover for $\iota \colon F \hookrightarrow \mathbb{C}$, the archimedean Langlands parameter*

$$\mathrm{rec}_v(\Pi_v)|_{\overline{F_v}^\times} \colon W_{\overline{F_v}} \to \mathrm{GL}_n(\mathbb{C}),$$

landing in a maximal torus T_n, is of the form $z \mapsto z^{\mu_\iota}\bar{z}^{\nu_\iota}$ for $\mu_\iota, \nu_\iota \in X_\bullet(T_n)$, and for any $\tau \colon F \subset F_v \hookrightarrow \overline{\mathbb{Q}}_\ell$, the Hodge-Tate co-character μ_τ of $\rho|_{\Gamma_{F_v}}$ is (conjugate to) $\mu_{\iota^*_{\infty,\ell}(\tau)}$.[8]

Conversely, given a cuspidal automorphic representation Π of $\mathrm{GL}_n(\mathbf{A}_F)$ with integral archimedean Langlands parameters, there exists an irreducible geometric Galois representation $\rho_\Pi \colon \Gamma_F \to \mathrm{GL}_n(\overline{\mathbb{Q}}_\ell)$ (depending on ι_∞, ι_ℓ) satisfying the above compatibilities.

REMARK 1.2.2. (1) Note that the Standard Conjectures are a special case of part 3 of Conjecture 1.2.1. We will rarely work directly with motives for homological equivalence in these notes, substituting instead André's category of motivated motives. Keeping one's faith in the Standard Conjectures, but wanting to assume less at the outset, one could substitute the algebra of motivated correspondences $C^0_{\mathrm{mot}}(X, X)$ (Definition 4.1.13 and following) for $C^0_{\mathrm{hom}}(X, X)$ in Conjecture 1.2.1.

(2) This is not a historically faithful presentation of these conjectures; for our purposes in these notes, however, this formulation is convenient.

(3) Part 4 implies a similar correspondence between semi-simple (not necessarily irreducible) geometric Galois representations and (suitably algebraic) isobaric automorphic representations of $\mathrm{GL}_n(\mathbf{A}_F)$. Alternatively, one can begin from this more general conjecture and deduce that cuspidal automorphic representations must correspond to irreducible Galois representations using standard properties of Rankin-Selberg L-functions.

The Fontaine-Mazur-Tate conjecture challenges us, given a geometric ℓ-adic representation ρ_ℓ, to produce a motive with ρ_ℓ as ℓ-adic realization, and in particular to produce a family of ℓ'-adic realizations (for all ℓ') that are 'compatible' with ρ_ℓ. It is often convenient to abstract this notion of compatible system of Galois representations, as an intermediary between the isolated ρ_ℓ and the robust motive.

DEFINITION 1.2.3. Let F and E be number fields, and let N be a positive integer. A rank N *weakly compatible system of* λ-adic (or, informally, ℓ-adic)

[7] Implicit in the conjecture is that the Satake parameters of Π are algebraic; of course this is not the case for general automorphic representations. A stronger version moreover asks that for all finite places v,

$$\mathrm{WD}(\rho|_{\Gamma_{F_v}})^{fr-ss} \cong \iota_\ell \circ \iota_\infty^{-1}(\mathrm{rec}_v(\Pi_v))$$

[8] $\iota^*_{\infty,\ell}(\tau)$ is the pullback of τ to an archimedean embedding via ι_ℓ, ι_∞; see §1.4.

representations of Γ_F *with coefficients in* E is a collection

$$\mathcal{R} = (\{\rho_\lambda\}_\lambda, S, \{Q_v(X)\}_{v \notin S}),$$

consisting of:

(1) for each finite place λ of E, a continuous semi-simple geometric representation
$$\rho_\lambda \colon \Gamma_F \to \mathrm{GL}_N(\overline{E_\lambda});$$
(2) a finite set of places S of F, containing the infinite places, such that for all $v \notin S$ and for all λ of different residue characteristic from v, $\rho_\lambda|_{\Gamma_{F_v}}$ is unramified;
(3) for all $v \notin S$, a polynomial $Q_v(X) \in E[X]$ such that for all λ of different residue characteristic from v, $Q_v(X)$ is the characteristic polynomial of $\rho_\lambda(\mathrm{fr}_v)$.

We will sometimes use a similar notion where GL_N is replaced by a, for simplicity, split connected reductive group H over the number field of coefficients E. In this case, repeat the above definition verbatim, except replace the characteristic polynomial of $\rho_\lambda(\mathrm{fr}_v)$ with the analogous point of the space of Weyl-invariant functions on a maximal torus in H, i.e. the image of $\rho_\lambda(\mathrm{fr}_v) \in H(\overline{E_\lambda})$ under the Chevalley map induced by the map on coordinate rings

$$E[H] \supset E[H]^H \xrightarrow[\text{res}]{\sim} E[T]^W,$$

where T is an E-split maximal torus, and W is the Weyl group of (H, T). That is, we ask that for all λ, the associated $\overline{E}_\lambda$-points of $E[T]^W$ in fact arise from a common $\overline{E}$-point.

Motivated again by [**Del74**], we will sometimes impose a *purity* hypothesis on $\mathcal{R}$:

DEFINITION 1.2.4. Fix an integer w. For any finite place v of F, let q_v denote the order of the residue field of F at v. We say that a weakly compatible system $\mathcal{R}$ is *pure* of weight $w \in \mathbb{Z}$ if for a density one set of places v of F, each root α of $Q_v(X)$ in $\overline{E}$ satisfies $|\iota(\alpha)|^2 = q_v^w$ for all embeddings $\iota \colon \overline{E} \hookrightarrow \mathbb{C}$.

Another standard variant of the definition of weakly compatible system would also specify that the ℓ-adic Hodge numbers are suitably 'independent of ℓ,' but we will not require this; see for instance [**BLGGT14**, §5.1]. Finally, an elementary argument (see, eg, [**CHT08**, Lemma 2.1.5]) shows that any continuous $\overline{E_\lambda}$-representation of Γ_F (a compact group) takes values in some $\mathrm{GL}_N(E')$ for some finite extension E' of E_λ.

1.3. Acknowledgments

Some of this work was carried out with the support of an NDSEG fellowship, and revisions were made with the support of NSF grant DMS-1303928. I thank the Institut Henri Poincaré, which I visited for the 2010 Trimestre Galois, and the Oxford University mathematics department, which I visited in 2011 – 2012, for their hospitality. I of course am grateful to the Princeton math department, both for its official support, and for its singular camaraderie.

The anonymous referees made many suggestions which have greatly improved the readability of this book; I heartily thank them, and also thank Kevin Buzzard for helpful comments and corrections. I am especially grateful to the referee who twice took the time to report on substantially different versions of this book.

Above all, it gives me great pleasure to thank the following people. Brian Conrad's paper [**Con11**] catalyzed much of the present work: indeed, I was stumbling towards the questions studied in [**Con11**] when Brian posted his paper, and the clarity and generality of the questions he raised put my ideas into the sharpest focus. Having posed such fruitful questions, Brian moreover generously stepped back to allow me the time to pursue my ideas. He further provided very helpful comments on an earlier draft. I am grateful to Peter Sarnak for his encouragement and for several enjoyable conversations about this material, and to Christopher Skinner for reading an earlier draft. Most of all, I thank Andrew Wiles, my advisor, for his ongoing support and patient encouragement during my graduate career.

1.4. Notation

We end this introductory chapter by collecting some of the most frequently-used notation in this text. The reader may also refer to the notation index to see where a given piece of notation is introduced. For a number field F, we always choose an algebraic closure $\overline{F}/F$ and let Γ_F denote $\mathrm{Gal}(\overline{F}/F)$. We write $C_F = \mathbf{A}_F^\times/F^\times$ for the idele class group of F.

If L/F is a finite extension inside $\overline{F}$ and W a representation of Γ_L $(= \mathrm{Gal}(\overline{F}/L))$, we abbreviate $\mathrm{Ind}_{\Gamma_L}^{\Gamma_F}(W)$ to $\mathrm{Ind}_L^F(W)$. For $g \in \Gamma_F$, we write $({}^gW)$ for the conjugate representation of $g\Gamma_L g^{-1}$.

We fix separable closures $\overline{\mathbb{Q}}$ and $\overline{\mathbb{Q}}_\ell$ of $\mathbb{Q}$ and $\mathbb{Q}_\ell$ for all ℓ. We fix throughout embeddings $\iota_\ell \colon \overline{\mathbb{Q}} \hookrightarrow \overline{\mathbb{Q}}_\ell$ and $\iota_\infty \colon \overline{\mathbb{Q}} \hookrightarrow \mathbb{C}$. An archimedean embedding $\iota \colon F \hookrightarrow \mathbb{C}$ thus induces an ℓ-adic embedding $\tau^*_{\ell,\infty}(\iota) = \iota_\ell \circ \iota_\infty^{-1} \circ \iota \colon F \hookrightarrow \overline{\mathbb{Q}}_\ell$; and similarly an ℓ-adic embedding $\tau \colon F \hookrightarrow \overline{\mathbb{Q}}_\ell$ induces $\iota^*_{\infty,\ell}(\tau) \colon F \hookrightarrow \mathbb{C}$. If there is no risk of confusion, we just write $\tau^*(\iota)$ and $\iota^*(\tau)$. These embeddings will be invoked (often implicitly) whenever we associate automorphic forms and Galois representations.

For a connected reductive F-group G, we construct dual and L-groups $G^\vee$ and ${}^LG$ over $\overline{\mathbb{Q}}$ (having chosen a maximal torus, Borel, and splitting, although we will only ever make the maximal torus explicit), and then use ι_ℓ and ι_∞ to regard the dual group over $\overline{\mathbb{Q}}_\ell$ or $\mathbb{C}$ as needed.

If v is a place of F, we denote by rec_v the local reciprocity map from irreducible admissible smooth representations (v finite) or irreducible admissible Harish-Chandra modules (v infinite) to (frobenius semi-simple) representations of the Weil(-Deligne) group of F_v. We use this in the unramified and archimedean cases, and for GL_n. For the group GL_1, this is local class field theory, normalized so that uniformizers correspond to geometric frobenii, which we denote fr_v.

It will be convenient to have a short-hand for the assertion that an automorphic representation and Galois representation 'correspond' under the conjectural global Langlands correspondence. We will elaborate on a number of related notions in §3.3.1, but for now we give two that will come up frequently. Recall that we have fixed embeddings ι_ℓ and ι_∞ of $\overline{\mathbb{Q}}$ into $\overline{\mathbb{Q}}_\ell$ and $\mathbb{C}$, respectively. If π is an automorphic representation of $G(\mathbf{A}_F)$, and $\rho \colon \Gamma_F \to {}^LG(\overline{\mathbb{Q}}_\ell)$ is a continuous, almost everywhere unramified, representation, then we write $\rho \sim_w \pi$ if for almost every unramified place v of F (at which we may assume ρ is unramified), $\mathrm{rec}_v(\pi_v) \colon W_{F_v} \to {}^LG(\mathbb{C})$

can be realized (up to $G^\vee$-conjugation) over $\overline{\mathbb{Q}} \xrightarrow{\iota_\infty} \mathbb{C}$, and that then

$$\iota_\ell \circ \iota_\infty^{-1} (\mathrm{rec}_v(\pi_v)) \sim \left(\rho|_{\Gamma_{F_v}}\right)^{\mathrm{fr-ss}},$$

where $\sim$ here denotes $G^\vee(\overline{\mathbb{Q}}_\ell)$-conjugacy, and 'fr-ss' means we replace $\rho(\mathrm{fr}_v)$ with its semi-simple part. More concretely, we restrict to places v at which both π and ρ are unramified, so that $\mathrm{rec}_v(\pi_v)$ and $\rho|_{\Gamma_{F_v}}$ are both just given by their evaluations at fr_v; after using $\iota_\ell \circ \iota_\infty^{-1}$ to regard them as defined over the same field, we ask that these two elements be $G^\vee(\overline{\mathbb{Q}}_\ell)$-conjugate.

As in the Fontaine-Mazur-Langlands Conjecture 1.2.1, we also often want to compare the archimedean component of π with the Hodge-Tate weights of ρ; when π is L-algebraic (see Definition 1.1.2), and ρ is de Rham, we may write, for all $\iota \colon F \hookrightarrow \mathbb{C}$, the archimedean Langlands parameter $\mathrm{rec}_v(\pi_v)|_{\overline{F}_v^\times}$ in the form

$$z \mapsto \iota(z)^{\mu_\iota} \bar{\iota}(z)^{\nu_\iota}$$

for $\mu_\iota, \nu_\iota \in X_\bullet(T^\vee)$. Then write $\rho \sim_{w,\infty} \pi$ if $\rho \sim_w \pi$, and moreover for all such $\iota \colon F \hookrightarrow \mathbb{C}$, and for all $\tau \colon F \subset F_v \hookrightarrow \overline{\mathbb{Q}}_\ell$, the τ-labeled Hodge-Tate co-character (see §2.2.2) of $\rho|_{\Gamma_{F_v}}$ is (conjugate to) $\mu_{\iota^*_{\infty,\ell}(\tau)}$.

By a CM field we mean as usual a quadratic totally imaginary extension of a totally real field; these, and their real subfields, are the number fields on which complex conjugation is well-defined, independent of the choice of complex embedding. For any number field, we write F_{cm} for the maximal subfield of F on which complex conjugation is well-defined; thus it is the maximal CM or totally real subfield, depending on whether F contains a CM subfield. We also write $\mathbb{Q}^{\mathrm{cm}}$ for the union of all CM extensions of $\mathbb{Q}$ inside $\overline{\mathbb{Q}}$.

For any topological group A, we denote by A^D the group $\mathrm{Hom}_{\mathrm{cts}}(A, \mathrm{S}^1)$ (S^1 is the unit circle); when A is abelian and locally compact, we topologize this as the Pontryagin dual.

For any ground field k (always characteristic zero for us) with a fixed separable closure k^s, and a (separable) algebraic extension K/k inside k^s, we write $\widetilde{K}$ for the normal (Galois) closure of K over k inside k^s.

We always denote base-changes of schemes by sub-scripts: thus, $X_{k'}$ is the base-change $X \times_{\mathrm{Spec}\, k} \mathrm{Spec}\, k'$ for a k-scheme X and a base-extension $\mathrm{Spec}\, k' \to \mathrm{Spec}\, k$. Notation such as $M \otimes_k k'$ will be reserved (see §4.1.4) to describe extension of scalars from an object M of a k-linear abelian category to one of a k'-linear abelian category; when there is no risk of confusion we will sometimes denote this also by $M_{k'}$.

Finally, we will signal either a significant change in running hypotheses, or an essential yet easily-overlooked point, with the 'dangerous bend' symbol ☡

CHAPTER 2

Foundations & examples

This chapter discusses the motivating examples and technical ingredients that underly the general arguments of Chapter 3. After (§2.1) recalling foundational lifting results of Tate, Wintenberger, and Conrad, and (§2.2) setting up the tools we will need from ℓ-adic Hodge theory, we undertake a detailed discussion (§2.3) of Hecke characters and ℓ-adic Galois characters, especially with reference to their possible infinity-types and Hodge-Tate weights. In §2.4.1 we first explain an abstract principle that is important for 'doing Hodge theory with coefficients;' in §2.4 we apply this to generalize Weil's result on descent of type A Hecke characters. Drawing further inspiration from [**Wei56**], we then discuss (§2.5) what seems to be the most useful generalization of the type A condition to higher-rank groups, which we call W-algebraic representations. The subsequent sections (§2.6 and §2.7) discuss the simplest non-abelian example, that of W-algebraic Hilbert modular forms, their associated Galois representations, and the first interesting cases of Conrad's geometric lifting question (Question 1.1.7); although certain results in these sections are superseded by the general theorems of Chapter 3, we can also prove much more refined statements in the Hilbert modular case, as well as some complementary results about the $\mathrm{GL}_2 \times \mathrm{GL}_2 \xrightarrow{\boxtimes} \mathrm{GL}_4$ functorial lift.[1] Another case where more precise results are accessible with the current technology is for certain automorphic representations on symplectic groups, and their associated (orthogonal and spin) Galois representations (§2.8).

2.1. Review of lifting results

In this section, we review the basic Galois lifting results due to Tate, Wintenberger, and Conrad. We then highlight some of the basic problems that remain unaddressed by these foundational results. The two main elements of Tate's proof will come up repeatedly throughout these notes, so they are worth making explicit: he requires information coming both from the 'automorphic-Galois correspondence' and from the bare automorphic theory, which in our work amounts to some question of what infinity-types automorphic representation can have. In Tate's proof, the former is global class field theory–in the form of the local-global structure of the Brauer group–and the latter takes the form of precise results about the structure of the connected component of the idele class group C_F. We will give a slightly different proof that emphasizes the continuity with some of our other arguments, especially Lemma 2.3.6.

THEOREM 2.1.1 (Tate; see [**Ser77**, Theorem 4]). *Let F be a number field. Then $H^2(\Gamma_F, \mathbb{Q}/\mathbb{Z})$ is zero.*

[1]Some of the results of §2.7, among other things, have been independently obtained by Tong Liu and Jiu-Kang Yu in their preprint [**LY**].

PROOF. It suffices to prove $H^2(\Gamma_F, \mathbb{Q}_p/\mathbb{Z}_p) = 0$ for all primes p, and then an easy inflation-restriction argument shows we may assume that F contains μ_p. Since $H^2(\Gamma_F, \mathbb{Q}_p/\mathbb{Z}_p)$ is p-power torsion, to show it is zero we may instead show that multiplication by p is injective, or equivalently that the boundary map

$$H^1(\Gamma_F, \mathbb{Q}_p/\mathbb{Z}_p) \xrightarrow{\delta} H^2(\Gamma_F, \mathbb{Z}/p) \cong \mathrm{Br}(F)[p]$$

is surjective (the identification with the Brauer group is possible since $\mu_p \subset F$). First we note that the corresponding claim holds for the completions F_v: since $\mathrm{Br}(F_v)[p] \cong \mathbb{Z}/p$ (for v finite; for v archimedean the argument is even simpler), we only need the corresponding (at v) boundary map $\delta_v \colon H^1(\Gamma_{F_v}, \mathbb{Q}_p/\mathbb{Z}_p) \to H^2(\Gamma_{F_v}, \mathbb{Z}/p)$ to be non-zero, and thus only that multiplication by p on $H^1(\Gamma_{F_v}, \mathbb{Q}_p/\mathbb{Z}_p)$ should not be surjective. But a character $F_v^\times \to \mathbb{Q}_p/\mathbb{Z}_p$ is a pth-power if and only if it is trivial on $\mu_p(F_v) \subset F_v^\times$, and there are obviously characters not satisfying this condition. Note also that for $\phi_v \in H^1(\Gamma_{F_v}, \mathbb{Q}_p/\mathbb{Z}_p)$, the image $\delta_v(\phi_v)$ depends only on the restriction $\phi_v|_{\mu_p(F_v)}$.

For global F, let α be a p-torsion element of $\mathrm{Br}(F) \subset \oplus_v \mathrm{Br}(F_v)$, with local components α_v. By the local theory, for all v we have characters $\phi_v \colon F_v^\times \to \mathbb{Q}_p/\mathbb{Z}_p$ such that $\delta_v(\phi_v) = \alpha_v$, and the collection of α_v is equivalent to the collection of restrictions $\phi_v|_{\mu_p(F_v)}$, i.e. to the corresponding character $\phi \colon \mu_p(\mathbf{A}_F) \to \mathbb{Q}_p/\mathbb{Z}_p$. The fact that the α_v arise from a global Brauer class α implies that ϕ factors through $\mu_p(F)\backslash\mu_p(\mathbf{A}_F)$, and we need only produce a finite-order extension to a Hecke character $\tilde{\phi} \colon C_F \to \mathbb{Q}/\mathbb{Z}$. The archimedean classes α_v are trivial for v complex, so Lemma 2.3.6 implies the existence of such an extension (note that it suffices to produce, as in the Lemma, an extension to a character $\tilde{\phi} \colon C_F \to \mathbb{Q}/\mathbb{Z}$, since we can then project down to $\mathbb{Q}_p/\mathbb{Z}_p$ and still have an extension of ϕ). □

EXAMPLE 2.1.2. As mentioned in the introduction, Tate's result certainly breaks down with finite coefficients. For example, consider the projectivization of the ℓ-adic Tate module of an elliptic curve over a totally real field F. The obstruction to lifting to $\mathrm{SL}_2(\overline{\mathbb{Q}}_\ell)$ lives in $H^2(\Gamma_F, \mathbb{Z}/2)$, and it vanishes if and only if the ℓ-adic cyclotomic character admits a square-root, which cannot happen for F totally real.

The papers [**Con11**] and [**Win95**] both study the problem of lifting Galois representation $\Gamma_F \to H(\overline{\mathbb{Q}}_\ell)$ through a surjection of linear algebraic groups $\widetilde{H} \to H$ with central kernel. The problem naturally breaks into two cases: that of finite kernel (an isogeny), and that of connected (torus) kernel. It is helpful to contrast these cases using an algebraic toy model:

EXAMPLE 2.1.3. Let $\widetilde{H} \twoheadrightarrow H$ be a surjection of *tori* over an algebraically closed field of characteristic zero, with kernel D, and suppose we are given an algebraic homomorphism $\rho \colon \mathbb{G}_m \to H$. Via the anti-equivalence of categories between diagonalizable algebraic groups and abelian groups (taking character groups), we see that the obstruction to lifting ρ as an algebraic morphism lives in $\mathrm{Ext}^1(X^\bullet(D), \mathbb{Z})$; in particular, when D is a torus, all such ρ lift, but they do not all lift when D is finite. For instance, we can fill in the dotted arrow in the diagram

$$\begin{array}{ccc} & & \mathbb{G}_m^2 \\ & \nearrow & \downarrow {\scriptstyle (w,z)\mapsto w^r z^s} \\ \mathbb{G}_m & \xrightarrow[z\mapsto z^n]{} & \mathbb{G}_m \end{array}$$

precisely when $gcd(r,s)|n$. The kernel of $\mathbb{G}_m^2 \to \mathbb{G}_m$ is connected precisely when $gcd(r,s) = 1$, so there is no obstruction in that case, and in general there is an explicit congruence obstruction.

Since it is the basis of all that follows, I recall the application of Theorem 2.1.1 to lifting through central torus quotients (see Proposition 5.3 of [**Con11**]).

PROPOSITION 2.1.4 (5.3 of [**Con11**]). *Let $\widetilde{H} \twoheadrightarrow H$ be a surjection of linear algebraic groups over $\overline{\mathbb{Q}}_\ell$ with kernel a central torus $S^\vee$.*[2] *Then any continuous representation $\rho\colon \Gamma_F \to H(\overline{\mathbb{Q}}_\ell)$ lifts to $\widetilde{H}(\overline{\mathbb{Q}}_\ell)$, i.e. we can fill in the diagram*

$$\begin{array}{ccc} & & \widetilde{H}(\overline{\mathbb{Q}}_\ell) \\ & \overset{\tilde{\rho}}{\nearrow} & \downarrow \\ \Gamma_F & \xrightarrow[\rho]{} & H(\overline{\mathbb{Q}}_\ell). \end{array}$$

PROOF. By induction, we may assume $S^\vee = \mathbb{G}_m$. There is an isogeny complement $\widetilde{H}_1$ in $\widetilde{H}$ to $S^\vee$: that is, we still have a surjection $\widetilde{H}_1 \twoheadrightarrow H$, but now $\widetilde{H}_1 \cap S^\vee$ is finite, and thus equal to $\mu_{n_0} \subset \mathbb{G}_m$ for some n_0. For any integer n divisible by n_0, we can enlarge $\widetilde{H}_1$ to $\widetilde{H}_n := \widetilde{H}_1 \cdot S^\vee[n]$, which now surjects onto H with kernel $S^\vee[n] \cong \mu_n$. These isogenies yield obstruction classes $c_n \in H^2(\Gamma_F, \mathbb{Z}/n)$ (as Γ_F-module, the $\mu_n \subset S^\vee[n]$ is of course trivial) that are compatible under the natural maps $\mathbb{Z}/n \to \mathbb{Z}/n'$ for $n|n'$. Tate's theorem (Theorem 2.1.1) tells us that

$$\varinjlim_n H^2(\Gamma_F, \mathbb{Z}/n) = H^2(\Gamma_F, \mathbb{Q}/\mathbb{Z}) = 0,$$

so for sufficiently large n, there exists a lift $\tilde{\rho}\colon \Gamma_F \to \widetilde{H}_n(\overline{\mathbb{Q}}_\ell)$. □

REMARK 2.1.5.
- Note that crucial to lifting here is the ability to enlarge the coefficients: if $\widetilde{H} \twoheadrightarrow H$ is a morphism of groups over $\mathbb{Q}_\ell$ and ρ lands in $H(\mathbb{Q}_\ell)$, then we only obtain a lift to $\widetilde{H}(\mathbb{Q}_\ell(\mu_n))$ for sufficiently large n.
- For example (and similarly in general; compare Lemma 3.2.2), in the case of $\mathrm{GL}_n \to \mathrm{PGL}_n$, this proof produces lifts with finite-order determinant. If ρ is Hodge-Tate, it is possible that no such lift is also Hodge-Tate. These non-Hodge-Tate Galois lifts have no parallel in the 'toy model' of Example 2.1.3, but we will use them as a stepping-stone toward finding geometric lifts, just as in the automorphic context we can choose an initial, possibly non-algebraic, lift, and then modify it to something algebraic (see Proposition 3.1.12). By contrast, in the 'motivic' version of this problem

[2] I use this awful notation to remain consistent with the 'dual picture' that we will later use to think about these questions.

(see §4.2), the motivic Galois group does not have the extra flexibility to admit such 'non-algebraic' lifts.

- If ρ is almost everywhere unramified, then it is easy to see that $\tilde{\rho}$ is almost everywhere unramified; see Lemma 5.2 of [**Con11**].

The problem of lifting through the isogeny $\widetilde{H}_1 \to H$ (or any isogeny $\widetilde{H} \to H$) is taken up in [**Win95**]. Before describing this work, we state results of Wintenberger and Conrad related to lifting p-adic Hodge theory properties through central quotients (Wintenberger treated isogenies; the general case is due to Conrad). Since all of our Galois representations will be ℓ-adic (rather than p-adic), from now on we will refer to "ℓ-adic Hodge theory," rather than p-adic Hodge theory. We will use the phrase "basic ℓ-adic Hodge theory property" to refer to any of the usual properties Hodge-Tate, de Rham, semistable, or crystalline. In the current subsection, we will only formally apply the following result; in the next subsection (§2.2), we will recall the more specific background from ℓ-adic Hodge theory needed later on in this book.

THEOREM 2.1.6 (Corollary 6.7 of [**Con11**]). *Let $\widetilde{H} \twoheadrightarrow H$ be a surjection of linear algebraic groups over $\overline{\mathbb{Q}}_\ell$ with central kernel of multiplicative type. For $K/\mathbb{Q}_\ell$ finite, let $\rho\colon \Gamma_K \to H(\overline{\mathbb{Q}}_\ell)$ be a representation satisfying a basic ℓ-adic Hodge theory property* **P**. *Provided that ρ admits a lift $\tilde{\rho}\colon \Gamma_K \to \widetilde{H}(\overline{\mathbb{Q}}_\ell)$, it has a lift satisfying* **P** *if and only if $\rho|_{I_K}$ has a Hodge-Tate lift $I_K \to \widetilde{H}(\overline{\mathbb{Q}}_\ell)$.*

REMARK 2.1.7. When the kernel of $\widetilde{H} \to H$ is a central torus, Proposition 2.1.4 shows that ρ always has some lift. We will see later (Corollary 3.2.12) that it even always has a Hodge-Tate lift, so there is no obstruction to finding a lift $\tilde{\rho}\colon \Gamma_K \to \widetilde{H}(\overline{\mathbb{Q}}_\ell)$ satisfying **P**. Even simple cases of this theorem yield very interesting results: for instance, applied to $\widetilde{H} = \mathrm{GL}_{n_1} \times \mathrm{GL}_{n_2} \to H \subset \mathrm{GL}_{n_1 n_2}$, for H the image of the exterior tensor product, it shows that if a tensor product of Γ_K-representations satisfies **P**, then up to twist they themselves satisfy **P**. We will use this example in Proposition 3.3.8.

We now sketch the proof of Wintenberger's main global result. This provides an occasion to simplify and generalize the arguments using subsequent progress in p-adic Hodge theory; it also serves to make clear what these methods do and do not prove, and to set up a contrast with the quite different techniques that we will use.

THEOREM 2.1.8 (2.1.4 of [**Win95**]). *Let $\widetilde{H} \xrightarrow{\pi} H$ be a central isogeny of linear algebraic groups over $\mathbb{Q}$. Let S be a finite set of non-archimedean places of F. Then there exist two extensions of number fields $F'' \supset F' \supset F$, and a finite set of finite places S' of F' such that for any prime number ℓ and any representation $\rho_\ell\colon \Gamma_F \to H(\mathbb{Q}_\ell)$ satisfying*

- *ρ_ℓ has "good reduction," i.e. is unramified for places not dividing ℓ and crystalline for places above ℓ, outside S;*
- *For all $v|\ell$, the one-parameter subgroups $\mu_v\colon \mathbb{G}_m \to H_{\mathbb{C}_{F_v}}$ giving the Hodge-Tate structure of $\rho_\ell|_{\Gamma_{F_v}}$ lift to $\widetilde{H}_{\mathbb{C}_{F_v}}$;*[3]

[3] When some local lift $I_{F_v} \to \widetilde{H}(\mathbb{Q}_\ell)$ exists, this condition is equivalent to the existence of a local Hodge-Tate lift. When $\rho_\ell|_{I_{F_v}}$ is moreover crystalline, or semistable, it is equivalent to the existence of a crystalline, or semistable, lift, by Theorem 2.1.6.

then the restriction $\rho_\ell|_{\Gamma_{F'}}$ *admits a geometric lift* $\tilde{\rho}_\ell$*, unramified outside* S'*, to* $\widetilde{H}(\mathbb{Q}_\ell)$*; if* ρ_ℓ *is crystalline (resp. semistable) at places above* ℓ*, then the lift may be chosen crystalline (resp. semistable). Moreover, the restriction* $\tilde{\rho}_\ell|_{\Gamma_{F''}}$ *is unique, i.e. independent of the initial choice of lift* $\tilde{\rho}_\ell$*.*

PROOF. We follow Wintenberger's arguments in detail, simplifying where the technology allows. First we explain some generalities: since the map on $\mathbb{Q}_\ell$-points $\pi_\ell \colon \widetilde{H}(\mathbb{Q}_\ell) \to H(\mathbb{Q}_\ell)$ need not be surjective, we have an initial obstruction $\mathcal{O}^1(\rho_\ell)$ in $\mathrm{Hom}(\Gamma_F, H^1(\Gamma_{\mathbb{Q}_\ell}, \ker \pi))$ given by the composition

$$\Gamma_F \xrightarrow{\rho_\ell} H(\mathbb{Q}_\ell)/\operatorname{im}(\pi_\ell) \hookrightarrow H^1(\Gamma_{\mathbb{Q}_\ell}, \ker \pi(\overline{\mathbb{Q}}_\ell)).$$

We have the choice of killing this obstruction by enlarging $\mathbb{Q}_\ell$ to some finite extension, or by restricting F; following Wintenberger, we will do the former. Having dealt with this obstruction, we will face the more serious lifting obstruction $\mathcal{O}^2(\rho_\ell)$ in $H^2(\Gamma_F, \ker \pi(\mathbb{Q}_\ell))$.

With these preliminaries aside, we construct the field F', first dealing with all but the (finite) set of ℓ lying below a place in S. Let $a(\pi)$ be the annihilator of $\ker \pi$. The field F' (and set of places S') will be defined, independent of ℓ, in the following three steps:

- Let F_1 be the maximal abelian extension of exponent $a(\pi)$, unramified outside S. For all ℓ and ρ_ℓ, the class $\mathcal{O}^1(\rho_\ell)$ dies after restriction to Γ_{F_1}: outside of $S_\ell = S \cup \{v|\ell\}$, it is of course unramified, and if $v|\ell$ but $v \notin S$ (i.e. $\rho_\ell|_{\Gamma_{F_v}}$ is crystalline), then Théorème 1.1.3 of [**Win95**]) shows the existence of a crystalline lift, valued in $\widetilde{H}(\mathbb{Q}_\ell)$, of $\rho_\ell|_{I_{F_v}}$. (Since weakly-admissible is now known to be equivalent to admissible, Wintenberger's argument applies without restriction on the Hodge-Tate weights, or the ramification of the local field $F_v/\mathbb{Q}_\ell$.)
- Let F_2/F_1 be a totally imaginary extension containing $\mathbb{Q}(\zeta_{a(\pi)})$ and such that for all places v' above a place v in S, $a(\pi)$ divides the local degree $[F_{2,v'} : F_{1,v}(\zeta_{a(\pi)})]$. If we write S_2 for the set of places of F_2 above S, ℓ, or $a(\pi_\ell)$ (the annihilator of $\ker \pi(\mathbb{Q}_\ell)$), then this implies local triviality of the obstruction classes, as long as we assume S contains no places above ℓ. That is,

$$\mathcal{O}^2(\rho_\ell) \in \ker\left(H^2(\Gamma_{F_2,S_2}, \ker \pi(\mathbb{Q}_\ell)) \to \bigoplus_{v \in S_2 \cup S_\infty} H^2(\Gamma_{F_v}, \ker \pi(\mathbb{Q}_\ell))\right).$$

  At v' above S, the fact that F_2 contains $\zeta_{a(\pi)}$ makes the local obstruction class a Brauer obstruction, which is killed over F_2 by the assumption $a(\pi)|[F_{2,v'} : F_{1,v}(\zeta_{a(\pi)})]$. At places not in S but above $a(\pi_\ell)$ there is no obstruction, since unramified reprsentations always lift (granted that $\mathcal{O}^1(\rho_\ell)$ is trivial). Finally, at places above ℓ, which we have assumed for now do not lie in S, Wintenberger's local result implies the existence of a crystalline lift: on the full decomposition group Γ_{F_v}, this is [**Win95**, Proposition 1.2]–note that for the extension from I_{F_v} to Γ_{F_v}, one must use the fact that a crystalline lift on inertia is unique.

- Let F_3 be the Hilbert class field of F_2. By global duality, the above kernel is Pontryagin dual to

$$\ker\left(H^1(\Gamma_{F_2,S_2}, \ker\pi(\mathbb{Q}_\ell)^*) \to \bigoplus_{v\in S_2\cup S_\infty} H^1(\Gamma_{F_2,S_2}, \ker\pi(\mathbb{Q}_\ell)^*)\right)$$

The Galois module $\ker\pi(\mathbb{Q}_\ell)^*$ is trivial since F_2 contains $\zeta_{a(\pi)}$, so this is just the space of homomorphisms $\Gamma_{F_2} \to \ker\pi(\mathbb{Q}_\ell)$ that are unramified everywhere and trivial at the places S_2. Restriction of the class $\mathcal{O}^2(\rho_\ell)$ to F_3 corresponds to (via global duality for F_3 now) the transfer on the Pontryagin dual of the H^1's, so by the Hauptidealsatz (trivialization of ideal classes upon restriction to the Hilbert class field), $\mathcal{O}^2(\rho_\ell)$ dies upon restriction to F_3.

We conclude that, independent of ℓ not below places of S, and of ρ_ℓ satisfying the hypotheses of the theorem, there exist lifts over the number field F_3. To deal with ℓ for which the local representations are not crystalline, we apply Proposition 0.3 of [**Win95**] to find a finite extension $F(\ell)/F$, unramified outside S_ℓ, and such that all $\rho_\ell|_{\Gamma_{F(\ell)}}$ will lift.[4] Enlarging F_3 by the composite of the $F(\ell)$ (for ℓ below places of S), we obtain a number field F_4 and a set of places S_4 equal to all places above S or $a(\pi)$, such that for *all* ℓ, any ρ_ℓ as in the theorem has a lift with good reduction outside S_4 after restriction to Γ_{F_4}.

The final step of the argument makes one final extension to show that these lifts can be taken crystalline (or semistable) whenever ρ_ℓ is, and even unramified outside all primes above S_ℓ (including those above $a(\pi)$, although this is not necessary for the theorem's conclusion); we omit the details, but see [**Win95**, Lemme 2.3.5ff.]. This will be the number field F', and the set of places S' can then be taken to be those above S. Finally, all such lifts $\tilde{\rho}_\ell$ (with good reduction outside S') are the same over the maximal abelian extension of exponent $a(\pi)$, unramified outside S', of F'; this is our F''. □

REMARK 2.1.9. We can apply the theorem to a system of representations ρ_ℓ all having good reduction outside S, and all having liftable Hodge-Tate cocharacters (which in the 'motivic' case we may assume all arise from a common underlying geometric Hodge structure), to obtain a common number field F', with finite set of places S', and a further extension F'' such that the whole system lifts to F' (with good reduction outside S'), and such that all the lifts are unique over F''. The crucial limitation of this result is that, even if the ρ_ℓ are assumed to be the system of ℓ-adic realizations of some motive (for absolute Hodge cycles, as in [**Win95**], or for motivated cycles), the proof of the theorem produce lifts $\tilde{\rho}_\ell$ that are neither (proven to be) motivic nor even weakly compatible (almost everywhere locally conjugate). So, even with the strong uniqueness properties ensured by Wintenberger's theorem, it is still not at all clear how to lift a compatible (or even motivic) system of ℓ-adic representations to another compatible (or motivic) system.

[4] This is a 'soft' result. It is easy to see that for fixed ℓ, all ρ_ℓ unramified outside S lift after restriction to some $F(\ell)$. The point of Wintenberger's theorem is that, with the ℓ-adic Hodge theory hypotheses, $F(\ell)$ can be taken independent of ℓ.

Here is the simplest cautionary example of how weakly compatible systems need not lift to weakly compatible systems; it arises from an example due to Serre (see [**Lar94**]) involving projective representations that are everywhere locally, but not globally, conjugate:

EXAMPLE 2.1.10. For any positive integer n, let H_n denote the Heisenberg group of upper-triangular unipotent elements in $\mathrm{GL}_3(\mathbb{Z}/n\mathbb{Z})$. This is a non-split extension

$$1 \to \mathbb{Z}/n \to H_n \to \mathbb{Z}/n \times \mathbb{Z}/n \to 1,$$

and writing A and B for lifts of the generators of the quotient copies of $\mathbb{Z}/n$, and Z for a generator of the center, we have the commutation relation $[A, B] = Z$. Now let ζ be a primitive n^{th} root of unity; for any $\alpha \in \mathbb{Z}/n$, we can then define a representation $\rho_\alpha \colon H_n \to \mathrm{GL}_n(\mathbb{C})$ by, in the standard basis e_i, $i = 1, \ldots, n$, and abusively letting e_0 denote e_n as well,

$$\begin{aligned} A(e_i) &= e_{i-1} \\ B(e_i) &= \zeta^{(i-1)\alpha} e_i \\ Z(e_i) &= \zeta^{\alpha} e_i \end{aligned}$$

for $i = 1, \ldots, n$. One can check that for all $\alpha \in (\mathbb{Z}/n\mathbb{Z})^\times$, the associated projective representations $\overline{\rho_\alpha} \colon H_n \to \mathrm{PGL}_n(\mathbb{C})$ are everywhere locally conjugate. Nevertheless, for distinct α, they are not globally conjugate: if they were, then the corresponding $\mathrm{GL}_n(\mathbb{C})$-representations would be twist-equivalent. The implications for our problem are the following: take $\overline{\rho_\alpha}$ and $\overline{\rho_\beta}$ for $\alpha \neq \beta$, and form a weakly-compatible system $\rho_\ell \colon \Gamma_F \twoheadrightarrow H_n \to \mathrm{PGL}_n(\overline{\mathbb{Q}}_\ell)$ (H_n certainly arises as Galois group of a finite extension of number fields.) in which some ρ_ℓ are formed from $\overline{\rho_\alpha}$ and others from $\overline{\rho_\beta}$. There is then no system of weakly-compatible lifts $\tilde{\rho}_\ell \colon \Gamma_F \to \mathrm{GL}_n(\overline{\mathbb{Q}}_\ell)$, since any character of Γ_F is trivial on Z (viewed as an element of $\mathrm{Gal}(F^{\mathrm{ab}}F'/F)$, where F'/F is the given H_n-extension; note that $\mathrm{Gal}(F^{\mathrm{ab}}F'/F^{\mathrm{ab}})$ is generated by Z). Also note that we can even produce such examples where the lifts to $\mathrm{GL}_n(\mathbb{C})$ have the same determinant:

$$\det(\rho_\alpha) \colon \begin{cases} A \mapsto (-1)^{n-1}; \\ B \mapsto 1 \text{ if } n \text{ is odd, or if } \alpha \text{ and } n \text{ are even; } -1 \text{ otherwise;} \\ Z \mapsto 1. \end{cases}$$

Thus, even weakly compatible systems $\Gamma_F \to \mathrm{PGL}_n(\overline{\mathbb{Q}}_\ell) \times \overline{\mathbb{Q}}_\ell^\times$ need not have compatible lifts through the isogeny $\mathrm{GL}_n \to \mathrm{PGL}_n \times \mathbb{G}_m$.

To connect this example more explicitly with endoscopy, note that $\mathrm{Cent}_{\mathrm{PGL}_n(\mathbb{C})}(\overline{\rho_\alpha})$ is the (cyclic order n) subgroup generated by $\overline{\rho_\alpha}(A)$, whereas ρ_α itself is irreducible. These centralizers partially govern the multiplicities of discrete automorphic representations in Arthur's conjectures. Explicitly, $\rho_\alpha(ABA^{-1}) = \zeta^\alpha B$.

2.2. ℓ-adic Hodge theory preliminaries

2.2.1. Basics. In this section, we recall some background and prove some simple lemmas in ℓ-adic Hodge theory. Throughout, let $K/\mathbb{Q}_\ell$ be a finite extension. Choose an algebraic closure $\overline{K}/K$, and let $\mathbb{C}_K$ denote the completion of $\overline{K}$. Since Γ_K acts by isometries on $\overline{K}$, $\mathbb{C}_K$ inherits a continuous Γ_K-action. We will mainly use only the simplest of Fontaine's period rings, $\mathrm{B_{HT}}$ and $\mathrm{B_{dR}}$, and even these only formally. Again, we refer the reader to the references in §1.2.2 for background, but here we quickly recall the little needed to follow future arguments. $\mathrm{B_{HT}}$ is the graded $\mathbb{C}_K$-algebra

$$\mathrm{B_{HT}} = \bigoplus_{i \in \mathbb{Z}} \mathbb{C}_K(i),$$

with the obvious continuous $\mathbb{C}_K$-semilinear action of Γ_K, $\mathbb{C}_K(i)$ denoting a twist of $\mathbb{C}_K$ by the ith power of the cyclotomic character. $\mathrm{B_{dR}}$ is the fraction field of a complete DVR and K-algebra $\mathrm{B_{dR}^+}$ whose residue field is $\mathbb{C}_K$. We do not recall the construction, but note that $\mathrm{B_{dR}}$ is a *filtered* K-algebra (the filtration associated to the maximal ideal of $\mathrm{B_{dR}^+}$) with a continuous K-linear action of Γ_K. Two of the fundamental results of the theory assert that $\mathrm{gr}^\bullet(\mathrm{B_{dR}}) \cong \mathrm{B_{HT}}$ (as graded $\mathbb{C}_K$-algebras with semi-linear Γ_K-action) and $\mathrm{B_{dR}^{\Gamma_K}} = \mathrm{B_{HT}^{\Gamma_K}} = K$. From this second point it follows that if V is a finite-dimensional representation of Γ_K over $\mathbb{Q}_\ell$, then

$$\mathrm{D_{HT}}(V) = (\mathrm{B_{HT}} \otimes_{\mathbb{Q}_\ell} V)^{\Gamma_K}$$

is a K-vector space of dimension at most $\dim_{\mathbb{Q}_\ell}(V)$, and similarly

$$\mathrm{D_{dR}}(V) = (\mathrm{B_{dR}} \otimes_{\mathbb{Q}_\ell} V)^{\Gamma_K}$$

is a K-vector space of dimension at most $\dim_{\mathbb{Q}_\ell}(V)$.

DEFINITION 2.2.1. Let V be a finite-dimensional representation of Γ_K over $\mathbb{Q}_\ell$. V is said to be *Hodge-Tate* if $\dim_K \mathrm{D_{HT}}(V) = \dim_{\mathbb{Q}_\ell}(V)$, and *de Rham* if $\dim_K \mathrm{D_{dR}}(V) = \dim_{\mathbb{Q}_\ell}(V)$.

If V is de Rham, then it is Hodge-Tate (see, eg, [**BC**, Proposition 6.3.2]), but not conversely.[5] However, the converse holds for characters ([**Ser98**, III-A6]), a fact to which we will frequently appeal.

We will usually consider Γ_K-representations V with coefficients in some extension $E/\mathbb{Q}_\ell$, typically either some unspecified finite extension, or $E = \overline{\mathbb{Q}}_\ell$. Then $\mathrm{D_{dR}}(V)$ (mutatis mutandis for $\mathrm{D_{HT}}$) is a filtered $K \otimes_{\mathbb{Q}_\ell} E$-module; if V, viewed as a $\mathbb{Q}_\ell$ representation by forgetting the E-linear structure, is de Rham, then $\mathrm{D_{dR}}(V)$ is a free $K \otimes_{\mathbb{Q}_\ell} E$-module. To see this, we may assume E is (finite-dimensional over $\mathbb{Q}_\ell$ and) large enough to contain all embeddings of K into $\overline{\mathbb{Q}}_\ell$. Then for all $\mathbb{Q}_\ell$-embeddings $\tau\colon K \hookrightarrow E$, we have orthogonal idempotents $e_\tau \in K \otimes_{\mathbb{Q}_\ell} E$, giving rise to an isomorphism

$$K \otimes_{\mathbb{Q}_\ell} E \xrightarrow[\sim]{(e_\tau)} \bigoplus_{\tau\colon K \hookrightarrow E} E$$
$$x \otimes \alpha \mapsto (\tau(x)\alpha)_\tau .$$

As $K \otimes_{\mathbb{Q}_\ell} E$-module,

$$\mathrm{D_{dR}}(V) \cong \bigoplus_\tau e_\tau \mathrm{D_{dR}}(V) \cong \bigoplus_\tau (\mathrm{B_{dR}} \otimes_{K,\tau} V)^{\Gamma_K}$$

[5] The standard example is a non-split extension of Γ_K- representations

$$0 \to \mathbb{Q}_\ell \to V \to \mathbb{Q}_\ell(1) \to 0.$$

Standard results on Galois cohomology of local fields show such a V exists. It's not so hard to show V is Hodge-Tate, but to show that it cannot be de Rham seems to require quite deep input; see for instance [**BC**, Example 6.3.5].

(compare Lemma 2.2.7), and each space on the right-hand side has E-dimension at most $\dim_E(V)$. The sum of their E-dimensions is then at most $\dim_{\mathbb{Q}_\ell}(K)\dim_E(V)$, so the $\mathbb{Q}_\ell$-dimension of the right-hand side is at most $\dim_{\mathbb{Q}_\ell}(K)\dim_{\mathbb{Q}_\ell}(V)$, which is, since V is de Rham, the $\mathbb{Q}_\ell$-dimension of the left-hand side. Therefore equality holds everywhere, and it easily follows that $D_{dR}(V)$ is free over $K \otimes_{\mathbb{Q}_\ell} E \cong \oplus_\tau E$.

EXAMPLE 2.2.2. Note that even when V is de Rham, other steps of the filtration need not be free over $K \otimes_{\mathbb{Q}_\ell} E$. We sketch a basic example, omitting the details because they require introducing more about B_{dR} than we will subsequently need. Take any extension to a character ψ of Γ_K of the character

$$I_K \xrightarrow{\mathrm{rec}_K} \mathcal{O}_K^\times \xrightarrow{\tau} \overline{\mathbb{Q}}_\ell^\times,$$

where $\tau\colon K \hookrightarrow \overline{\mathbb{Q}}_\ell$ is any fixed $\mathbb{Q}_\ell$-embedding. If $K = \mathbb{Q}_\ell$, so that there is only one such embedding, then $\psi|_{I_K}$ is just the restriction to inertia of the cyclotomic character, and $D_{dR}(\psi)$ has filtration concentrated in degree -1. If $K \neq \mathbb{Q}_\ell$, then for all embeddings $\tau' \neq \tau$, the filtered one-dimensional $\overline{\mathbb{Q}}_\ell$-vector space $e_{\tau'}D_{dR}(\psi)$ has its filtration concentrated in degree zero, whereas the τ piece is concentrated in degree -1.

It will be crucial for us to have a refined notion of Hodge-Tate weight that takes into account the different embeddings $\tau\colon K \hookrightarrow \overline{\mathbb{Q}}_\ell$:

DEFINITION 2.2.3. Let V be a de Rham representation of Γ_K on a $\overline{\mathbb{Q}}_\ell$-vector space of dimension d, so that $D_{dR}(V)$ is a free $K \otimes_{\mathbb{Q}_\ell} \overline{\mathbb{Q}}_\ell$-module of rank d. Then for each $\tau\colon K \hookrightarrow \overline{\mathbb{Q}}_\ell$, we define the d-element multi-set of *τ-labeled Hodge-Tate weights*, $\mathrm{HT}_\tau(D)$, to be the multi-set of integers h such that

$$\mathrm{gr}^h(e_\tau(D_{dR}(V))) \neq 0,$$

where h has multiplicity $\dim_{\overline{\mathbb{Q}}_\ell} \mathrm{gr}^h(e_\tau D_{dR}(V))$.

We could also make this definition for V having coefficients in a subfield $E \subset \overline{\mathbb{Q}}_\ell$, possibly then having to enlarge E in order to contain the embeddings of K (and thus define the labeled Hodge-Tate weights). For example, for any embedding $\tau\colon K \hookrightarrow \overline{\mathbb{Q}}_\ell$, the cyclotomic character $\kappa\colon \Gamma_K \to \mathbb{Q}_\ell^\times$ has a single Hodge-Tate weight of -1. In §2.4.1, we will discuss the elementary but significant implications of E *not* containing all embeddings of K (which in the cyclotomic example just described, is that $\mathrm{HT}_\tau(\kappa)$ is independent of τ).

2.2.2. Labeled Hodge-Tate-Sen weights. Later we will work with Γ_K-representations that are not even Hodge-Tate and will need a generalized notion of (labeled) Hodge-Tate weight. Moreover, we will need an analogue of the sets HT_τ for representations of Γ_K valued in linear algebraic groups other than GL_N. Both of these flexibilities will be required to address Question 1.1.7. First, again let V be a de Rham representation of Γ_K on a $\overline{\mathbb{Q}}_\ell$-vector space. The identification of $\mathrm{gr}^\bullet(B_{dR})$ with B_{HT} implies that the multiplicity of q in $\mathrm{HT}_\tau(V)$ also equals $\dim_{\overline{\mathbb{Q}}_\ell}(V \otimes_{\tau,K} \mathbb{C}_K(-q))^{\Gamma_K}$. We use this formulation to extend the definition of τ-labeled weights to the Hodge-Tate case. Note that if K'/K is any finite extension, and $\tau'\colon K' \hookrightarrow \overline{\mathbb{Q}}_\ell$ is any embedding extending $\tau\colon K \hookrightarrow \overline{\mathbb{Q}}_\ell$, then $\mathrm{HT}_{\tau'}(V|_{\Gamma_{K'}}) = \mathrm{HT}_\tau(V)$.

Now we drop the assumption that V is Hodge-Tate. To analyze such V, we use a construction of Sen ([**Sen81**]): for Sen's theory, [**Ber04**, II.1.2] is a very brief, and very illuminating, overview; see the references there or [**BC**, Part IV] for a detailed introduction. Sen's theory gives much more precise results than what we use here, and we will not need any details of the construction. We will therefore only summarize the properties of the construction that we will formally apply:

THEOREM 2.2.4 (Sen). *For any $\mathbb{C}_K$-semilinear representation V of Γ_K, there is a $\mathbb{C}_K$-linear endomorphism Θ_V, the 'Sen operator' of V, satisfying the following functorial properties for a pair V_1, V_2 of $\mathbb{C}_K$-semilinear representations:*

- *If $V_1 \xrightarrow{T} V_2$ is a Γ_K-equivariant map of $\mathbb{C}_K$-semilinear representations, then* $\mathrm{Lie}(T) \circ \Theta_{V_1} = \Theta_{V_2} \circ \mathrm{Lie}(T)$.
- $\Theta_{V_1 \oplus V_2} = \Theta_{V_1} \oplus \Theta_{V_2}$.
- $\Theta_{V_1 \otimes V_2} = \Theta_{V_1} \otimes \mathrm{id}_{V_2} + \mathrm{id}_{V_1} \otimes \Theta_{V_2}$.

The representation V is is Hodge-Tate[6] *if and only if Θ_V is semi-simple with integer eigenvalues. In general, we define the 'Hodge-Tate-Sen' weights of V to be the eigenvalues of Θ_V.*

PROOF. See [**Sen81**] or, more specifically, [**BC**, Corollary 15.1.6, Theorem 15.1.7, Exercise 15.5.4]. □

We will now be more precise about how to work with Γ_K-representations with coefficients in $\overline{\mathbb{Q}}_\ell$. We will also describe Sen operators for representations valued in a general linear algebraic group. For some finite extension $E/\mathbb{Q}_\ell$ inside $\overline{\mathbb{Q}}_\ell$, our (arbitrary) Γ_K-representation V descends to an E-linear representation V_E. For any embedding $\iota\colon \overline{\mathbb{Q}}_\ell \hookrightarrow \mathbb{C}_K$, we have the Sen operator $\Theta_{\rho,\iota} \in \mathrm{End}_{\mathbb{C}_K}(V \otimes_{\overline{\mathbb{Q}}_\ell, \iota} \mathbb{C}_K)$. More precisely, let $K' = \iota(E)K$, so that $V_E \otimes_{E,\iota} \mathbb{C}_K$ is a $\mathbb{C}_K$-semilinear representation of $\Gamma_{K'}$, to which we can in the usual way associate the Sen operator $\Theta_{\rho,\iota}$. Note that the Sen operator of a $\mathbb{C}_K$ semi-linear representation is insensitive to finite restriction, so this construction is independent of the choice of (sufficiently large) K'. Moreover, choosing a model $V_{E'}$ of V with coefficients in some finite extension E'/E inside $\overline{\mathbb{Q}}_\ell$ also yields the same $\Theta_{\rho,\iota}$, since we don't change the $\mathbb{C}_K$-semilinear representation of $\Gamma_{K'}$ (this may be a bigger K', having enlarged E). This allows the following Tannakian observation about Sen operators (§6 of [**Con11**]).

LEMMA 2.2.5. *Let G be a linear algebraic group over $\overline{\mathbb{Q}}_\ell$ with Lie algebra $\mathfrak{g}$ and $\rho\colon \Gamma_K \to G(\overline{\mathbb{Q}}_\ell)$ a (continuous) representation. Then for each embedding $\iota\colon \overline{\mathbb{Q}}_\ell \hookrightarrow \mathbb{C}_K$, there exists a unique Sen operator $\Theta_{\rho,\iota} \in \mathfrak{g} \otimes_{\overline{\mathbb{Q}}_\ell, \iota} \mathbb{C}_K$ such that for all linear representations $r\colon G \underset{/\overline{\mathbb{Q}}_\ell}{\longrightarrow} \mathrm{GL}_V$, $\mathrm{Lie}(r)(\Theta_{\rho,\iota})$ is equal to the previously-defined $\Theta_{r \circ \rho, \iota}$.*

PROOF. This follows from Tannaka duality for Lie algebras (a precise reference is [**HC50**]). Namely, for each $\overline{\mathbb{Q}}_\ell$-representation $G \to \mathrm{GL}_V$ we get an element $\Theta_{V,\iota} \in \mathfrak{gl}_V \otimes_\iota \mathbb{C}_K$, and these satisfy the functorial properties

- If $V_1 \xrightarrow{T} V_2$ is a G-morphism, then $\mathrm{Lie}(T) \circ \Theta_{V_1,\iota} = \Theta_{V_2,\iota} \circ \mathrm{Lie}(T)$;
- $\Theta_{V_1 \oplus V_2, \iota} = \Theta_{V_1,\iota} \oplus \Theta_{V_2,\iota}$;
- $\Theta_{V_1 \otimes V_2, \iota} = \Theta_{V_1,\iota} \otimes \mathrm{id}_{V_2} + \mathrm{id}_{V_1} \otimes \Theta_{V_2,\iota}$.

---

[6] Strictly speaking, in §2.2.1 we only defined Hodge-Tate representations on finite-dimensional $\mathbb{Q}_\ell$-vector spaces. This is a special case of a more general notion for $\mathbb{C}_K$-semi-linear Γ_K-representations: see [**BC**, Definition 2.3.4].

(If different fields E and K' as above are needed to define the $\Theta_{V_i,\iota}$, we can pass to a common extension and apply the remarks preceding the lemma.) Tannaka duality then implies that all $\Theta_{V,\iota}$ arise from a unique element $\Theta_{\rho,\iota} \in \mathfrak{g} \otimes_\iota \mathbb{C}_K$. $\square$

REMARK 2.2.6. The necessary functorial properties of the Sen operators are not automatic. All the relevant statements are in §15 of [**BC**].

As mentioned previously, $V \otimes_{\overline{\mathbb{Q}}_\ell,\iota} \mathbb{C}_K$ is Hodge-Tate if and only if $\Theta_{\rho,\iota}$ is semi-simple with integer eigenvalues. We will need the following comparison:

LEMMA 2.2.7. *Suppose $\rho\colon \Gamma_K \to \mathrm{Aut}_{\overline{\mathbb{Q}}_\ell}(V)$ is Hodge-Tate. Any embedding $\iota\colon \overline{\mathbb{Q}}_\ell \hookrightarrow \mathbb{C}_K$ induces $\tau_\iota\colon K \hookrightarrow \overline{\mathbb{Q}}_\ell$, and we then have*

$$\mathrm{HT}_{\tau_\iota}(\rho) = \{\textit{eigenvalues of } \Theta_{\rho,\iota}\}.$$

PROOF. Let E and K' be as above. The Γ_K-representation V_E is Hodge-Tate, and there is a natural isomorphism of graded $E \otimes_{\mathbb{Q}_\ell} K'$-modules

$$(V_E \otimes_{\mathbb{Q}_\ell} \mathrm{B}_{HT})^{\Gamma_K} \otimes_K K' \xrightarrow{\sim} (V_E \otimes_{\mathbb{Q}_\ell} \mathrm{B}_{HT})^{\Gamma_{K'}}.$$

Restricting to the qth graded pieces, we get an $E \otimes_{\mathbb{Q}_\ell} K'$-isomorphism

$$(V_E \otimes_{\mathbb{Q}_\ell} \mathbb{C}_K(-q))^{\Gamma_K} \otimes_K K' \xrightarrow{\sim} (V_E \otimes_{\mathbb{Q}_\ell} \mathbb{C}_K(-q))^{\Gamma_{K'}},$$

which in turn is an $E \otimes_{\mathbb{Q}_\ell} K'$-isomorphism

$$\bigoplus_{\tau\colon K \hookrightarrow E} (V_E \otimes_{\tau,K} \mathbb{C}_K(-q))^{\Gamma_K} \otimes_K K' \xrightarrow{\sim} \bigoplus_{\iota\colon E \hookrightarrow \mathbb{C}_K} (V_E \otimes_{E,\iota} \mathbb{C}_K(-q))^{\Gamma_{K'}}.$$

Projecting to the τ-component, we obtain an $E \otimes_{\tau,K} K'$-isomorphism

$$(V_E \otimes_{\tau,K} \mathbb{C}_K(-q))^{\Gamma_K} \otimes_K K' \xrightarrow{\sim} \bigoplus_{\iota\colon E \hookrightarrow \mathbb{C}_K : \tau_\iota = \tau} (V_E \otimes_{E,\iota} \mathbb{C}_K(-q))^{\Gamma_{K'}}.$$

Writing $m_{q,\tau}$ for the multiplicity of q in $\mathrm{HT}_\tau(V)$, the left-hand side is a free $E \otimes_{\tau,K} K'$-module of rank $m_{q,\tau}$. Meanwhile, the K'-dimension of the ι-factor of the right-hand side is by definition the multiplicity of q as a Hodge-Tate weight of $V_E \otimes_{E,\iota} \mathbb{C}_K$, hence is q's multiplicity as an eigenvalue of $\Theta_{\rho,\iota}$. We deduce (applying the ι-projection in $E \otimes_{\tau,K} K' \cong \prod_{\iota:\tau_\iota=\tau} K'$) that for all ι the eigenvalues of $\Theta_{\rho,\iota}$ match the multi-set $\mathrm{HT}_{\tau_\iota}(V)$.

REMARK 2.2.8. We will only apply this when V is in fact de Rham, with the exception of Corollary 3.2.12.

2.2.3. Induction and ℓ-adic Hodge theory. We will later need a result on compatibility of Fontaine's functors with induction. The following lemma is certainly well-known, but I don't know of a proof in the literature, so I record some details. For a finite extension $K/\mathbb{Q}_\ell$, we let (as is standard notation in the subject) $K_0 \subset K$ denote the maximal unramified sub-extension of K.

LEMMA 2.2.9. *Let $L/K/\mathbb{Q}_\ell$ be finite, and let W be a de Rham representation of Γ_L. Then $V = \mathrm{Ind}_L^K W$ is also de Rham, and $\mathrm{D}_{\mathrm{dR}}(V)$ is the image under the forgetful functor $\mathrm{Fil}_L \to \mathrm{Fil}_K$ of $\mathrm{D}_{\mathrm{dR}}(W)$. Moreover, $\mathrm{Ind}_L^K W$ is crystalline if and only if W is crystalline and L/K is unramified.*

PROOF. This follows almost immediately from Frobenius reciprocity if one uses contravariant Fontaine functors:

$$D^*_{dR}(V) := \mathrm{Hom}_{\Gamma_K}(V, B_{dR}) \cong \mathrm{Hom}_{\Gamma_L}(W, B_{dR}|_{\Gamma_L}) = D^*_{dR}(W).$$

Since $D^*_{dR}(V) \cong D_{dR}(V^*)$, and $\mathrm{Ind}_L^K(W^*) \cong \mathrm{Ind}_L^K(W)^*$, this shows that $D_{dR}(V)$ is simply the filtered K-vector space underlying $D_{dR}(W)$. Comparing dimensions, it is clear that V is de Rham if and only if W is. In the crystalline case, the same observation yields $D^*_{cris}(V) \cong D^*_{cris}(W)$, so

$$\dim_{K_0} D_{cris}(V) = \dim_{K_0} D_{cris}(W) = \dim_{L_0} D_{cris}(W)[L_0 : K_0],$$

and comparing dimensions we see that V is crystalline if and only if W is crystalline and $[L_0 : K_0] = [L : K]$, i.e. L/K is unramified. □

2.3. GL$_1$

A detailed analysis of the GL$_1$ theory provides both motivation and technical tools for addressing the more general Galois and automorphic lifting problems.

2.3.1. The automorphic side. We begin by reviewing some basic facts about Hecke characters $\psi\colon \mathbf{A}_F^\times/F^\times \to \mathbb{C}^\times$. Any such ψ is the twist by $|\cdot|^r_{\mathbf{A}_F}$, for some $r \in \mathbb{R}$, of a unitary Hecke character. At each place v of F, $\psi_v = \psi|_{F_v^\times}$ can be decomposed via projection to the maximal compact subgroup in $F_v^\times$; at $v|\infty$, this lets us write any unitary ψ_v in terms of a given embedding $\iota_v\colon F_v \hookrightarrow \mathbb{C}$ as

$$\psi_v(x_v) = (\iota_v(x_v)/|\iota_v(x_v)|)^{m_v}|\iota_v(x_v)|_{\mathbb{C}}^{it_v},$$

with $m_v = m_{\iota_v}$ an integer, and t_v a real number. To avoid cumbersome notation, we will sometimes omit reference to the embedding ι_v. Hecke characters encode finite Galois-theoretic information as well as rather subtle archimedean information; the latter will be particularly relevant in what follows, and the basic observation (**[Wei56]**) is:

LEMMA 2.3.1. *There is a (unitary, say) Hecke character ψ of F with archimedean parameters $\{m_v, t_v\}_{v|\infty}$, as above, if and only if for some positive integer M, all global units $\alpha \in \mathcal{O}_F^\times$ satisfy*

$$\left(\prod_{v|\infty}\left(\frac{\iota_v(\alpha)}{|\iota_v(\alpha)|}\right)^{m_v}|\iota_v(\alpha)|^{it_v}\right)^M = 1.$$

Equivalently, the inside product is trivial for all α in some finite-index subgroup of $\mathcal{O}_F^\times$.

It follows that if F is a CM field, and $\{m_v\}_{v|\infty}$ is a set of integers indexed by the archimedean places of F , then there is a unitary Hecke character ψ of F with archimedean components given by $\psi_v(x_v) = (\iota_v(x_v)/|\iota_v(x_v)|)^{m_v}$, for all $v|\infty$.

Using the notation we have just established, we recall Weil's notions of type A_0 and type A Hecke characters:

DEFINITION 2.3.2. A Hecke character ψ is said to be of type A if $r \in \mathbb{Q}$ and $t_v = 0$ for all $v|\infty$. ψ is said to be type A_0 if $t_v = 0$ and $\frac{m_v}{2} + r$ is an integer for all $v|\infty$. Equivalently, ψ is type A_0 if for all $v|\infty$, there exist integers p_v and q_v such that $\psi_v(z) = \iota_v(z)^{p_v}\bar{\iota}_v(z)^{q_v}$.

In particular, note that Lemma 2.3.1 implies the existence of all the type A Hecke characters over CM fields that one could hope for. The situation is more interesting over general number fields, as we will see in Lemma 2.3.4. First we note another result of Weil ([**Wei56**]), which reveals the central role in the algebraic theory of automorphic forms (for GL_1, here) played by the notion of 'type A.'

LEMMA 2.3.3 (Weil). *Let χ be a type A Hecke character locally trivialized by the modulus $\mathfrak{m}$ (i.e. unramified outside $\mathfrak{m}$, and, if $\mathfrak{p}_v^{f_v}$ is the exact power of a prime ideal $\mathfrak{p}_v$ dividing $\mathfrak{m}$, trivial on $1+\mathfrak{p}_v^{f_v}$). The character χ then induces a character $\chi^{\mathfrak{m}}$ of $I^{\mathfrak{m}}$, the group of non-zero, prime to $\mathfrak{m}$, fractional ideals, whose values are all algebraic numbers.*

Later we will be interested in a higher-rank version of the type A condition; in current discussions of algebraicity of automorphic representations, people tend to focus on analogues of type A_0. We recall a basic result of Weil-Artin ([**Wei56**]), which will later inspire much of our discussion in higher-rank as well. Recall that for Hecke characters, base-change (in the sense of Langlands functoriality) from a field F to a field L containing F is simply precomposition with the norm $N_{L/F}$; we denote the base-change of a Hecke character ψ of F to one of L by $\mathrm{BC}_{L/F}(\psi)$.

LEMMA 2.3.4 (Weil-Artin). *Let ψ be a type-A Hecke character of a number field F. Write F_{cm} for the maximal CM subfield of F. Then there exists a finite-order Hecke character χ_0 of F and a type-A Hecke character ψ' of F_{cm} such that $\psi=\chi_0\cdot \mathrm{BC}_{F/F_{\mathrm{cm}}}(\psi')$. If F_{cm} is totally real, then ψ is a finite-order twist of some power $|\cdot|^r$, $r\in\mathbb{Q}$, of the absolute value.*

REMARK 2.3.5. We emphasize that the true content of this result, which should fully generalize to higher-rank (see Proposition 2.4.8), is that the infinity-type of a type-A Hecke character descends to the maximal CM subfield.

FIRST PROOF: Weil leaves it as an exercise, so we give a proof. We then indicate a second proof, which we will apply more generally in Proposition 2.4.8. By twisting by a rational power of the absolute value, and then if necessary squaring, we may assume that ψ is type A_0. First assume $F/\mathbb{Q}$ is Galois, and set $G=\mathrm{Gal}(F/\mathbb{Q})$. We may regard F as a subfield of $\mathbb{C}$ (i.e., choose an embedding), so the above relation becomes

$$\prod_{g\in G} g(\alpha)^{m_g}=1$$

for all α in some finite-index subgroup of $\mathcal{O}_F^\times$ and some integers m_g. Now take any embedding $\iota\colon F\hookrightarrow\mathbb{C}$, using it to define a complex conjugation c_ι on F, and thus on G. In particular, when F is totally imaginary we can rewrite the above as

$$\prod_{G/c_\iota} g(\alpha)^{m_g}(c_\iota g)(\alpha)^{m_{c_\iota g}}=1.$$

Applying $\log|\iota(\cdot)|$, we get (for all α)

$$\sum_{G/c_\iota}(m_g+m_{c_\iota g})\log|\iota(g(\alpha))|_{\mathbb{C}}=0,$$

which by the unit theorem is only possible when $m_g+m_{c_\iota g}$ equals a constant w_ι independent of g; the analogous argument in the totally real case shows that the m_g themselves are independent of g, concluding that case of the proof. But w_ι is independent of ι as well,[7] hence $m_g + m_{c_\iota g}$ is independent both of g and the choice of conjugation. This implies that for any two ι, ι', $m_g = m_{c_\iota c_{\iota'} g}$, and since F_{cm} is precisely the fixed field (in F) of the subgroup generated by all products $c_\iota c_{\iota'}$, we are done.

Now let F be arbitrary, and let ψ be a type A Hecke character of F. As usual letting $\widetilde{F}$ denote the Galois closure, we know from the Galois case that $\psi|_{\widetilde{F}}$ has infinity-type descending to $(\widetilde{F})_{\mathrm{cm}}$. Setting $H_1 = \mathrm{Gal}(\widetilde{F}/F)$ and $H_2 = \mathrm{Gal}(\widetilde{F}/(\widetilde{F})_{\mathrm{cm}})$, the values m_{ι_v} (for embeddings $\iota_v \colon \widetilde{F} \hookrightarrow \mathbb{C}$) are constant on H_1-orbits (by construction) and H_2-orbits (by the Galois case), hence on H_1H_2-orbits. But $F \cap (\widetilde{F})_{\mathrm{cm}} = F_{\mathrm{cm}}$, so $H_1 H_2 = \mathrm{Gal}(\widetilde{F}/F_{\mathrm{cm}})$, and the infinity-type descends all the way down to F_{cm}, proving the lemma in general. □

Second Proof: Later, we will generalize the following argument (see Proposition 2.4.8 for more details) in the totally imaginary case: for all $\sigma \in \mathrm{Aut}(\mathbb{C})$, there is a Hecke character ${}^\sigma\psi$ whose finite part is ${}^\sigma\psi_f(x) = \sigma(\psi_f(x))$ and whose infinity-type is given by the integers $m_{\sigma^{-1}\iota_v}$ (the key but easy check is $F^\times$-invariance). For some integer r, the twist $|\cdot|^{r/2}\psi$ is unitary, so ${}^c\psi = |\cdot|^{-r}\psi^{-1}$, and therefore the fixed field $\mathbb{Q}(\psi_f) \subset \overline{\mathbb{Q}}$ of all $\sigma \in \mathrm{Aut}(\mathbb{C})$ such that ${}^\sigma\psi_f \cong \psi_f$ is contained in $\mathbb{Q}^{\mathrm{cm}}$. The result follows from some Galois theory. □

As pointed out to me by Brian Conrad, a third, more algebraic, approach is possible, where infinity-types of algebraic Hecke characters are related to (algebraic) characters of the connected Serre group; compare the discussion of potentially abelian motives in §4.1.30. Of course, in all arguments, the unit theorem is the essential ingredient.

We will have to analyze Hecke characters in the following setting: An automorphic representation π of an F-group G has a central character $\omega_\pi \colon Z_G(F)\backslash Z_G(\mathbf{A}) \to \mathbb{C}^\times$, and we will want to understand the possible infinity-types of extensions

$$\begin{array}{ccc} Z_G(F)\backslash Z_G(\mathbf{A}) & \xrightarrow{\ \omega_\pi\ } & \mathbb{C}^\times \\ \downarrow & \nearrow_{\tilde{\omega}} & \\ \widetilde{Z}(F)\backslash \widetilde{Z}(\mathbf{A}) & & \end{array}$$

where $\widetilde{Z}$ is an F-torus containing Z_G. The basic case to which this is reduced, for split groups at least, is where $Z_G = \mu_n$ and $\widetilde{Z} = \mathbf{G}_m$, where we have to extend a character from $\mu_n(F)\backslash\mu_n(\mathbf{A})$ to the full idele class group $F^\times\backslash\mathbf{A}^\times$.

Lemma 2.3.6. *Let F be any number field. Given a continuous character $\omega \colon \mu_n(F)\backslash\mu_n(\mathbf{A}) \to \mathrm{S}^1$, with archimedean component ω_∞, fix embeddings $\iota_v \colon F_v \hookrightarrow \mathbb{C}$ and write*

$$\omega_\infty \colon (x_v)_{v|\infty} \mapsto \prod_v \iota_v(x_v)^{m_v}$$

for some set of residue classes $m_v \in \mathbb{Z}/n\mathbb{Z}$. Then:

- *There exist type A (unitary) Hecke character extensions $\tilde{\omega} \colon F^\times\backslash\mathbf{A}^\times \to \mathrm{S}^1$ if and only if the images $m_v \in \mathbb{Z}/n\mathbb{Z}$ depend only on the restriction $v|_{F_{\mathrm{cm}}}$. There exists a finite-order extension $\tilde{\omega}$ if and only if for all complex $v|\infty$, the classes $m_v \in \mathbb{Z}/n\mathbb{Z}$ are all zero.*

[7] For instance, whatever ι, $\frac{1}{2}|G|w_\iota = \sum_{g \in G} m_g$.

- *The extension $\tilde{\omega}$ is unique up to the nth power of another type A Hecke character, except in the Grunwald-Wang special case (see the proof), where one first has two choices of extension to $C_F[n]$, and then the nth power ambiguity.*
- *In particular, if F is totally real or CM, then (unitary) type A extensions of ω always exist. If F is totally real, they are finite-order, and if F is CM, a finite-order extension can be chosen if and only if $\omega|_{F_\infty^\times}$ is trivial.*

PROOF. For the time being, let F be arbitrary. The cokernel of $\mu_n(F)\backslash\mu_n(\mathbf{A}) \to C_F[n]$ is either trivial or order 2, the latter case being the Grunwald-Wang special case (see [**Con11**, Appendix A]), so we first choose an extension ω' to $C_F[n]$ (we will use this flexibility in Proposition 3.1.4). Then Pontryagin duality gives an isomorphism

$$C_F^D/nC_F^D \xrightarrow{\sim} (C_F[n])^D,$$

so there exists some $\tilde{\omega} \in C_F^D$ extending ω', and it is unique up to n^{th} powers. Restricted to $F_\infty^\times$, we can write (implicitly invoking ι_v)

$$\tilde{\omega}_\infty \colon (x_v)_{v|\infty} \to \prod_{v|\infty} (x_v/|x_v|)^{m_v} |x_v|_{\mathbb{C}}^{it_v}$$

for integers m_v and real numbers t_v. Then a type A lift of ω' exists if and only if there is a Hecke character ψ with infinity-type

$$\psi_\infty \colon (x_v) \mapsto \prod_{v|\infty} (x_v/|x_v|)^{f_v} |x_v|_{\mathbb{C}}^{it_v/n}$$

for some integers f_v (the desired type A character is then $\tilde{\omega}\psi^{-n}$). Following Weil (Lemma 2.3.1), the existence of a Hecke character $\tilde{\omega}$ corresponding to the infinity-data $\{m_v, t_v\}_v$ is equivalent to the existence of some integer M such that for all $\alpha \in \mathcal{O}_F^\times$,

$$\prod_{v|\infty} (\alpha/|\alpha|)^{m_v M} |\alpha|_v^{it_v M} = 1,$$

and, similarly, such a ψ can exist if and only if there exists $M' \in \mathbb{Z}$ such that

$$\prod_{v|\infty} (\alpha/|\alpha|)^{f_v MM'n} |\alpha|_v^{it_v MM'} = 1$$

for all $\alpha \in \mathcal{O}_F^\times$. Substituting, the left-hand side is $\prod_{v|\infty}(\alpha/|\alpha|)^{(f_v n - m_v)MM'}$, which can equal 1 for all $\alpha \in \mathcal{O}_F^\times$ if and only if there exists a type A Hecke character with infinity-type

$$(x_v) \mapsto \prod_{v|\infty} (x_v/|x_v|)^{f_v n - m_v}.$$

By Weil's classification of type A Hecke characters, this is possible if and only if $f_v n - m_v$ depends only on $v|_{F_{\mathrm{cm}}}$. This in turn is possible (for some choice of f_v) if and only if $m_v \in \mathbb{Z}/n\mathbb{Z}$ depends only on $v|_{F_{\mathrm{cm}}}$. In particular, over CM and totally real fields F, there is no obstruction, so a type A lift of ω always exists.

The claim about existence of finite-order extensions is a simple variant: if all m_v (v complex) are zero in $\mathbb{Z}/n\mathbb{Z}$, then there is a type A lift $\tilde{\omega}$ with infinity-type

$$\tilde{\omega}_\infty \colon (x_v)_{v|\infty} \to \prod_{v|\infty} (x_v/|x_v|)^{m_v}$$

for some integers m_v, all divisible by n and only depending on $\iota_v|_{F_{\mathrm{cm}}}$. But then there is also a type A Hecke character ψ of F with infinity-type

$$\psi_\infty \colon (x_v)_{v|\infty} \to \prod_{v|\infty} (x_v/|x_v|)^{m_v/n},$$

so $\tilde{\omega}\psi^{-n}$ is a finite-order extension of ω.

As for uniqueness, once a type A lift of ω' is chosen, any other is (by the discussion at the start of the proof) a twist by the nth power of another Hecke character, which must itself clearly be of type A. This is then the ambiguity in extending ω itself, except in the Grunwald-Wang special case, in which there is the additional $\mathbb{Z}/2\mathbb{Z}$ ambiguity noted above. □

REMARK 2.3.7.
- Of course any quasi-character of $\mu_n(\mathbf{A})$ is unitary, and to understand all extensions it suffices (twisting by powers of $|\cdot|_{\mathbf{A}}$) to understand unitary extensions–hence the restriction to S^1 instead of $\mathbb{C}^\times$.
- This raises a tantalizing question: certainly over a non-CM field F we can produce characters ω that have no type A extensions (although they certainly all have some extension to a Hecke character, by Pontryagin duality). What if ω actually arises as the central character of a suitably algebraic cuspidal automorphic representation (on $\mathrm{SL}_n(\mathbf{A})$, for instance)? We return to this in §2.4.

A few consequences, either of the result or the method of proof, will follow; first, though, let us make a definition:

DEFINITION 2.3.8. Let F be any number field, and let $\psi \in C_F^D$ be a unitary Hecke character. We say that ψ is *of Maass type* if for all $v|\infty$, the restriction $\psi_v \colon F_v^\times \to \mathrm{S}^1$ has the form $x \mapsto |x|^{it_v}$ for real numbers t_v.

COROLLARY 2.3.9. *Let $\psi \colon C_F \to \mathbb{C}^\times$ be a Hecke character of a number field F.*

- *If F is totally real or CM, then ψ can be decomposed as*

$$\psi = \psi_{\mathrm{alg}}\psi_{\mathrm{Maass}}\psi_{\mathrm{fin}}|\cdot|^w,$$

where w is the unique real number twisting ψ to a unitary character, ψ_{alg} is unitary type A, ψ_{Maass} is of Maass type, and ψ_{fin} is finite order. The last three characters are all unique up to finite-order characters.
- *If ψ is of Maass type (after twisting to a unitary character), with F arbitrary, then ψ is 'nearly divisible': for any $n \in \mathbb{Z}$ there exist Hecke characters χ of Maass type and χ_0 of finite order such that $\chi^n\chi_0 = \psi$. In particular, after a finite base-change ψ is n-divisible.*
- *Write $\mathcal{A}_F(1)$ for the space (topological group) of all Hecke characters of F. Suppose F is CM, of degree $2s$ over $\mathbb{Q}$. Then there is an exact sequence*

$$1 \to \Gamma_F^D \to \mathcal{A}_F(1) \to \mathbb{R} \times \mathbb{Z}^s \times \mathbb{Q}^{s-1} \to 1.$$

(As always, Γ_F^D denotes $\mathrm{Hom}_{\mathrm{cts}}(\Gamma_F, \mathrm{S}^1)$; it is the space of Dirichlet characters.) For the totally real subfield F_+ of F, we have a similar sequence

$$1 \to \Gamma_{F_+}^D \to \mathcal{A}_{F_+}(1) \to \mathbb{R} \times \mathbb{Q}^{s-1} \to 1.$$

It is also possible (via the unit theorem) to compute all possible infinity-types of Hecke characters of any number field, but we will make no use of this calculation. Roughly speaking, the new transcendentals that arise in the 'mixed' case where algebraic and Maass parts cannot be separated are the arguments (angles) of fundamental units. It is very tempting to ask whether for CM (or totally real fields) algebraic and spherical ('Maass') infinity-types on higher-rank groups can twist together in a non-trivial way. If this question is too naïve, is there any other higher-rank generalization of part 1 of Corollary 2.3.9?

We include a couple other related useful results. For the first, note that in contrast with the situation for ℓ-adic Galois characters (see Lemma 2.3.15 below), a general Hecke character cannot be written as an nth-power up to a finite-order twist.[8]

LEMMA 2.3.10. *Let ψ be a Hecke character of F, and suppose that L/F is a finite Galois extension over which $\mathrm{BC}_{L/F}(\psi) = \psi \circ N_{L/F}$ is an nth-power. Then up to a finite-order twist, ψ itself is an nth-power.*

PROOF. Let ω be a Hecke character of L such that $\psi \circ N_{L/F} = \omega^n$. Then for any $\sigma \in \mathrm{Gal}(L/F)$, $\omega^{-1} \cdot (\omega \circ \sigma)$ has finite order. After passing to a further finite extension $\widetilde{L}$ that kills all these characters, we see that $\mathrm{BC}_{\widetilde{L}/F}\psi = \mathrm{BC}_{\widetilde{L}/L}(\omega)^n$, where $\mathrm{BC}_{\widetilde{L}/L}(\omega)$ is now $\mathrm{Gal}(\widetilde{L}/F)$-invariant. The next lemma shows that ω descends to F up to a finite-order twist. $\square$

The next lemma completes the previous one, and also enables a slight refinement of a result of Rajan (see Remark 2.3.12):

LEMMA 2.3.11. *Let L/F be a Galois extension of number fields, and let ψ be a $\mathrm{Gal}(L/F)$-invariant Hecke character of L. Then there exists a Hecke character ψ_F of F and a finite-order Hecke character ψ_0 of L such that $\psi = (\psi_F \circ N_{L/F}) \cdot \psi_0$.*

PROOF. We may assume ψ is unitary, and we may choose, for all infinite places w of L, embeddings $\iota_w \colon L_w \to \mathbb{C}$ such that all ι_w for w above a fixed place v of F restrict to the same embedding $\iota_v \colon F_v \to \mathbb{C}$. $\mathrm{Gal}(L/F)$ acts transitively on the places $w|v$, so when we write

$$\psi_w(x_w) = \left(\frac{\iota_w(x_w)}{|\iota_w(x_w)|}\right)^{m_w} \cdot |\iota_w(x_w)|^{it_w},$$

$\mathrm{Gal}(L/F)$-invariance implies that the m_w and t_w depend only on the place v below w (from now on in the proof, the embeddings ι_w will be implicit). We therefore denote these by m_v and t_v. Lemma 2.3.1 implies there is a Hecke character of F with infinity-type given by the data $\{m_v, t_v\}_v$. Namely, there is an integer M such that for all $\alpha \in \mathcal{O}_L^\times$,

$$1 = \left(\prod_{w|\infty} \left(\frac{\alpha}{|\alpha|}\right)^{m_w} |\alpha|_{\mathbb{C}}^{it_w}\right)^M.$$

[8] Consider, for example, a type A Hecke character whose integral parameters at infinity are not divisible by n.

Restricting to $\alpha \in \mathcal{O}_F^\times$, this becomes

$$1 = \left(\prod_{v|\infty}\prod_{w|v}\left(\frac{\alpha}{|\alpha|}\right)^{m_v}|\alpha|_{\mathbb{C}}^{it_v}\right)^M = \left(\prod_{v|\infty}\left(\frac{\alpha}{|\alpha|}\right)^{m_v}|\alpha|_{\mathbb{C}}^{it_v}\right)^{M\#\{w|v\}},$$

($\#\{w|v\}$ is independent of v) which is simply the criterion for there to be a Hecke character of F with infinity-type given by the collection $\{m_v, t_v\}_v$. Any such character has base-change differing from ψ by a finite-order character, so we are done. □

REMARK 2.3.12. A theorem of Rajan describes the image of solvable base-change: precisely, for a solvable extension L/F of number fields and a $\mathrm{Gal}(L/F)$-invariant cuspidal automorphic representation π of $\mathrm{GL}_n(\mathbf{A}_L)$, Theorem 1 of [**Raj02**] asserts that there is a cuspidal representation π_F of $\mathrm{GL}_n(\mathbf{A}_F)$ and a $\mathrm{Gal}(L/F)$-invariant Hecke character ψ of L such that

$$\mathrm{BC}_{L/F}(\pi_0) \otimes \psi = \pi.$$

Lemma 2.3.11 shows that this ψ may be chosen to have finite-order; in particular, for some cyclic extension L'/L (so L'/F is still solvable), $\mathrm{BC}_{L'/L}(\pi)$ descends to F.

2.3.2. Galois GL_1**.** We now discuss (continuous) Galois characters $\hat{\psi}\colon \Gamma_F \to \overline{\mathbb{Q}}_\ell^\times$. For the time being, F is any number field. These are necessarily almost everywhere unramified, and we focus on ℓ-adic Hodge theory aspects. The following is well-known, and is proven in [**Ser98**, Chapter III]:

THEOREM 2.3.13. *Suppose that $\hat{\psi}$ is de Rham. Then for all places v not dividing ℓ, $\hat{\psi}|_{\Gamma_{F_v}}$ assumes algebraic values in $\overline{\mathbb{Q}} \xrightarrow{\iota_\ell} \overline{\mathbb{Q}}_\ell$, and there exists a type A_0 Hecke character $\psi\colon \mathbf{A}_F/F^\times \to \mathbb{C}^\times$ corresponding to $\hat{\psi}$ (via $\iota_\infty\colon \overline{\mathbb{Q}} \hookrightarrow \mathbb{C}$). Moreover, $\hat{\psi}$ is motivic: the Fontaine-Mazur conjecture holds for GL_1/F.*

PROOF. We indicate how this follows from the arguments of [**Ser98**, III]; see Remark 2.3.14 for the explanation of a simple case. Since $\hat{\psi}$ is Hodge-Tate at all $v|\ell$, it is (by a theorem of Tate: see [**Ser98**, III-A6]) locally algebraic in the sense of [**Ser98**, III-1.1] (and therefore de Rham in the language that post-dates [**Ser98**]). Then the argument of [**Ser98**, III.2.3 Theorem 2] implies that, for a suitable modulus $\mathfrak{m}$, $\hat{\psi}$ is the ℓ-adic Galois representation associated to an algebraic homomorphism $\mathcal{S}_{\mathfrak{m},\overline{\mathbb{Q}}_\ell} \to \mathbb{G}_{m,\overline{\mathbb{Q}}_\ell}$, where $\mathcal{S}_{\mathfrak{m}}$ is the $\mathbb{Q}$-torus of [**Ser98**, II-2.2]. Up to isomorphism, this algebraic representation can be realized over some finite extension E of $\mathbb{Q}$ inside $\overline{\mathbb{Q}}_\ell$. Taking the 'archimedean realization' as in [**Ser98**, II-2.7], we obtain the desired type A_0 Hecke character. It then follows by Lemma 2.3.3 that $\hat{\psi}|_{\Gamma_{F_v}}$ assumes algebraic values for $v \nmid \ell$. The Fontaine-Mazur conjecture for $\hat{\psi}$ follows from a theorem of Deligne ([**DMOS82**, Proposition IV.D.1]), which shows the somewhat stronger statement that $\hat{\psi}$ is the ℓ-adic realization of a motive for absolute Hodge cycles. □

REMARK 2.3.14. Consider the simple case in which $F = \mathbb{Q}$. Then $\hat{\psi}|_{\Gamma_{\mathbb{Q}_\ell}}$ has a single labeled Hodge-Tate weight; it therefore has the same (labeled) Hodge-Tate weights as an integer power ω_ℓ^r of the ℓ-adic cyclotomic character. It follows then from the above-cited theorem of Tate ([**Ser98**, III-A6]) and global class field theory that $\hat{\psi}\omega_\ell^{-r}$ is finite-order, hence is the ℓ-adic realization (via a fixed isomorphism

$\mathbb{C} \xrightarrow{\sim} \overline{\mathbb{Q}}_\ell$) of some finite-order Hecke character χ. The Hecke character corresponding to $\hat{\psi}$ is then $\chi| \cdot |^r_{\mathbf{A}_\mathbb{Q}}$. It is also easy to show that $\hat{\psi}$ is motivic: ω_ℓ^r is the ℓ-adic realization of the Tate motive $\mathbb{Q}(r)$, and if $\hat{\psi}\omega_\ell^{-r}$ factors through a finite quotient $\mathrm{Gal}(F/\mathbb{Q})$, then it is a suitable direct factor of $H^0(X_{\overline{\mathbb{Q}}}, \overline{\mathbb{Q}}_\ell)$, for the zero-dimensional variety $X = \mathrm{Res}_{F/\mathbb{Q}}(\mathrm{Spec}\, F)$ whose $\overline{\mathbb{Q}}$-points are naturally indexed by embeddings $\tau\colon F \hookrightarrow \overline{\mathbb{Q}}$, with the $\Gamma_\mathbb{Q}$-action on H^0 arising from permutation of the set of embeddings (for more details, see the discussion of Artin motives in §4.1.2).

More generally, the idea behind establishing the Fontaine-Mazur conjecture in the abelian case is to find a sub-quotient of the cohomology of a CM abelian variety having 'labeled Hodge numbers' matching those of $\hat{\psi}$; that this is possible essentially follows from the constraints on the ∞-type of the type A_0 Hecke character underlying $\hat{\psi}$ (as in Lemma 2.3.4). If the CM abelian variety were actually defined over F, this would realize $\hat{\psi}$ inside some finite-order twist of its cohomology. The subtle part of the theorem is the need to have some control over the field of definition of the CM abelian variety.

We make repeated use of the following well-known observation (see for instance Lemma 3.1 of [**Con11**]).

LEMMA 2.3.15. *Let $\chi\colon \Gamma_F \to \overline{\mathbb{Q}}_\ell^\times$ be an ℓ-adic Galois character. For any non-zero integer m, there are characters $\chi_1, \chi_0\colon \Gamma_F \to \overline{\mathbb{Q}}_\ell^\times$, with χ_0 finite-order, such that $\chi = (\chi_1)^m\chi_0$.*

We will need to consider Galois characters somewhat outside the algebraic range. As is customary, we write $\mathrm{HT}_\tau(\rho)$ to indicate the $\tau\colon F \hookrightarrow \overline{\mathbb{Q}}_\ell$-labeled Hodge-Tate weights of an ℓ-adic Galois representation ρ; see §2.2 for details, as well as for what we mean when we write non-integral τ-labeled Hodge-Tate-Sen weights.

COROLLARY 2.3.16. *Let F be a totally imaginary field, and n a non-zero integer. For all $\tau\colon F \hookrightarrow \overline{\mathbb{Q}}_\ell$, fix an integer k_τ. Then there exists a Galois character $\hat{\psi}\colon \Gamma_F \to \overline{\mathbb{Q}}_\ell^\times$ with $\mathrm{HT}_\tau(\hat{\psi}) = \frac{k_\tau}{n}$ (respectively, with $\mathrm{HT}_\tau(\hat{\psi}) \equiv \frac{k_\tau}{n} \mod \mathbb{Z}$) if and only if*

(1) *k_τ depends only on $\tau_0 := \tau|_{F_{\mathrm{cm}}}$ (respectively, only depends modulo n on τ_0);*
(2) *and there exists an integer w such that $k_{\tau_0} + k_{\tau_0 \circ c} = w$ (respectively, $k_{\tau_0} + k_{\tau_0 \circ c} \equiv w \mod n$) for all τ, and c the complex conjugation on F_{cm}.*[9]

PROOF. Suppose such a $\hat{\psi}$ exists, with weights $\frac{k_\tau}{n}$. Then $\hat{\psi}^n$ is geometric (the de Rham and Hodge-Tate conditions are equivalent for characters, as noted in Theorem 2.3.13), and so (via ι_ℓ and ι_∞) there exists a type A_0 Hecke character ψ of F corresponding to $\hat{\psi}^n$. For $v|\infty$ and $\iota_v\colon F_v \xrightarrow{\sim} \mathbb{C}$, $\psi|_{F_v^\times}$ is given by

$$\psi_v(x_v) = \iota_v(x_v)^{k_{\tau^*(\iota_v)}}\bar{\iota}_v(x_v)^{k_{\tau^*(\bar{\iota}_v)}},$$

where $\tau^*(\iota_v) = \tau^*_{\ell,\infty}(\iota_v)$ denotes the embedding $F \hookrightarrow \overline{\mathbb{Q}}_\ell$ induced by ι_v, ι_ℓ, and ι_∞. By purity for Hecke characters–immediate from the description of characters of $\mathbb{R}^\times$ and $\mathbb{C}^\times$–there is an integer w such that $k_{\tau^*(\iota_v)} + k_{\tau^*(\bar{\iota}_v)} = w$; moreover, by

[9] If F is Galois, we can rephrase this as $k_\tau + k_{\tau\circ c} = w$ for all choices of complex conjugation c on F.

Weil's descent result, $k_{\tau^*(\iota_v)}$ depends only on $\iota_v|_{F_{\mathrm{cm}}}$. The constraint on the weights follows.

Conversely, given a set of weights satisfying the purity constraint, we form a putative infinity-type $p_{\iota_v} = k_{\tau^*(\iota_v)}$ for a Hecke character of F; that this is in fact an achievable infinity-type follows from Weil. We form the associated geometric Galois character and then use the fact that, up to a finite-order twist, we can always extract n^{th} roots of ℓ-adic characters.

The mod n statement is a simple modification. For instance, to construct a character with given weights satisfying the purity constraint modulo n, proceed as follows. Choose a maximal set modulo c of embeddings $\tau_0 \colon F_{\mathrm{cm}} \hookrightarrow \mathbb{C}$. Choose lifts to $\mathbb{Q}$ of the congruence classes $\frac{k_{\tau_0}}{n}$ mod $\mathbb{Z}$, and declare $k_\tau = k_{\tau_0}$ for all τ lying above these τ_0. Then choose $w \in \mathbb{Z}$ lifting $k_{\tau_0} + k_{\tau_0 \circ c} \pmod{n}$, which is possible by hypothesis, and set $k_{\tau_0 \circ c} = w - k_{\tau_0}$. □

The same technique yields an easy example of the constraints on Galois characters over totally real fields:

LEMMA 2.3.17. *Suppose F is totally real, and $\hat{\psi} \colon \Gamma_F \to \overline{\mathbb{Q}}_\ell^\times$ is a character with all Hodge-Tate-Sen weights in $\mathbb{Q}$. Then all of these weights are equal. Conversely, for $x, d \in \mathbb{Z}$, there are global characters with all HTS weights equal to $\frac{x}{d}$.*

PROOF. For some integer d, all the HTS weights lie in $\frac{1}{d}\mathbb{Z}$, so $\hat{\psi}^d$ is geometric, and therefore corresponds to a type A_0 Hecke character. F is totally real, so $\hat{\psi}^d$ must be, up to a finite-order twist, an integer power of the cyclotomic character. □

It is worth remarking that the most general answer to the question 'what Galois characters $\hat{\psi} \colon \Gamma_F \to \overline{\mathbb{Q}}_\ell^\times$ exist' is essentially Leopoldt's conjecture.

Finally, we will need a lemma refining the construction of certain Galois characters over CM fields:

LEMMA 2.3.18. *Let F be a totally real field, and $\ell \neq 2$ a prime unramified in F. Let $L = KF$ be its composite with an imaginary quadratic field K in which ℓ is inert, and let ψ be any unitary type A Hecke character of L. Then:*

(1) *There exists a Galois character $\hat{\psi} \colon \Gamma_L \to \overline{\mathbb{Q}}_\ell^\times$ such that $\hat{\psi}^2$ corresponds to ψ^2 (which is type A_0).*
(2) *Moreover, the frobenius eigenvalues $\hat{\psi}(\mathrm{fr}_v)$ at all unramified v lie in $\mathbb{Q}^{\mathrm{cm}}$.*

PROOF. First, let m_{ι_v} as before denote the integers giving the infinity-type of ψ. Using the known algebraicity of ψ (Lemma 2.3.3) and the fixed embeddings ι_∞, ι_ℓ, we can define the ℓ-adic representation associated to ψ^2:

$$(\widehat{\psi^2})_\ell \circ \mathrm{rec}_L^{-1} \colon \mathbf{A}_L^\times / (\overline{L^\times L_\infty^\times}) \to \overline{\mathbb{Q}}_\ell$$

$$(x_w) \mapsto \prod_{w \nmid \ell\infty} \psi_w(x_w)^2 \prod_{w | \ell} \left(\psi_w(x_w)^2 \prod_{\tau \colon L_w \hookrightarrow \overline{\mathbb{Q}}_\ell} \tau(x_w)^{m_{\iota_\infty^* ,\ell(\tau)}} \right).$$

By the dual Grunwald-Wang theorem ([**Con11**, Remark 1.1]), to show that $(\widehat{\psi^2})_\ell$ is the square of some character $\Gamma_L \to \overline{\mathbb{Q}}_\ell^\times$, it suffices to check that for all places w of L, the above character is locally on $L_w^\times$ a square. This in turn immediately reduces to seeing whether, for each $w|\ell$, the character $\chi_w \colon L_w^\times \to \overline{\mathbb{Q}}_\ell^\times$ given by

$$\chi_w(x_w) = \prod_{\tau\colon L_w \hookrightarrow \overline{\mathbb{Q}}_\ell} \tau(x_w)^{m_{\iota^*_{\infty,\ell}(\tau)}}$$

is a square. Writing

$$L_w^\times = \langle \ell \rangle \times \mu_\infty(L_w^\times) \times (1 + \ell\mathcal{O}_w),$$

it suffices to check on each component of this factorization. On the $\langle \ell \rangle$ factor, this is clear (choose a square root of $\chi_w(\ell)$). On the $1 + \ell\mathcal{O}_w$ factor, the ℓ-adic logarithm, using our hypotheses that $\ell \neq 2$ is unramified, lets us define a single-valued square-root function, and thus extract a square root of $\chi_w|_{1+\ell\mathcal{O}_w}$. Now note that the product over $\tau\colon L_w \hookrightarrow \overline{\mathbb{Q}}_\ell$ is a product over pairs of complex-conjugate embeddings extending a given $F_v \hookrightarrow \overline{\mathbb{Q}}_\ell$, and $m_{\iota^*(\bar{\tau})} = m_{\overline{\iota^*(\tau)}} = -m_{\iota^*(\tau)}$ (ψ is unitary). $\mu_\infty(L_w^\times)$ is isomorphic to $\mathbb{Z}/(q-1)\mathbb{Z}$, where q is the order of the residue field at w. Let $\zeta \mapsto 1$ give the isomorphism, for ζ a primitive $(q-1)^{st}$ root of unity. Complex conjugation must identify to multiplication by an element $r \in \mathbb{Z}/(q-1)$ satisfying $r^2 \equiv 1 \mod q-1$; in particular, r is an odd residue class. Then $x \mapsto \tau(x)^{m_{\iota^*(\tau)}} \cdot (\tau \circ c)(x)^{-m_{\iota^*(\tau)}}$ takes ζ to an even power of $\tau(\zeta)$, hence $\chi_w|_{\mu_\infty(L_w^\times)}$ is a square.[10]

The second part of the lemma follows from the construction of $\hat{\psi}$ and the corresponding statement (Lemma 2.3.3) for the type A Hecke character ψ. □

2.4. Coefficients: Generalizing Weil's CM descent of type A Hecke characters

We now begin to pursue higher-rank analogues of two aspects of Weil's paper [**Wei56**]. This section extends to higher rank the important observation that type A Hecke characters of a number field F descend, up to a finite-order twist, to the maximal CM subfield F_{cm} (see Lemma 2.3.4). Of interest in its own right, this generalization also provides some of the intuition necessary for a general solution to Conrad's lifting question (Question 1.1.7).

2.4.1. Coefficients in Hodge theory. The guiding principle that allows us to reinterpret, and correspondingly generalize, Weil's result is that careful attention to the 'field of coefficients' of an arithmetic object can yield non-trivial information about its 'field of definition,' or that of certain of its invariants. We begin by recording in an abstract setting a lemma whose motivation is 'doing Hodge theory with coefficients.' Let k be a field of characteristic zero, and let F and E be extensions of k with F/k finite. Let D be a filtered $F \otimes_k E$-module that is free of rank d. Let E' be a Galois extension of E large enough to split F over k. For all (k-embeddings) $\tau\colon F \hookrightarrow E'$, we define as in Definition 2.2.3 the τ-labeled Hodge-Tate weights $\mathrm{HT}_\tau(D)$ as follows: for such τ we have orthogonal idempotents $e_\tau \in F \otimes_k E'$ giving rise to projections

$$F \otimes_k E' \xrightarrow[\sim]{(e_\tau)} \prod_\tau E'$$
$$x \otimes \alpha \mapsto (\tau(x)\alpha)_\tau .$$

[10]Note also that if ℓ is split in $L/\mathbb{Q}$, χ_w cannot be a square when $m_{\iota^*_{\infty,\ell}(\tau)}$ is odd.

The projection $e_\tau(D \otimes_E E')$ is then a filtered E'-vector space of dimension d, and we define $\mathrm{HT}_\tau(D)$ to be the collection of integers h (with multiplicity) such that

$$\mathrm{gr}^h(e_\tau(D \otimes_E E')) \neq 0.$$

In §2.2.1 we applied this formalism to the filtered $K \otimes_{\mathbb{Q}_\ell} \overline{\mathbb{Q}}_\ell$-module $\mathrm{D}_{\mathrm{dR}}(V)$, when V was a representation of Γ_K ($K/\mathbb{Q}_\ell$ finite) on a $\overline{\mathbb{Q}}_\ell$-vector space. In this section, however, we emphasize the general formalism: one should really keep in mind not ℓ-adic Hodge theory, but rather the de Rham realization of a motive over F with coefficients in E.

LEMMA 2.4.1. *Let D as above be a filtered $F \otimes_k E$-module that is free of some rank d. Then the set of integers $\mathrm{HT}_\tau(D)$ depends only on the $\mathrm{Gal}(E'/E)$-orbit of $\tau\colon F \hookrightarrow E'$. If E/k is Galois, then $\mathrm{HT}_\tau(D)$ depends only on the restriction $\tau|_{\tau^{-1}(E)}$.*

PROOF. We decompose $F \otimes_k E$ into a product of fields $\prod E_i$, writing $q_i\colon F \otimes_k E \twoheadrightarrow E_i$ for the quotient map. This yields filtered E_i-vector spaces D_i for all i. Any E-algebra homomorphism $\tau\colon F \otimes_k E \to E'$ factors through $q_{i(\tau)}$ for a unique index $i(\tau)$, and then $\mathrm{HT}_\tau(D)$ is simply the multi-set of weights of $D_{i(\tau)}$. This implies the first claim, since the $\mathrm{Gal}(E'/E)$-orbit of τ is simply all embeddings τ' for which $i(\tau') = i(\tau)$.

Now we assume E/k Galois and address the second claim. Having fixed a 'reference' embedding $\tau_0\colon F \hookrightarrow E'$, it makes sense to speak of $F \cap E := F \cap \tau_0^{-1}(E)$ inside E. Obviously E splits $F \cap E$ over k, so we can write

$$F \otimes_k E \cong F \otimes_{F\cap E} (F \cap E \otimes_k E) \cong \prod_\sigma F \otimes_{F\cap E,\sigma} E,$$

where the σ range over all embeddings $F \cap E \hookrightarrow E$. These factors $F \otimes_{F\cap E,\sigma} E$ are themselves fields, since E/k is Galois, and so this decomposition realizes explicitly the decomposition of $F \otimes_k E$ into a product of fields. In particular, by the first part of the Lemma, $\mathrm{HT}_\tau(D)$ depends only on $i(\tau)$, i.e. only on the $\sigma\colon F \cap E \hookrightarrow E$ to which τ restricts. □

DEFINITION 2.4.2. We call a filtered $F \otimes_k E$-module D *regular* if the multi-sets $\mathrm{HT}_\tau(D)$ are multiplicity-free.

A very simple application that we will use later is:

COROLLARY 2.4.3. *Let D be a filtered $L \otimes_k E$-module (globally free) for some finite extension L/F, and suppose L does not embed in $\widetilde{E}$. Then the restriction of scalars (image under the forgetful functor) $\mathrm{Res}_{L/F}(D)$ is not regular.*

2.4.2. CM descent. We will see that Weil's result (Lemma 2.3.4) is the conjunction of Lemma 2.4.1 with the fact that algebraic Hecke characters have CM 'fields of coefficients.' This observation will lead us naturally to the desired higher-rank generalization.

We recall some notation from §1.4. Let G be a connected reductive F-group. For each $v|\infty$ fix an isomorphism $\iota_v\colon \overline{F}_v \xrightarrow{\sim} \mathbb{C}$. For π an automorphic representation of $G(\mathbf{A}_F)$, we can write (in Langlands' normalization) the restriction to $W_{\overline{F}_v}$ of its L-parameter as

$$\mathrm{rec}_v(\pi_v)\colon z \mapsto \iota_v(z)^{\mu_{\iota_v}} \bar{\iota}_v(z)^{\nu_{\iota_v}} \in T^\vee(\mathbb{C}).$$

with $\mu_{\iota_v}, \nu_{\iota_v} \in X^\bullet(T)_{\mathbb{C}}$ and $\mu_{\iota_v} - \nu_{\iota_v} \in X^\bullet(T)$. For v imaginary, $\mu_{\bar{\iota}_v} = \nu_{\iota_v}$. Unless there is risk of confusion, we will omit reference to the embedding ι_v, writing $\mu_v = \mu_{\iota_v}$, etc. We also recall the following terminology, introduced in [**BG11**]:

DEFINITION 2.4.4. In the above notation, the automorphic representation π is L-algebraic if for all $v|\infty$, μ_v and ν_v lie in $X^\bullet(T)$; it is C-algebraic if μ_v and ν_v lie in $\rho + X^\bullet(T)$, where ρ denotes the half-sum of the positive roots (with respect to our fixed Borel containing T used to define a based root datum of G).

Our starting point is a result (and, more generally, conjecture) of Clozel. We take $G = \mathrm{GL}_n/F$ and collect the data of π's archimedean L-parameters as $M = \{\mu_\iota\}_\iota$, which we will loosely refer to as the 'infinity-type' of π. For $\sigma \in \mathrm{Aut}(\mathbb{C})$, define the action ${}^\sigma M = \{\mu_{\sigma^{-1}\iota}\}_\iota$. Recall that π is said to be *regular* if all roots of GL_n are non-vanishing on all of the co-characters μ_ι.

THEOREM 2.4.5 (Théorème 3.13 of [**Clo90**]). *Let F be any number field, and suppose π is a cuspidal, C-algebraic automorphic representation of $\mathrm{GL}_n(\mathbf{A}_F)$ that is moreover regular. Then π_f has a model over the fixed field $\mathbb{Q}(\pi_f) \subset \overline{\mathbb{Q}} \subset \mathbb{C}$ of all automorphisms $\sigma \in \mathrm{Aut}(\mathbb{C})$ such that ${}^\sigma\pi_f \cong \pi_f$, and $\mathbb{Q}(\pi_f)$ is in fact a number field. For each $\sigma \in \mathrm{Aut}(\mathbb{C})$ there is a cuspidal representation ${}^\sigma\pi$ with finite part ${}^\sigma\pi_f = \pi_f \otimes_{\mathbb{C},\sigma} \mathbb{C}$ and with infinity-type ${}^\sigma M$.*

More generally, Clozel conjectures this for any C-algebraic isobaric automorphic representation of $\mathrm{GL}_n(\mathbf{A}_F)$. Note that for GL_n, the notions of C- and L-algebraic are equivalent when n is odd, but they differ by a twist when n is even. I'm grateful to Kevin Buzzard for pointing out to me that this twist suffices to make Theorem 2.4.5 fail for L-algebraic representations; our interest, however, is a consequence (Proposition 2.4.8) with which this twist does not interfere.

HYPOTHESIS 2.4.6. *Throughout the rest of this section, we will assume that π is a cuspidal C- or L-algebraic automorphic representation of $\mathrm{GL}_n(\mathbf{A}_F)$ whose C-algebraic twists satisfy the conclusion of Clozel's theorem (if one such twist has this property, then all do).*

We first note an elementary but crucial refinement of Hypothesis 2.4.6:

COROLLARY 2.4.7. *Assume π satisfies Hypothesis 2.4.6. Continue to denote by $\mathbb{Q}(\pi_f) \subset \overline{\mathbb{Q}} \subset \mathbb{C}$ the fixed field of all $\sigma \in \mathrm{Aut}(\mathbb{C})$ such that ${}^\sigma\pi_f = \pi_f \otimes_{\mathbb{C},\sigma} \mathbb{C} \cong \pi_f$ (this may no longer be a number field). Then the field $\mathbb{Q}(\pi_f)$ is contained in $\mathbb{Q}^{\mathrm{cm}}$, the union of all CM extensions of $\mathbb{Q}$ inside $\overline{\mathbb{Q}}$.*

PROOF. The conclusion of the corollary remains unchanged if we replace π by its twist by $|\cdot|^{r/2}$ for some integer r, since for all primes p, $\mathbb{Q}(\sqrt{p})$ is a CM field. Therefore we may assume that π is C-algebraic, and that either π or $\pi|\cdot|^{1/2}$ is unitary. Using the L^2 inner product we deduce that in the former case, ${}^c\pi \cong \pi^\vee$, and in the latter case ${}^c\pi|\cdot| \cong \pi^\vee$ (note that for any automorphic representation Π, ${}^c\Pi$ makes sense as an automorphic representation–in contrast to twists by more general $\sigma \in \mathrm{Aut}(\mathbb{C})$, which seem only to exist for C-algebraic Π). Either way, we see that for all $\sigma \in \mathrm{Aut}(\mathbb{C})$, ${}^{c\sigma}\pi \cong {}^{\sigma c}\pi$, and therefore that the fixed field $\mathbb{Q}(\pi_f)$ is contained in $\mathbb{Q}^{\mathrm{cm}}$. □

We can now formulate (and prove under Hypothesis 2.4.6) the appropriate higher-rank generalization of Weil's result that type A Hecke characters of F, up to a finite-order twist, descend to F_{cm}.

PROPOSITION 2.4.8. *Assume π satisfies Hypothesis 2.4.6. Then the infinity-type of π descends to F_{cm}.*

PROOF. This statement too is invariant under twisting by powers of the absolute value, so we may assume π is C-algebraic. Let E denote the Galois closure (in $\mathbb{C}$) of $\mathbb{Q}(\pi_f)$; E is also a CM field. Extend each $\iota\colon F \hookrightarrow \mathbb{C}$ to an embedding $\tilde{\iota}\colon \widetilde{F} \hookrightarrow \mathbb{C}$ of the Galois closure $\widetilde{F}$. The image $\tilde{\iota}(\widetilde{F}) \subset \mathbb{C}$ does not depend on the extension. By the corollary, $\tilde{\iota}(\widetilde{F})$ is linearly disjoint from E over $\tilde{\iota}((\widetilde{F})_{\mathrm{cm}})$, and therefore we can find $\sigma \in \mathrm{Aut}(\mathbb{C}/E)$ restricting (via $\tilde{\iota}$) to any element we like of $\mathrm{Gal}(\widetilde{F}/(\widetilde{F})_{\mathrm{cm}})$; the collection of such σ acts transitively on the set of embeddings $F \hookrightarrow \mathbb{C}$ lying above a fixed $F \cap (\widetilde{F})_{\mathrm{cm}} = F_{\mathrm{cm}} \hookrightarrow \mathbb{C}$.[11] For any two such $\iota, \iota'\colon F \hookrightarrow \mathbb{C}$ (related by $\sigma \in \mathrm{Aut}(\mathbb{C}/E)$), we deduce, since ${}^{\sigma}\pi \cong \pi$ and hence ${}^{\sigma}M = M$, that $\mu_\iota = \mu_{\sigma^{-1}\iota} = \mu_{\iota'}$. □

Again, we remark that Hypothesis 2.4.6 is expected to hold for all cuspidal C- or L-algebraic automorphic representations of $\mathrm{GL}_n(\mathbf{A}_F)$. Any study of algebraic automorphic forms over non-CM fields will need to take this result into account.

REMARK 2.4.9.
- If π is regular, this yields an unconditional descent result for its infinity-type.
- Let F be a CM field and π be a regular L or C-algebraic cuspidal representation of $\mathrm{GL}_n(\mathbf{A}_F)$. Suppose that $\pi = \mathrm{Ind}_L^F(\pi_0)$ for some extension L/F. Then L is CM. This modest consequence of the proposition suggests that in the study of regular automorphic representations/motives/Galois representations over CM fields, we will never have to grapple with the (less well-understood) situation over non-CM fields. In §3.4.2 we discuss an abstract Galois-theoretic analogue.
- If we know more about $\mathbb{Q}(\pi_f)$ than that it is CM (an extreme case: $\mathbb{Q}(\pi_f) = \mathbb{Q}$), the proof of the proposition yields a correspondingly stronger result. For the related formalism, see §2.4.1.

REMARK 2.4.10. Let us also note that Proposition 2.4.8 is a 'seed' result (under Hypothesis 2.4.6); if we moreover assume functoriality (in a form that requires functorial transfers to be strong transfers–i.e. of archimedean L-packets–at infinity), then it implies:
- Let G be a connected reductive group over a totally imaginary field F, and let π be a cuspidal tempered L-algebraic automorphic representation of $G(\mathbf{A}_F)$. Then the infinity-type of π descends to F_{cm}.
- Let G and π be as above. Let $\widetilde{G} \supset G$ be a connected reductive F-group obtained by enlarging the center of G to a torus $\widetilde{Z}$ (see §3.1). Then the central character $\omega_\pi\colon Z_G(F)\backslash Z_G(\mathbf{A}_F) \to \mathbb{C}^\times$ extends to an L-algebraic Hecke character of $\widetilde{Z}(\mathbf{A}_F)$. (Compare Proposition 13.3.1.)

2.5. W-algebraic representations

The current section generalizes a second aspect of [**Wei56**], discussing a higher-rank analogue of the type A, but not necessarily A_0, condition, and begins to

---

[11] $G = \mathrm{Gal}(\widetilde{F}/F_{\mathrm{cm}})$ is generated by $H = \mathrm{Gal}(\widetilde{F}/F)$ and $H' = \mathrm{Gal}(\widetilde{F}/\widetilde{F}_{\mathrm{cm}})$, with H' normal. The set $\mathrm{Hom}_{F_{\mathrm{cm}}}(F, \mathbb{C})$ is permuted transitively by G, with H acting trivially, so for any such embedding x, $H'x = H'Hx = Gx = \mathrm{Hom}_{F_{\mathrm{cm}}}(F, \mathbb{C})$.

motivate its arithmetic significance. As always, let G be a connected reductive group over a number field F. We continue with the infinity-type notation of §2.4.

For many interesting questions about algebraicity of automorphic representations, and especially the interaction of algebraicity and functoriality, the framework of C and L-algebraic representations does not suffice. Motivated initially by Weil's study of type-A Hecke characters, we make the following definition:

DEFINITION 2.5.1. Let π be an automorphic representation of $G(\mathbf{A}_F)$. We say that π is W-algebraic if for all $v|\infty$, μ_{ι_v} and ν_{ι_v} in fact lie in $\frac{1}{2}X^\bullet(T)$.

EXAMPLE 2.5.2. For GL_1, a unitary W-algebraic representation is precisely a unitary Hecke character of type A in the sense of Weil ([**Wei56**]). Weil's type A characters also include arbitrary twists $|\cdot|^r$ for $r \in \mathbb{Q}$, since these also yield L-series with algebraic coefficients. The L- and C-algebraic representations are the Hecke characters of type A_0.

EXAMPLE 2.5.3. Consider a Hilbert modular form f on GL_2/F with (classical) weights $\{k_\tau\}_{\tau \colon F \hookrightarrow \mathbb{R}}$, where the positive integers k_τ are not all congruent modulo 2; these are called 'mixed-parity'. f then gives rise to an automorphic representation (in the unitary normalization, say) that is W-algebraic but neither L- nor C-algebraic. The previous example yields a special case: choose a quadratic CM (totally imaginary) extension L/F, and let ψ be a unitary Hecke character of L that is type A but not type A_0. Then the automorphic induction $\mathrm{Ind}_L^F \psi$ yields an example on GL_2/F. We will elaborate on the case of mixed-parity Hilbert modular forms in §2.6.

We are led to ask (compare §2.4 and Conjectures 3.1.5 and 3.1.6 of [**BG11**]):

QUESTION 2.5.4.

- Let π be a W-algebraic automorphic representation on G. Does the $G(\mathbf{A}_{F,f})$-module π_f have a model over $\overline{\mathbb{Q}}$ (or $\mathbb{Q}^{\mathrm{cm}}$)?. Note that we do not seek a model over a fixed number field. Alternatively, is π_v defined over $\overline{\mathbb{Q}}$ for almost all finite places v? By analogy with the terminology of [**BG11**], let us call the latter condition W-arithmetic, and ask whether some condition (which will not quite be W-algebraicity!) on infinity-types characterizes W-arithmeticity.
- Similarly, we can ask the CM descent question of §2.4 for W-algebraic representations. Example 2.5.6 below will be a cautionary tale: if non-induced cuspidal Π with infinity-types as in that example exist, then there is no evidence that they would satisfy the analogue of Clozel's algebraicity conjecture. If they do not exist, or if they miraculously still satisfy the conclusion of Clozel's theorem, then we could confidently extend the CM descent conjectures to the W-algebraic case.

For tori, the most optimistic conjecture holds:

LEMMA 2.5.5. *For arbitrary F-tori, W-algebraic implies W-arithmetic.*

PROOF. After squaring, this follows from the corresponding result for L-algebraic/L-arithmetic (§4 of [**BG11**]). □

Continuing with Example 2.5.2, let us emphasize that the W-algebraic condition does not capture all automorphic representations with algebraic Satake parameters; we nevertheless want to make a case for isolating this condition, rather than allowing arbitrary rational parameters $\mu_v, \nu_v \in X^\bullet(T)_\mathbb{Q}$, which would be the naïve analogue of type A. First, note that if a unitary π is tempered at infinity with real parameters $\mu_v, \nu_v \in X^\bullet(T)_\mathbb{R}$, then these parameters in fact lie in $\frac{1}{2}X^\bullet(T)$. In particular, the Ramanujan conjecture implies that all cuspidal unitary π on GL_n/F with real infinity-type are in fact W-algebraic. It is easy to construct non-cuspidal automorphic representations with rational Satake parameters that do not twist to W-algebraic representations, but the point is that all such examples will be degenerate, so W-algebraicity is the condition of basic importance. A more interesting question, concerning the difference between W, L, and C-algebraicity, is the following:

EXAMPLE 2.5.6. Let $G = \mathrm{GL}_4/\mathbb{Q}$ (for example; there are obvious analogues for any totally real field). Let $F/\mathbb{Q}$ be real quadratic, and let π as in Example 2.5.3 be a mixed parity (unitary) Hilbert modular representation. It cannot be isomorphic to its $\mathrm{Gal}(F/\mathbb{Q})$-conjugate, so $\Pi = \mathrm{Ind}_F^\mathbb{Q}(\pi)$ is cuspidal automorphic on $\mathrm{GL}_4/\mathbb{Q}$, and at infinity its Langlands parameter (restricted to $\mathbb{C}^\times$) looks like

$$z \mapsto \begin{pmatrix} (z/\bar{z})^{\frac{k_1-1}{2}} & 0 & 0 & 0 \\ 0 & (z/\bar{z})^{\frac{1-k_1}{2}} & 0 & 0 \\ 0 & 0 & (z/\bar{z})^{\frac{k_2-1}{2}} & 0 \\ 0 & 0 & 0 & (z/\bar{z})^{\frac{1-k_2}{2}}, \end{pmatrix}$$

where k_1 and k_2 are the classical weights of π at the two infinite places of F. This exhibits the 'parity-mixing' within a single infinite place, which implies the following:

LEMMA 2.5.7. *For any functorial transfer* ${}^L\mathrm{GL}_4 \xrightarrow{r} {}^L\mathrm{GL}_N$ *arising from an irreducible representation* r, *not equal to any power of the determinant, of* $\mathrm{GL}_4(\mathbb{C})$, *the transfer* $\mathrm{Lift}_r(\Pi)$ *cannot be* L*-algebraic.*

PROOF. We have stated the lemma globally, and therefore conjecturally, but of course the analogous archimedean statement (which is well-defined since the local transfer of Π_∞ via r is known to exist) is all we are really interested in. The proof is elementary highest-weight theory. □

We will see (in §2.6) that mixed-parity Hilbert modular representations are W-arithmetic, using the fact that they have L-algebraic functorial transfers; by contrast, if a representation Π with the infinity-type of the above example is W-arithmetic, it may be harder to establish. Of course, in the above example, Π is automorphically induced, so its W-arithmeticity is an immediate consequence of that of π. We are led to ask whether there exist non-automorphically induced examples of such Π, or more generally of cuspidal, non-induced Π on GL_n/F that exhibit the 'parity-mixing' within a single infinite place. The trace formula does not easily yield them, since such a Π is not the transfer from a classical group of a form that is discrete series at infinity (in these cases there is a parity constraint on regular elliptic parameters).

One further motivation for considering W-algebraic representations comes from studying the fibers of functorial lifts, and their algebraicity properties. For instance, given two mixed-parity Hilbert modular forms π_1 and π_2, with the (classical) weight of $\pi_{1,v}$ and $\pi_{2,v}$ having the same parity for each $v|\infty$, the tensor product $\Pi = \pi_1 \boxtimes \pi_2$

is L-algebraic. The π_i are not themselves twists of L-algebraic automorphic representations, so Π cannot be expressed as a tensor product of L-algebraic representations (this is proven in Corollary 2.6.3). This is, however, the 'farthest' from L-algebraic that the π_i's can be:

PROPOSITION 2.5.8. *Let Π be an L-algebraic cuspidal automorphic representation of $\mathrm{GL}_{nn'}(\mathbf{A}_F)$ for some number field F. Assume that $\Pi = \pi \boxtimes \pi'$ is in the image of $\mathrm{GL}_n \times \mathrm{GL}_{n'} \xrightarrow{\boxtimes} \mathrm{GL}_{nn'}$ for cuspidal automorphic representations π and π' of $\mathrm{GL}_n(\mathbf{A}_F)$ and $\mathrm{GL}_{n'}(\mathbf{A}_F)$.*[12] *If F is totally real or CM, then there are quasi-tempered (i.e., tempered up to a twist) W-algebraic automorphic representations π of $\mathrm{GL}_n(\mathbf{A}_F)$ and π' of $\mathrm{GL}_{n'}(\mathbf{A}_F)$ such that $\Pi = \pi \boxtimes \pi'$.*

PROOF. The first part of the argument is similar to, and will make use of, Lemme 4.9 of [**Clo90**] (Clozel's 'archimedean purity lemma'), which shows that Π_∞ itself is quasi-tempered. First assume F is totally imaginary. Let w_Π (the 'motivic weight') be the integer such that $\Pi|\cdot|^{-w_\Pi/2}$ is unitary, and write $z^{\mu_v}\bar{z}^{\nu_v}$ and $z^{\mu'_v}\bar{z}^{\nu'_v}$ for the L-parameters at $v|\infty$ of π and π'. Temperedness of Π implies that for all $i, j = 1, \ldots, n$,

$$\mathrm{Re}(\mu_{v,i} + \mu'_{v,j} + \nu_{v,i} + \nu'_{v,j}) = w_\Pi.$$

Fixing j and varying i, we find that $\mathrm{Re}(\mu_{v,i} + \nu_{v,i})$ is independent of i–call it w_{π_v}. Similarly, for $w_{\pi'_v} = w_\Pi - w_{\pi_v}$, we have $\mathrm{Re}(\mu'_{v,i} + \nu'_{v,i}) = w_{\pi'_v}$ for all i. For each of π and π' there is a unique real number r, r' such that each of $\pi|\cdot|^{-r/2}$, $\pi'|\cdot|^{-r'/2}$ is unitary. Thus, $z^{\mu_v - r/2}\bar{z}^{\nu_v - r/2}$ (likewise for π'_v) is a generic (by cuspidality) unitary parameter.[13] This implies (see the proof of Lemme 4.9 of [**Clo90**]) that these parameters are sums of the parameters of *unitary* characters and 2×2 complementary series blocks (what Clozel denotes $J(\chi, 1)$ and $J(\chi, \alpha, 1)$, where $\alpha \in (0, 1/2)$ is the complementary series parameter). Since $\mathrm{Re}(\mu_{v,i} + \nu_{v,i})$ is independent of i, we can immediately rule out any complementary series factors, and we deduce that each $z^{\mu_v - r/2}\bar{z}^{\nu_v - r/2}$ is a direct sum of (parameters of) unitary characters, and that $w_{\pi_v} = r$ is independent of v (and $w_{\pi'_v} = r'$), and $r + r' = w_\Pi \in \mathbb{Z}$. We may replace π by $\pi|\cdot|^{\frac{r'-r}{2}}$ and π' by $\pi'|\cdot|^{\frac{r-r'}{2}}$, so that still $\pi \boxtimes \pi' = \Pi$ but now each has half-integral 'motivic weight' (namely, $\frac{1}{2}w_\Pi \in \frac{1}{2}\mathbb{Z}$). In particular, $\mathrm{Re}(\mu_v)$, $\mathrm{Re}(\nu_v)$, $\mathrm{Re}(\mu'_v)$, and $\mathrm{Re}(\nu'_v)$ all now consist of half-integers (it is easily seen that all of these half-integers are moreover congruent modulo $\mathbb{Z}$). Now, $\mathrm{Im}(\mu_{v,i} + \mu'_{v,j}) = 0$ for all i, j, so there exists a $t_v \in \mathbb{R}$ such that

$$\mathrm{Im}(\mu_{v,i}) = \mathrm{Im}(\nu_{v,i}) = -\mathrm{Im}(\mu'_{v,i}) = -\mathrm{Im}(\nu'_{v,i}) = t_v$$

for all i. To conclude the proof, note that if there exists a Hecke character of F with infinity-componenets $|z|_{\mathbb{C}}^{it_v}$ for all $v|\infty$, then we can twist π and π' to arrange that both be W-algebraic (in fact, either both L-algebraic or both C-algebraic). Looking at central characters, we are handed a Hecke character with infinity-type

$$z \mapsto z^{\det(\mathrm{Re}(\mu_v)) + int_v}\bar{z}^{\det(\mathrm{Re}(\mu_v)) + int_v},$$

and by Corollary 2.3.9, the desired Hecke character exists if and only if the set of half-integers $\det(\mathrm{Re}(\mu_v)) := \sum_i \mathrm{Re}(\mu_{v,i})$ is the infinity-type of a type A Hecke

[12] To be precise, we want Π to be a weak lift that is also a strong lift at archimedean places.

[13] On GL_n, L-packets are singletons, and we can identify which parameters correspond to generic representations.

character of F, i.e. descends to F_{cm}. Of course, when F itself is CM, this is no obstruction, and the proof is complete.

Now assume F is totally real. By a global base-change argument (as in Clozel's Lemme 4.9), we may deduce temperedness of π and π' from the totally imaginary (or even CM) case, and to determine whether there are W-algebraic descents π and π', as usual it suffices to look at the archimedean L-parameters restricted to $W_{\overline{F}_v}$. The purity constraint (see Corollary 2.3.9) on the half-integers $\det(\mathrm{Re}(\mu_v))$ forces these to be independent of v, and we can argue as in the CM case to finish the proof. □

In the non-CM case, we can describe the infinity-types of possible Hecke characters; this gives rise to an explicit relation between the $\det(\mathrm{Re}(\mu_v))$ and t_v that needs to be satisfied if π and π' are to have W-algebraic twists. Conversely, note that the tensor product transfer (only known to exist locally, of course) $\mathrm{GL}_n \times \mathrm{GL}_m \xrightarrow{\boxtimes} \mathrm{GL}_{nm}$ clearly preserves L- or W-algebraicity (but not C-algebraicity).

In §3.1, we take up in detail the (algebraicity of the) fibers of the functorial transfer ${}^L\widetilde{G} \to {}^L G$, where $G \subset \widetilde{G}$ with torus quotient.

2.6. Further examples: The Hilbert modular case and $\mathrm{GL}_2 \times \mathrm{GL}_2 \xrightarrow{\boxtimes} \mathrm{GL}_4$

In this section we will elaborate on the example of mixed-parity Hilbert modular forms: we discuss W-arithmeticity in this context and make some initial forays on the Galois side. First, however, we recall known results on the automorphic tensor product $\mathrm{GL}_2 \times \mathrm{GL}_2 \xrightarrow{\boxtimes} \mathrm{GL}_4$ and provide a description of its fibers.

2.6.1. General results on the $\mathrm{GL}_2 \times \mathrm{GL}_2 \xrightarrow{\boxtimes} \mathrm{GL}_4$ functorial transfer. Ramakrishnan ([**Ram00**]) has proven the existence of the automorphic tensor product transfer in the case $\mathrm{GL}_2 \times \mathrm{GL}_2 \xrightarrow{\boxtimes} \mathrm{GL}_4$. Moreover, he establishes a cuspidality criterion:

THEOREM 2.6.1 (Ramakrishnan, [**Ram00**, Theorem M]). *Let π_1 and π_2 be cuspidal automorphic representations on GL_2/F. Then $\pi_1 \boxtimes \pi_2$ is automorphic. The cuspidality criterion divides into two cases:*

- *If neither π_i is an automorphic induction, $\pi_1 \boxtimes \pi_2$ is cuspidal if and only if π_1 is not equivalent to $\pi_2 \otimes \chi$ for some Hecke character χ of F.*
- *If $\pi_1 = \mathrm{Ind}_L^F(\psi)$ for a Hecke character ψ of a quadratic extension L/F, then $\pi_1 \boxtimes \pi_2$ is cuspidal if and only if $\mathrm{BC}_L(\pi_2)$ is not isomorphic to its own twist by $\psi^\theta\psi^{-1}$, where θ is the non-trivial automorphism of L/F.*

Here is the Galois-theoretic heuristic for the first part of the cuspidality criterion. Assume V_1 and V_2 are two-dimensional, irreducible, non-dihedral representations. Non-dihedral implies that each $\mathrm{Sym}^2 V_i$ is irreducible. Suppose that $V_1 \otimes V_2 \cong W_1 \oplus W_2$. Taking exterior squares, we get

$$(\mathrm{Sym}^2 V_1 \otimes \det V_2) \oplus (\det V_1 \otimes \mathrm{Sym}^2 V_2) \cong \wedge^2 W_1 \oplus (W_1 \otimes W_2) \oplus \wedge^2 W_2.$$

We may therefore assume $\dim W_1 = 3$, $\dim W_2 = 1$ (so rename W_2 as a character ψ), and thus

$$V_1 \otimes V_2(\psi^{-1}) \cong W_1(\psi^{-1}) \oplus 1,$$

which implies that V_1 and V_2 are twist-equivalent.

We now describe the fibers of this tensor product lift, beginning with a Galois heuristic. Suppose we have four irreducible 2-dimensional representations (over $\overline{\mathbb{Q}}_\ell$, say) satisfying

$$V_1 \otimes V_2 \cong W_1 \otimes W_2.$$

Taking exterior squares as above, we get

$$(\mathrm{Sym}^2 V_1 \otimes \det V_2) \oplus (\det V_1 \otimes \mathrm{Sym}^2 V_2) \cong (\mathrm{Sym}^2 W_1 \otimes \det W_2) \oplus (\det W_1 \otimes \mathrm{Sym}^2 W_2)$$

Assume V_1 is non-dihedral, so $\mathrm{Sym}^2 V_1$ is irreducible. Then we may assume

$$\mathrm{Sym}^2 V_1 \otimes \det V_2 \cong \mathrm{Sym}^2 W_1 \otimes \det W_2$$

and

$$\det V_1 \otimes \mathrm{Sym}^2 V_2 \cong \det W_1 \otimes \mathrm{Sym}^2 W_2.$$

Comparing determinants in this and the initial isomorphism, we find $\det V_1 \det V_2 = \det W_1 \det W_2$, and so (for $i = 1, 2$)

$$\mathrm{Ad}^0(V_i) \cong \mathrm{Ad}^0(W_i).$$

V_i and W_1 therefore give rise to isomorphic two-dimensional projective representations, and so, by the (not very) long exact sequence in continuous cohomology associated to

$$1 \to \overline{\mathbb{Q}}_\ell^\times \to \mathrm{GL}_2(\overline{\mathbb{Q}}_\ell) \to \mathrm{PGL}_2(\overline{\mathbb{Q}}_\ell) \to 1, \text{[14]}$$

V_i and W_i are twist-equivalent: we find a pair of characters ψ_i such that $V_1 \cong W_1(\psi_1)$ and $V_2 \cong W_2(\psi_2)$. If V_2 is also non-induced, then $\psi_2 = \psi_1^{-1}$, else $V_1 \otimes V_2$, and hence one of V_1 or V_2, is an induction.

Because the functorial transfers $\mathrm{GL}_2 \xrightarrow{\mathrm{Sym}^2} \mathrm{GL}_3$ ([**GJ78**]) and $\mathrm{GL}_4 \xrightarrow{\wedge^2} \mathrm{GL}_6$ ([**Kim03**]) are known, this argument works automorphically until the last step. To conclude a purely automorphic argument, we have to know the fibers of Ad^0. These were determined by Ramakrishnan as a consequence of his construction of the tensor product lift (and his cuspidality criterion):

THEOREM 2.6.2 (Ramakrishnan, [**Ram00**] Theorem 4.1.2). *Let π and π' be unitary cuspidal automorphic representations on GL_2/F such that $\mathrm{Ad}^0 \pi \cong \mathrm{Ad}^0 \pi'$. Then there exists a Hecke character χ of F such that $\pi \cong \pi' \otimes \chi$.*

It would be remiss not to mention the special nature of this fiber description: by [**LL79**], it is essentially multiplicity one for SL_2. In any case, from this and the preceding calculation, we conclude:

COROLLARY 2.6.3. *Let $\pi_1, \pi_2, \pi_1', \pi_2'$ be unitary cuspidal automorphic representations on GL_2/F satisfying $\pi_1 \boxtimes \pi_2 \cong \pi_1' \boxtimes \pi_2'$ (it suffices to assume this almost everywhere locally). Assume that $\pi_1 \boxtimes \pi_2$ is cuspidal (equivalently, satisfies Ramakrishnan's criterion). Up to reordering, we have $\mathrm{Ad}^0 \pi_i \cong \mathrm{Ad}^0 \pi_i'$ for $i = 1, 2$, and there are Hecke characters χ_i of F such that $\pi_i \cong \pi_i' \otimes \chi_i$ $(i = 1, 2)$. Moreover, $\chi_1 = \chi_2^{-1}$, so the fibers of the lift are just twists by Hecke characters.*

PROOF. If at least one, say π_1, of the π_i is non-dihedral, then it only remains to check $\chi_1 = \chi_2^{-1}$. Since $\pi_1 \boxtimes \pi_2 \cong (\pi_1 \boxtimes \pi_2) \otimes (\chi_1\chi_2)$, we deduce from a theorem of Lapid-Rogawski–which we elaborate on in Lemma 2.6.4 below–that either $\chi_1\chi_2 = 1$, or $\pi_1 \boxtimes \pi_2$ is an automorphic induction from the (quadratic or quartic) cyclic extension of F cut out by $\chi_1\chi_2$. In the latter case, there is a cyclic base change

[14]Note that the surjection admits a topological section.

L/F such that $\mathrm{BC}_L(\pi_1 \boxtimes \pi_2) = \mathrm{BC}_L(\pi_1) \boxtimes \mathrm{BC}_L(\pi_2)$ is non-cuspidal. If both π_1 and π_2 are non-dihedral, or if $\pi_2 = \mathrm{Ind}_K^F(\psi_2)$ with K a quadratic extension of F not contained in L, then the cuspidality criterion implies there exists a Hecke character ψ of L such that $\mathrm{BC}_L(\pi_1) \otimes \psi \cong \mathrm{BC}_L(\pi_2)$. Since π_1 is non-dihedral, ψ is invariant under $\mathrm{Gal}(L/F)$; for cyclic extensions, there is no obstruction to descending invariant Hecke characters, so ψ descends to a character of F that we also write as ψ, i.e. $\mathrm{BC}_L(\pi_1 \otimes \psi) \cong \mathrm{BC}_L(\pi_2)$. By cyclic descent (also Theorem B of [**LR98**]), π_1 and π_2 are twist-equivalent, contradicting cuspidality of $\pi_1 \boxtimes \pi_2$. We conclude that $\chi_1\chi_2 = 1$.

We resume the above argument when π_2 is dihedral with $K \subset L$. If $L = K$, then $\chi_1\chi_2$ is the quadratic character χ of $\mathrm{Gal}(K/F)$, so $\pi_2 = \pi_2' \otimes (\chi_1^{-1}\chi)$, and thus $\pi_2 = \pi_2' \otimes \chi_1^{-1}$ since $\pi_2 \otimes \chi = \pi_2$. If L/F is quartic, then the result of Lapid-Rogawski still implies $\pi_1 \boxtimes \pi_2 = \mathrm{Ind}_L^F(\psi)$, where ψ is now a Hecke character of L. Base-changing to L and comparing isobaric constituents, we again contradict the assumption that π_1 is non-dihedral.

Finally, we treat the case where both π_1 and π_2 are dihedral. Base-change implies that both π_1' and π_2' are dihedral as well, and moreover that we may assume that π_i and π_i' are both induced from the same field L_i, say by characters ψ_i and ψ_i'. Write σ_i for the non-trivial element of $\mathrm{Gal}(L_i/F)$. Then

$$\mathrm{Ind}_{L_1}^F \left(\psi_1 \otimes \mathrm{BC}_{L_1}(\mathrm{Ind}_{L_2}^F(\psi_2))\right) = \mathrm{Ind}_{L_1}^F \left(\psi_1' \otimes \mathrm{BC}_{L_1}(\mathrm{Ind}_{L_2}^F(\psi_2'))\right),$$

and up to replacing ψ_1 by $\psi_1^{\sigma_1}$, we may assume that on GL_2/L_1 we have

$$\frac{\psi_1}{\psi_1'} \otimes \mathrm{BC}_{L_1}(\mathrm{Ind}_{L_2}^F(\psi_2)) = \mathrm{BC}_{L_1}(\mathrm{Ind}_{L_2}^F(\psi_2')).$$

If $L_1 \neq L_2$, then we find

$$\mathrm{Ind}_{L_1L_2}^{L_1}(\frac{\psi_1}{\psi_1'} \cdot \psi_2) = \mathrm{Ind}_{L_1L_2}^{L_1}(\psi_2'),$$

and possibly replacing ψ_2 by its conjugate $\psi_2^{\sigma_2}$, we may assume that as Hecke characters of L_1L_2, $\psi_1/\psi_1' = \psi_2'/\psi_2$. Let α be the Hecke character of L_1 given by ψ_1/ψ_1'. This equality shows α is σ_1-invariant, and therefore descends to a Hecke character of F. Consequently, $\pi_1 = \pi_1' \otimes \alpha$, and $\pi_2 = \pi_2' \otimes \alpha^{-1}$ as automorphic representations on GL_2/F. The case $L_1 = L_2$ is similar, but easier, and we omit the details. □

The above corollary made use of a characterization of the image of cyclic automorphic induction (in the quadratic and quartic case). This characterization in the case of non-prime-degree does not follow from the results of [**AC89**], but it is now known thanks to work of Lapid-Rogawski. The following is stated without proof in [**LR98**], as an easy application of their Statement B[15]. We fill in the details:

PROPOSITION 2.6.4. *[Lapid-Rogawski] Let* Π *be a cuspidal automorphic representation on* GL_n/F*, and let* L/F *be a cyclic extension. Then* Π *is automorphically induced from* L *if (and only if)* $\Pi \cong \Pi \otimes \omega$ *for a Hecke character* ω *of* F *that cuts out the extension* L/F.

[15]The proof of which was conditional on versions of the fundamental lemma that are now known.

PROOF. Lapid-Rogawski establish the following, which implies the Proposition:[16]

THEOREM 2.6.5 (Statement B, part (*a*) of [**LR98**]). *Let E/F be a cyclic extension of number fields, with σ a generator of* $\mathrm{Gal}(E/F)$, *and let ω be a Hecke character of E. Denote by ω_F its restriction to $\mathbf{A}_F^\times \subset \mathbf{A}_E^\times$. Suppose Π is a cuspidal automorphic representation on GL_n/E satisfying $\Pi^\sigma \cong \Pi \otimes \omega$. Let K/F be the extension (necessarily of order dividing n) of F cut out by ω_F, and let $L = KE$. Then $E \cap K = F$, and $[K : F]$ divides n.*

Suppose ω cuts out L/F, cyclic of order m, and suppose m is divisible by a prime ℓ (and $m \neq \ell$). We see that $\Pi \cong \Pi \otimes \omega^{m/l}$ as well, and letting E/F be the (cyclic degree ℓ) extension cut out by $\omega^{m/l}$, the case of prime degree implies that $\Pi = \mathrm{Ind}_E^F(\pi)$ for some cuspidal representation π on GL_n/E. Moreover,

$$\mathrm{Ind}_E^F(\pi) \cong \mathrm{Ind}_E^F(\pi \otimes (\omega \circ N_{E/F})),$$

so the description of the fibers in the prime case implies π and $\pi \otimes (\omega \circ N_{E/F})$ are Galois-conjugate: writing τ for a generator of $\mathrm{Gal} L/F$ (or for its restriction to E), there exists an integer i such that

$$\pi^{\tau^i} \cong \pi \otimes (\omega \circ N_{E/F}).$$

Part (a) of Theorem 2.6.5 implies that the restriction of $\omega \circ N_{E/F}$ to $\mathbf{A}_F^\times/F^\times$, which is just ω^ℓ, cuts out an extension K/F that is linearly disjoint from E/F,[17] so in particular $(\omega \circ N_{E/F})^\ell$ is a non-trivial Hecke character of E. But now, since τ^ℓ is trivial on E, we can iterate the previous conjugation relation to deduce

$$\pi \cong \pi \otimes (\omega \circ N_{E/F})^\ell.$$

It now follows from induction on the degree $[L : F]$ that $\pi \cong \mathrm{Ind}_L^E(\pi_0)$ for some cuspidal π_0 on $\mathrm{GL}_{\frac{n}{[L:F]}}/L$, so we're done.

□

2.6.2. The Hilbert modular case. Now we take up the case of Hilbert modular forms, starting with an observation of Blasius-Rogawski ([**BR93**]): while a 'mixed parity' Hilbert modular representation π (Definition 2.6.6) does not itself twist to an L-algebraic representation, its base changes to every CM field will.[18] Since we work up to twist, we may assume the parameters of π at places $v|\infty$ have the form (restricted to $\overline{F}_v^\times \subset W_{F_v}$ and implicitly invoking ι as above)

$$z \mapsto \begin{pmatrix} (z/\bar{z})^{\frac{k_v-1}{2}} & 0 \\ 0 & (z/\bar{z})^{\frac{1-k_v}{2}} \end{pmatrix}.$$

DEFINITION 2.6.6. In the above notation, π is *mixed parity* if as v varies over all $v|\infty$, k_v takes both even and odd values.

---

[16]Which is in turn used to prove the more refined parts (*b*) and (*c*) of Statement B of [**LR98**].

[17]The point is that Π is cuspidal. For instance, in the simple case $\ell^2 \nmid m$, there's no need to appeal to the Lapid-Rogawski result: ℓ divides $\frac{m}{\ell} i$, so $\ell | i$, and then $\pi \cong \pi \otimes (\omega \circ N_{E/F})$.

[18]This observation of Blasius-Rogawski is the only time I have seen substantive use made of type A but not A_0 Hecke characters.

For a quadratic CM extension L/F, let $\psi\colon \mathbf{A}_L^\times/L^\times \to \mathbf{S}^1$ be a Hecke character whose infinity type at v is given by

$$z \mapsto (z/|z|)^{k_v-1}$$

for all v. Then $\mathrm{BC}_L(\pi)\otimes\psi$ is L-algebraic, with infinity type

$$z \mapsto \begin{pmatrix} (z/\bar{z})^{k_v-1} & 0 \\ 0 & 1 \end{pmatrix}$$

Given two such π_1 and π_2, both W- but not L-algebraic, and with $\Pi = \pi_1 \boxtimes \pi_2$ L-algebraic, Blasius-Rogawski (see Theorem 2.6.2 and Corollary 2.6.3 of [**BR93**]) can then associate an ℓ-adic Galois representation $\rho_{\Pi,\ell}$ to Π via the identity

$$\mathrm{BC}_L(\Pi) \cong (\mathrm{BC}_L(\pi_1)(\psi_1)) \boxtimes (\mathrm{BC}_L(\pi_2)(\psi_2)) \otimes (\psi_1\psi_2)^{-1},$$

since all three tensor factors on the right-hand side are L-algebraic with associated Galois representations. (To get a Galois representation for Π over F itself, they vary L and use the now-famous patching argument.) We pursue this a little farther, emphasizing the 'W-algebraic' and non-de Rham/geometric aspects of this construction.

PROPOSITION 2.6.7. *Let Π be the tensor product $\pi_1 \boxtimes \pi_2$ for π_i as above corresponding to mixed parity Hilbert modular forms (but π_1 and π_2 having common weight-parity at each infinite place). Assume that Π is cuspidal, and for simplicity assume that the π_i are non-dihedral.*

(i) *The ℓ-adic Galois representation $\rho_{\Pi,\ell}$ is Lie irreducible.*
(ii) *$\rho_{\Pi,\ell}$ is isomorphic to a tensor product $\rho_{1,\ell}\otimes\rho_{2,\ell}$ of two-dimensional continuous, almost everywhere unramified, representations of Γ_F. No twist of either $\rho_{i,\ell}$ is de Rham (or even Hodge-Tate), and $\rho_{\Pi,\ell}$ is not a tensor product of geometric Galois representations (although it is after every CM base change, by construction).*[19]
(iii) *$\mathrm{Ad}^0(\rho_{i,\ell})$ is geometric, corresponding to the L-algebraic $\mathrm{Ad}^0(\pi_i)$ (suitably ordering the representations). In particular, a mixed-parity Hilbert modular representation π gives rise to a geometric projective Galois representations $\bar{\rho}_{\pi,\ell}\colon \Gamma_F \to \mathrm{PGL}_2(\overline{\mathbb{Q}}_\ell)$; these are the representations predicted by Conjecture 3.2.2 of [**BG11**] for irreducible $\mathrm{SL}_2(\mathbf{A}_F)$-constituents of $\pi|_{\mathrm{SL}_2(\mathbf{A}_F)}$.*
(iv) *With the π_i normalized to be W-algebraic, they are also W-arithmetic (see Question 2.5.4).*

PROOF. To show $\rho_{\Pi,\ell}$ is a tensor product, observe that it is essentially self-dual, since

$$\Pi^\vee \cong \pi_1^\vee \boxtimes \pi_2^\vee \cong (\pi_1 \otimes \omega_{\pi_1}^{-1}) \boxtimes (\pi_2 \otimes \omega_{\pi_2}^{-1}) \cong \Pi \otimes (\omega_{\pi_1}\omega_{\pi_2})^{-1}.$$

On the Galois side, we deduce that $\rho_{\Pi,\ell}^\vee \cong \rho_{\Pi,\ell}\otimes\mu_\Pi^{-1}$, where μ_Π is the (geometric) Galois character corresponding to the (L-algebraic) Hecke character $\omega_{\pi_1}\omega_{\pi_2}$. Since after one of these quadratic base changes L/F $\rho_{\Pi,\ell}|_{\Gamma_L}$ is irreducible and orthogonal (rather than symplectic), we see that it (as Γ_F-representation) is orthogonal, i.e.

$$\rho_{\Pi,\ell}\colon \Gamma_F \to \mathrm{GO}_4(\overline{\mathbb{Q}}_\ell).$$

[19]Compare Proposition 3.3.8.

Writing μ for the multiplier character on GO_4, and observing that $\det^2 = \mu^4$ on this group, we find an exact sequence of algebraic groups

$$1 \to \mathbb{G}_m \to \mathrm{GL}_2 \times \mathrm{GL}_2 \xrightarrow{\boxtimes} \mathrm{GO}_4 \xrightarrow{\det \cdot \mu^{-2}} \{\pm 1\} \to 1.$$

The image of $\rho_{\Pi,\ell}$ is contained in the kernel GSO_4 of $\det \cdot \mu^{-2}$: for this it suffices to know that $\rho_{\Pi,\ell}$ is isomorphic to a tensor product after two disjoint quadratic base changes, for then $(\det \cdot \mu^{-2}) \circ \rho_{\Pi,\ell}$ is a character of Γ_F that is trivial on a set of primes of density strictly greater than $1/2$, and therefore it is trivial. We can then apply Proposition 2.1.4 to deduce that $\rho_{\Pi,\ell}$ lifts across $\mathrm{GL}_2(\overline{\mathbb{Q}}_\ell) \times \mathrm{GL}_2(\overline{\mathbb{Q}}_\ell) \to \mathrm{GSO}_4(\overline{\mathbb{Q}}_\ell)$, i.e. it is isomorphic to a tensor product $\rho_{1,\ell} \otimes \rho_{2,\ell}$. The same exact sequence implies that any expression of $\rho_{\Pi,\ell}$ as a tensor product has the form $(\rho_{1,\ell} \otimes \chi) \otimes (\rho_{2,\ell} \otimes \chi^{-1})$ for some continuous character $\chi \colon \Gamma_F \to \overline{\mathbb{Q}}_\ell^\times$. Lie-irreducibility also follows since the Zariski closure of the image of each $\rho_{i,\ell}$ contains SL_2 (otherwise $\rho_{\Pi,\ell}$, and hence Π, is automorphically induced), and thus the Zariski closure of the image of $\rho_{\Pi,\ell}$ contains SO_4, which acts Lie-irreducibly in its standard 4-dimensional representation.

Now we show all twists $\rho_{i,\ell} \otimes \chi$ are almost everywhere unramified, but none are geometric (they fail to be de Rham). This follows from the Galois-theoretic arguments of the next section, but here we give a more automorphic argument. Note that both the de Rham and almost everywhere unramified conditions can be checked after a finite extension. By construction, for any $L = FK$ for $K/\mathbb{Q}$ imaginary quadratic, there exist (L-algebraic) cuspidal automorphic representations τ_i ($i = 1, 2$) on GL_2/L with associated geometric Galois representations $\sigma_i \colon \Gamma_L \to \mathrm{GL}_2(\overline{\mathbb{Q}}_\ell)$ such that

$$\rho_{1,\ell}|_{\Gamma_L} \otimes \rho_{2,\ell}|_{\Gamma_L} \cong \rho_{\Pi,\ell}|_{\Gamma_L} \cong \sigma_1 \otimes \sigma_2.$$

This implies there is a Galois character $\chi \colon \Gamma_L \to \overline{\mathbb{Q}}_\ell^\times$ such that

$$\sigma_1(\chi) \cong \rho_{1,\ell}|_{\Gamma_L},$$
$$\sigma_2(\chi^{-1}) \cong \rho_{2,\ell}|_{\Gamma_L},$$

and every abelian ℓ-adic representation is unramified almost everywhere (easy), so each $\rho_{i,\ell}$ is almost everywhere unramified. We then have the equivalences

$$\rho_{i,\ell} \text{ is geometric} \iff \rho_{i,\ell}|_{\Gamma_L} \text{ is geometric} \iff \chi \text{ is geometric.}$$ [20]

But if χ is geometric, then (Theorem 2.3.13) there exists a type A_0 Hecke character $\chi_{\mathbf{A}}$ of L such that

$$\tau_1 \otimes \chi_{\mathbf{A}} \sim_{w,\infty} \rho_{1,\ell}$$
$$\tau_2 \otimes \chi_{\mathbf{A}}^{-1} \sim_{w,\infty} \rho_{2,\ell},$$

where $\sim_{w,\infty}$ is the notation of §1.4. $\mathrm{Gal}(L/F)$-invariance and cyclic base change then imply there are cuspidal automorphic representations $\tilde{\tau}_i$ on GL_2/F lifting $\tau_1 \otimes \chi_{\mathbf{A}}$ and $\tau_2 \otimes \chi_{\mathbf{A}}^{-1}$, and therefore $\mathrm{BC}_L(\tilde{\tau}_1 \boxtimes \tilde{\tau}_2) \cong \mathrm{BC}_L(\Pi)$. This implies $\tilde{\tau}_1 \boxtimes \tilde{\tau}_2$ and Π are twist-equivalent, and Corollary 2.6.3 ensures that (up to relabeling) π_i and $\tilde{\tau}_i$ are twist-equivalent. But $\tilde{\tau}_i$ is L-algebraic since τ_i and $\chi_{\mathbf{A}}$ are, yet it is easy to see (because π is mixed-parity and F is totally real) that no twist of π_i can be L-algebraic.

[20] Moreover, $\rho_{i,\ell}$ is Hodge-Tate $\iff$ $\rho_{i,\ell}|_{\Gamma_L}$ is Hodge-Tate $\iff$ χ is Hodge-Tate $\iff$ χ is geometric!

That, suitably ordered, $\mathrm{Ad}^0(\rho_{i,\ell})$ is a geometric Galois representation whose local factors match those of the L-algebraic cuspidal representation $\mathrm{Ad}^0(\pi_i)$ follows easily from the earlier $\wedge^2$ argument (see §2.6.1)and the fact that our forms (representations) are all non-dihedral.

Part (iv): Normalize π_i to be W-algebraic. Then $\pi_i \boxtimes \pi_i = \mathrm{Sym}^2(\pi_i) \boxplus \omega_{\pi_i}$ is L-algebraic, and by [**BR93**] it has Satake parameters in $\overline{\mathbb{Q}}$ at unramified primes. Writing $\{\alpha_{i,v}, \beta_{i,v}\}$ for the parameters of $\pi_{i,v}$ (when unramified), we conclude that $\alpha_{i,v}^2, \beta_{i,v}^2 \in \overline{\mathbb{Q}}$, hence $\alpha_{i,v}, \beta_{i,v} \in \overline{\mathbb{Q}}$. This implies $\pi_{i,v}$ has a model over $\overline{\mathbb{Q}}$ for all such v,[21] answering the weaker version of Question 2.5.4. In this case the stronger version (that $\pi_{i,f}$ is defined over $\overline{\mathbb{Q}}$) follows as well, by a check (omitted) at the ramified primes. □

Now we prove a more refined result in the case $\pi_1 = \pi_2$:

PROPOSITION 2.6.8. *Let π be a unitary, cuspidal, non-induced, mixed-parity Hilbert modular representation, so that $\Pi = \pi \boxtimes \pi$ has an associated Galois representation, the sum of Galois representations associated to the L-algebraic representations $\mathrm{Sym}^2(\pi)$ and ω_π. As before, there are ℓ-adic representations $\rho_{i,\ell}$ such that $\rho_{\Pi,\ell} = \rho_{1,\ell} \otimes \rho_{2,\ell}$.*

- *We can normalize the $\rho_{i,\ell}$ so that $\rho_{i,\ell}(\mathrm{fr}_v)$ has characteristic polynomial in $\overline{\mathbb{Q}}[X]$ for all unramified v. Moreover, its eigenvalues in fact lie in $\mathbb{Q}^{\mathrm{cm}}$, and are pure of some weight, which we may take to be zero.*
- *For quadratic CM extensions $L = KF$ with $K/\mathbb{Q}$ a quadratic imaginary field in which ℓ is inert, there is a $\rho_\ell \colon \Gamma_F \to \mathrm{GL}_2(\overline{\mathbb{Q}}_\ell)$ for which $\mathrm{Sym}^2(\rho_\ell) \sim_{w,\infty} \mathrm{Sym}^2(\pi)$. Later on (Corollary 2.7.8), we will see this is not possible over a totally real field.*
- *Nevertheless, we can choose ρ_ℓ and a finite-order character χ such that $\mathrm{Sym}^2(\rho_\ell) \colon \Gamma_F \to \mathrm{GO}_3(\overline{\mathbb{Q}}_\ell)$ is, viewing $\mathrm{GO}_3(\overline{\mathbb{Q}}_\ell)$ as the dual group $(\mathbb{G}_m \times \mathrm{SL}_2)^\vee(\overline{\mathbb{Q}}_\ell)$, associated to an automorphic representation $(\omega_\pi\chi, \pi_0)$ of $(\mathbb{G}_m \times \mathrm{SL}_2)(\mathbf{A}_F)$, as in 3.2.2 of [**BG11**], with π_0 any irreducible constituent of $\pi|_{\mathrm{SL}_2(\mathbf{A}_F)}$.*

PROOF. If we take $\pi_1 = \pi_2 = \pi$ in the unitary normalization (this will ensure $\pi \boxtimes \pi$ has 'motivic weight zero,' with an implicit–and provable–application of Ramanujan at infinity) and decompose the Galois representation $\rho_{\Pi,\ell}$ associated to $\Pi = \pi \boxtimes \pi$ as $\rho_1 \otimes \rho_2$, then the now-familiar $\wedge^2$ argument shows that $\mathrm{Ad}^0(\rho_1) \cong \mathrm{Ad}^0(\rho_2)$. We can therefore write $\rho_{\Pi,\ell} \cong \rho \otimes (\rho \otimes \chi)$ for some Galois character χ. This χ need not be a square, but we can normalize ρ so that χ is finite order (by Lemma 2.3.15). Now, the Ramanujan conjecture is known for Π[22], and of course for χ, so $\rho \otimes \rho$ is pure of weight zero (at unramified primes, say). Let α_v and β_v be frobenius eigenvalues of $\rho(\mathrm{fr}_v)$ at an unramified prime v, so that α_v^2 and β_v^2 (being eigenvalues of $\mathrm{Sym}^2(\rho)$) have absolute value one under any embedding $\overline{\mathbb{Q}}_\ell \hookrightarrow \mathbb{C}$. The same then holds for α_v and β_v, so ρ itself is pure of weight zero. Therefore the frobenius eigenvalues of ρ lie in $\mathbb{Q}^{\mathrm{cm}}$.

We conclude by showing that, after certain CM base-changes, we can take the finite-order character χ to be trivial. Restricting to $L = KF$ as above, we can

[21] Note that the fields of definition of π_v and its Satake parameters may be different; but if one is algebraic, then both are.

[22] Of course, Π is not cuspidal, so literally this is for $\mathrm{Sym}^2(\pi)$ and ω_π.

find a type A Hecke character ψ of L such that $\pi \otimes \psi$ is L-algebraic, and then letting $\widehat{(\psi^2)}_\ell$ be the geometric ℓ-adic representation associated to ψ^2, we claim that, possibly replacing ψ by ψ^{-1}, there exists a Galois character λ of L such that

$$\rho|_{\Gamma_L} \otimes \lambda \cong \rho|_{\Gamma_L} \otimes (\chi \widehat{(\psi^2)}_\ell \lambda^{-1}).$$

Over L, $\pi \cdot \psi^{-1}$ has associated Galois representation r, and then $r \cdot \widehat{(\psi^2)}_\ell$ corresponds to $\pi \cdot \psi$. Then there exists a λ such that either $r = \rho \cdot \lambda$ and $r \cdot \widehat{(\psi^2)}_\ell = \rho(\chi\lambda^{-1})$, or $r = \rho \cdot (\chi\lambda^{-1})$ and $r \cdot \widehat{(\psi^2)}_\ell = \rho \cdot \lambda$. Either way, plugging one equation into the other we get the claim.

ρ is not an induction, so $\chi|_{\Gamma_L} = \lambda^2 \widehat{(\psi^2)}_\ell^{-1}$. In particular, $\chi|_{\Gamma_L}$ is a square if and only if $\widehat{(\psi^2)}_\ell$ admits a Galois-theoretic square-root. This was discussed in Lemma 2.3.18 above: it need not always be the case (for instance, if ℓ is split in $L/\mathbb{Q}$), but it is if ℓ is inert in K. Choosing such an $L = KF$, we can now find $\chi_L \colon \Gamma_L \to \overline{\mathbb{Q}}_\ell$ such that

$$\rho_{\Pi,\ell}|_{\Gamma_L} \cong (\rho|_{\Gamma_L} \otimes \chi_L)^{\otimes 2}.$$

Here $\chi_L = \lambda\hat{\psi}$ for $\hat{\psi}$ a square-root of $\widehat{(\psi^2)}_\ell$, and we're done.

For the final point, note that the inclusion $\mathbb{G}_m \times \mathrm{SO}_3 \to \mathrm{GO}_3$ is an isomorphism, with inverse map $g \mapsto (\frac{\det}{\mu}(g), \frac{\det}{\mu}(g)^{-1}g)$. In particular, $\mathrm{Sym}^2 \colon \mathrm{GL}_2 \to \mathrm{GO}_3$ becomes $g \mapsto (\det(g), \mathrm{Ad}^0(g)$ in these coordinates. The $\mathrm{SL}_2(\mathbf{A}_F)$-constituents of π have Langlands parameters corresponding to the projectivization of those of π, and since $\mathrm{Ad}^0 \colon \mathrm{GL}_2 \to \mathrm{SO}_3 \cong \mathrm{PGL}_2$ is just the quotient map $\mathrm{GL}_2 \twoheadrightarrow \mathrm{PGL}_2$, the claim follows. □

REMARK 2.6.9. In particular, the Proposition tells us that there are pure ℓ-adic Galois representations no twist of which are geometric. They do not, however, live in compatible systems.

These propositions would have no content if the only examples of mixed parity Hilbert modular forms were the inductions described in Example 2.5.3; we end this section by showing

LEMMA 2.6.10. *Over any totally real field, non-CM mixed parity Hilbert modular forms exist.*

PROOF. Considering the (semi-simple, connected) $\mathbb{Q}$-group

$$G = \mathrm{Res}_{F/\mathbb{Q}}(\mathrm{SL}_2/F),$$

we can apply Theorem 1B from §4 of Clozel's paper [**Clo86**]. We fix a discrete series representation π_∞ of $G(\mathbb{R}) \cong \prod_{\tau \colon F \hookrightarrow \mathbb{R}} \mathrm{SL}_2(\mathbb{R})$ corresponding to the desired mixed-parity infinity-type. Then fix an auxiliary supercuspidal level $K_{p_0} \subset G(\mathbb{Q}_{p_0})$ for some prime p_0. At some other finite prime p, let π_p be a fixed (twist of) Steinberg representation. Finally, let S be a set of finite primes (disjoint from p_0, p) at which we will let the level go to infinity in the manner of Clozel's paper (written $K_S \to 1$). Letting K be any fixed compact open subgroup away from S, p, p_0, ∞, Clozel proves that

$$\liminf_{K_S \to 1} \left[\mathrm{vol}(K_S) \cdot \mathrm{mult}\left(\pi_\infty \otimes \pi_p, L^2_{\mathrm{cusp}}(G(\mathbb{Q})/G(\mathbf{A}_F))^{K_{p_0} \times K_S \times K}\right)\right] > 0.$$

(He also determines an explicit but non-optimal constant.) This suffices for the Lemma: it implies the existence of (infinitely many) cuspidal automorphic representations on G that are Steinberg at p, and therefore not automorphically induced, and have the desired mixed parity at infinity. □

REMARK 2.6.11. In [**Shi12**], Shin derives, as a corollary of very general existence results for automorphic representations, an exact limit multiplicity formula for *cohomological* Hilbert modular forms. Unfortunately, this excludes precisely the mixed-parity forms we are interested in. It seems, however, that his methods should extend to cover all discrete series infinity types.

2.6.3. A couple of questions. In Proposition 2.6.7, the key point was an understanding of which type A Hecke characters–or their Galois analogues–could exist. I would like to raise here a couple questions in higher rank about the existence of automorphic forms with certain infinity-types.

QUESTION 2.6.12. Are there automorphic representations π on GL_2/F (F totally real) such that at two infinite places v_1 and v_2,

- π_{v_1} is discrete series and π_{v_2} is limit of discrete series, but excluding the easily-constructed dihedral cases, as in Example 2.5.3? Clozel's result does not apply to this case; but after the initial writing of this paper, Kevin Buzzard informed me that one such example, for $F = \mathbb{Q}(\sqrt{5})$, has been produced by computer calculation in [**MS14**];
- π_{v_1} is discrete series (or limit of discrete series), and π_{v_2} is principal series (looks like a Maass form)?
- as in the previous item, but with algebraic infinity-type (so a 'Plancherel density'-type result will not suffice)?

2.7. Galois lifting: Hilbert modular case

2.7.1. Outline. We continue to elaborate on the examples of the previous section, now turning to some preliminary cases of a problem raised by Conrad in [**Con11**]. Recall from the introduction that he addresses lifting problems

$$\begin{array}{ccc} & & \widetilde{H}(\overline{\mathbb{Q}}_\ell) \\ & \overset{\tilde{\rho}}{\nearrow} & \downarrow \\ \Gamma_F & \xrightarrow{\rho} & H(\overline{\mathbb{Q}}_\ell) \end{array},$$

where $\widetilde{H} \twoheadrightarrow H$ is a surjection of linear algebraic groups with central kernel, and the key remaining question is:

QUESTION 2.7.1. Suppose that the kernel of $\widetilde{H} \twoheadrightarrow H$ is a torus. If ρ is geometric, when does there exist a geometric lift $\tilde{\rho}$?

We will address this question in stages. Here is an outline of the coming sections:

- First we address, in the regular case, with F totally real, Conrad's question for lifting across $\mathrm{GL}_2(\overline{\mathbb{Q}}_\ell) \to \mathrm{PGL}_2(\overline{\mathbb{Q}}_\ell)$. Modulo the difference between potential automorphy and automorphy, we find that all examples of ρ not having geometric lifts are accounted for by mixed-parity Hilbert modular forms (a converse to Propositions 2.6.7 and 2.6.8).

- To give a higher-rank example in which potential automorphy theorems still allow us to link the automorphic and Galois sides, we then carry out in §2.8 an analogous discussion for $\mathrm{GSpin}_{2n+1}(\overline{\mathbb{Q}}_\ell) \to \mathrm{SO}_{2n+1}(\overline{\mathbb{Q}}_\ell)$ (§2.2 introduces the necessary background about Sen operators and 'labeled Hodge-Tate-Sen weights' in ℓ-adic Hodge theory; in the current section, we use more elementary terminology, although some manipulations are justified by the general theory).
- Before proceeding to the general solution (§3.2) of Conrad's question, we (§3.1) discuss the automorphic analogue. This is in some sense more intuitive, and it motivates the Galois-theoretic solution. Note that in general we have neither constructions of automorphic Galois representations nor (potential) automorphy theorems, so we cannot unconditionally bridge the automorphic-Galois chasm. Proving a 'modular lifting theorem' from H to $\widetilde{H}$ in this context is an important remaining question.

2.7.2. $\mathrm{GL}_2(\overline{\mathbb{Q}}_\ell) \to \mathrm{PGL}_2(\overline{\mathbb{Q}}_\ell)$. In this subsection we prove:

THEOREM 2.7.2. *Let F be totally real. Suppose $\rho\colon \Gamma_F \to \mathrm{PGL}_2(\overline{\mathbb{Q}}_\ell)$ is a geometric representation with no geometric lift to $\mathrm{GL}_2(\overline{\mathbb{Q}}_\ell)$. Then*

- *There exists a lift $\tilde{\rho}\colon \Gamma_F \to \mathrm{GL}_2(\overline{\mathbb{Q}}_\ell)$ such that for all CM L/F, $\tilde{\rho}|_{\Gamma_L}$ is the twist of a geometric Galois representation (in particular, $\rho|_{\Gamma_L}$ has a geometric lift, which ought to be automorphic).*

Now assume moreover that $\mathrm{Ad}(\rho)\colon \Gamma_F \to \mathrm{SO}_3(\overline{\mathbb{Q}}_\ell) \subset \mathrm{GL}_3(\overline{\mathbb{Q}}_\ell)$ satisfies the hypotheses of (the potential automorphy theorem) Corollary 4.5.2 of [**BLGGT14**][23] *Then:*

- *After a totally real base-change F'/F, $\mathrm{Ad}(\rho)$ is automorphic, and more precisely*
- *we may normalize $\tilde{\rho}$ such that $\mathrm{Sym}^2(\tilde{\rho})|_{\Gamma_{F'}}$ is a geometric Galois representation corresponding to $\mathrm{Sym}^2(\pi) \otimes \chi$ for some mixed parity Hilbert modular representation π on GL_2/F and* non-trivial *finite order character χ. Furthermore, $\tilde{\rho}$ is totally odd.*

REMARK 2.7.3. If $\mathrm{Ad}^0(\rho)$ is not regular, there are two possibilities: either it fails to be regular everywhere, in which case Fontaine-Mazur conjecture that ρ (up to twist) has finite image. Such ρ always have finite-image (hence geometric) lifts by Tate's theorem (Proposition 2.1.4). If ρ is totally odd, and $\mathrm{Ad}^0(\rho)$ exhibits a mixture of regular and irregular behavior, it should be related to the existence of a mixed-parity Hilbert modular form that is limit of discrete series at some infinite places, and genuine discrete series at others. I know of no such examples not arising from Hecke characters. The question of whether there exist non-dihedral ρ that are even at some and odd at other infinite places is related to Question 2.6.12.

We now make our first use of non-Hodge-Tate Galois representations; recall that in §2.2 we have described the theory of (non-integral) Hodge-Tate-Sen weights. In fact, a reader unfamiliar with this general theory will have no difficulty following the arguments of the current section.

---

[23]Essentially: there is a lift $\tilde{\rho}$ of ρ such that $(\tilde{\rho} \mod \ell)|_{\Gamma_{F(\zeta_\ell)}}$ is irreducible and $\tilde{\rho}$ is regular (if true for one lift, this is true for all), and ℓ is sufficiently large.

LEMMA 2.7.4. *For any number field F and any geometric $\rho\colon \Gamma_F \to \mathrm{PGL}_2(\overline{\mathbb{Q}}_\ell)$, there is a lift $\tilde{\rho}\colon \Gamma_F \to \mathrm{GL}_2(\overline{\mathbb{Q}}_\ell)$ of ρ whose Hodge-Tate-Sen (HTS) weights at all $v|\ell$ belong to $\frac{1}{2}\mathbb{Z}$. In this case, $\mathrm{Sym}^2(\tilde{\rho})$ and $\det(\tilde{\rho})$ are geometric.*

PROOF. Proposition 2.1.4 ensures that we can find some lift $\tilde{\rho}$. By assumption, $\mathrm{Ad}^0(\tilde{\rho})$ is geometric, so

$$\tilde{\rho} \otimes \tilde{\rho} \otimes \det(\tilde{\rho})^{-1} \cong \mathrm{Ad}^0(\tilde{\rho}) \oplus \mathbf{1}$$

is geometric. We ask whether $\det(\tilde{\rho})$ is a square. There at least exists a Galois character $\psi\colon \Gamma_F \to \overline{\mathbb{Q}}_\ell^\times$ such that $\det(\tilde{\rho}) = \psi^2\chi$ with χ finite order (Lemma 2.3.15). Thus

$$[\tilde{\rho} \otimes \psi^{-1}]^{\otimes 2} \cong (\mathrm{Ad}^0(\tilde{\rho}) \otimes \chi) \oplus \chi$$

is also geometric, and in particular Hodge-Tate. Replacing $\tilde{\rho}$ with the new lift $\tilde{\rho} \otimes \psi^{-1}$ we conclude that $\mathrm{Sym}^2(\tilde{\rho})$ and $\det\tilde{\rho}$ are Hodge-Tate. The theorem of Wintenberger and Conrad (Theorem 2.1.6) then proves they must be de Rham, since the original projective ρ was. Any lift $\tilde{\rho}$ is almost everywhere unramified by Lemma 5.3 of [**Con11**], so the proof is complete. □

We now proceed in the same spirit as in the construction of W-algebraic forms via type A Hecke characters. The following lemma is an *ad hoc* version of Theorem 3.2.7; it may help in navigating that rather formal proof:

LEMMA 2.7.5. *Consider $\tilde{\rho}\colon \Gamma_F \to \mathrm{GL}_2(\overline{\mathbb{Q}}_\ell)$ as produced by Lemma 2.7.4. Let $L/\mathbb{Q}$ be a quadratic CM extension. Then for all such L, the restriction of $\tilde{\rho}$ to Γ_L is the twist of a geometric Galois representation.*

PROOF. Recall that our fixed embeddings $\overline{\mathbb{Q}} \overset{\iota_\infty}{\hookrightarrow} \mathbb{C}$ and $\overline{\mathbb{Q}} \overset{\iota_\ell}{\hookrightarrow} \overline{\mathbb{Q}}_\ell$ yield a map

$$\iota_{\infty,\ell}^*\colon \bigcup_{v|\ell} \mathrm{Hom}_{\mathbb{Q}_\ell}(F_v, \overline{\mathbb{Q}}_\ell) \to \mathrm{Hom}_{\mathbb{Q}}(F, \mathbb{C}).$$

$\mathrm{Sym}^2(\tilde{\rho})$ is de Rham, so for each place $v|\ell$ of F and each $\tau\colon F_v \hookrightarrow \overline{\mathbb{Q}}_\ell$, it has τ-labeled Hodge-Tate weights[24]

$$\mathrm{HT}_\tau\left(\mathrm{Sym}^2(\tilde{\rho}|_{\Gamma_{F_v}})\right) = \{A_\tau, \frac{A_\tau + B_\tau}{2}, B_\tau\},$$

where the A_τ and B_τ are integers, necessarily having the same parity. We can distinguish the subset

$$\{A_\tau, B_\tau\} \subset \mathrm{HT}_\tau\left(\mathrm{Sym}^2(\tilde{\rho}|_{\Gamma_{F_v}})\right)$$

(trivially in the non-regular case, easily otherwise). It follows that $\tilde{\rho}|_{\Gamma_{F_v}}$ is Hodge-Tate (hence de Rham) if and only if all A_τ and B_τ are even. Consider the map $\iota_{\infty,\ell}^*$ for L as well as F; the two versions are compatible, so conjugate archimedean embeddings $\iota, \bar{\iota}\colon L \hookrightarrow \mathbb{C}$ pull back to embeddings $L \hookrightarrow \overline{\mathbb{Q}}_\ell$ that lie above a common $\tau\colon F \subset F_v \hookrightarrow \overline{\mathbb{Q}}_\ell$. To each τ we then unambiguously associate the parity $\epsilon_\tau \in \{0,1\}$ of A_τ (and B_τ), and at the unique archimedean place $v_{\infty,\tau}$ of L above $\iota_{\infty,\ell}^*(\tau)$ we define a character

$$\begin{aligned}\psi_{v_{\infty,\tau}}\colon L_{v_{\infty,\tau}}^\times &\to \mathbb{C}^\times\\ z &\mapsto \iota(z)^{\frac{k_\tau}{2}}\bar{\iota}(z)^{-\frac{k_\tau}{2}},\end{aligned}$$

where k_τ is any integer with parity ϵ_τ.

[24] By duality, or by the Hodge-Tate-Sen theory.

L is a CM field, so Weil ([**Wei56**]) tells us that there exists a type A Hecke character ψ of L with components at archimedean places given by these $\psi_{v_{\infty,\tau}}$. We wish to twist $\tilde{\rho}|_{\Gamma_L}$ by a Galois realization of ψ, shifting all of its HTS weights to be integers (recall this can be checked by looking at $\mathrm{Sym}^2(\tilde{\rho}|_{\Gamma_L})$), but care is needed: there is no Galois representation canonically associated to a type A (not A_0) Hecke character. Instead, we associate (canonically, having specified ι_∞ and ι_ℓ) an ℓ-adic representation $(\widehat{\psi^2})_\ell$ to the type A_0 Hecke character ψ^2, and then we non-canonically extract, up to a finite-order character, a (Galois-theoretic, non-geometric) square root $\hat{\psi}$. Then $\tilde{\rho}|_{\Gamma_L} \otimes \hat{\psi}$ is Hodge-Tate, since the analogues of the A_τ and B_τ, but now for $\mathrm{Sym}^2(\tilde{\rho}|_{\Gamma_L} \otimes \hat{\psi}) = \mathrm{Sym}^2(\tilde{\rho}|_{\Gamma_L}) \otimes (\widehat{\psi^2})_\ell$, are all even. Thus $\tilde{\rho}|_{\Gamma_L} \otimes \hat{\psi}$ is geometric (again by Theorem 2.1.6). □

This completes the proof of the first part of Theorem 2.7.2. We need deeper tools to make further progress. Let $\tilde{\rho}$ be the lift as above, with both $\mathrm{Sym}^2(\tilde{\rho})$ and $\det(\tilde{\rho})$ geometric. $\mathrm{Sym}^2(\tilde{\rho})$ factors through $\mathrm{GO}_3(\overline{\mathbb{Q}}_\ell)$ with totally even multiplier character $\det(\tilde{\rho})^2$. We now make further assumptions on $\tilde{\rho}$ (or ρ) in order to apply a potential automorphy theorem:

HYPOTHESIS 2.7.6. (1) *Assume* $\ell > 7$.

(2) *Assume* $\mathrm{Sym}^2(\tilde{\rho})$ *is 'potentially diagonalizable' and regular at all* $v|\ell$. *For instance, we require that for each* $v|\ell$ *it be crystalline with* τ*-labeled Hodge-Tate weights distinct and falling in the Fontaine-Laffaille range for all* $\tau\colon F_v \hookrightarrow \overline{\mathbb{Q}}_\ell$.

(3) *Assume that the reduction mod* ℓ *of* $\mathrm{Sym}^2(\tilde{\rho})|_{\Gamma_{F(\zeta_\ell)}}$ *is irreducible. (We have taken* $\ell \neq 2$, *so this is equivalent to the analogous assumption for* $\tilde{\rho}$.)

Then we can can apply Theorem C of [**BLGGT14**] to deduce that after base-change to some totally real field F', $\mathrm{Sym}^2(\tilde{\rho})|_{\Gamma_{F'}}$ is automorphic, corresponding to a RAESDC automorphic representation Π on GL_3/F'. Let us abusively denote by $\det(\tilde{\rho})$ the Hecke character (of type A_0) corresponding to $\det(\tilde{\rho})$. Then $\Pi \otimes \det(\tilde{\rho})^{-1}$ is self-dual with trivial central character, and (by, for instance, [**Art13**]–details, in greater generality, appear in Lemma 2.8.6–although in this case the result is older) there exists a cuspidal π on GL_2/F' such that $\mathrm{Ad}^0(\pi) \cong \Pi \otimes \det(\tilde{\rho})^{-1}$. We may assume that ω_π has finite order (applying Lemma 2.3.6 and Proposition 3.1.4, since F' is totally real), because the descent is originally to SL_2/F', and we then have control over the choice of extension to GL_2/F'.[25]

LEMMA 2.7.7. *Such a* π *on* GL_2/F' *satisfying* $\mathrm{Ad}^0(\pi) \cong \Pi \otimes \det(\tilde{\rho})^{-1}$ *is necessarily a mixed parity Hilbert modular representation.*

PROOF. It is clear from the archimedean L-parameters that π is W-algebraic. If it were L-algebraic, there would be a corresponding Galois representation ρ_π[26] with $\mathrm{Ad}^0(\rho_\pi) \cong \mathrm{Ad}^0(\tilde{\rho})$, and thus ρ_π and $\tilde{\rho}|_{\Gamma_{F'}}$ would be twist-equivalent. Lemma 2.3.17 then implies $\tilde{\rho}$ has all integral or all half-integral Hodge-Tate weights, which contradicts the assumption that ρ has no geometric lift (note that both F' and F are totally real). Similarly, if π were C-algebraic, then there would be a geometric representation corresponding to $\pi|\cdot|^{1/2}$, and we can use the same argument. We conclude that π must be mixed parity. □

[25] We only sketch this here, because these matters will be discussed in greater detail and generality in §3.1.

[26] By [**BR93**].

COROLLARY 2.7.8. *For any* $\rho\colon \Gamma_F \to \mathrm{PGL}_2(\overline{\mathbb{Q}}_\ell)$ *as in Theorem* 2.7.2, *and satisfying the auxiliary potential automorphy Hypothesis* 2.7.6, *there exists a lift* $\tilde{\rho}\colon \Gamma_F \to \mathrm{GL}_2(\overline{\mathbb{Q}}_\ell)$ *and, after a totally real base change* F'/F, *a mixed-parity Hilbert modular representation* π *on* GL_2/F', *such that*

$$\mathrm{Sym}^2(\tilde{\rho})|_{\Gamma_{F'}} \sim_{w,\infty} \mathrm{Sym}^2(\pi) \otimes \chi$$

for some finite order Hecke character χ *of* F'. *Moreover, any lift* $\tilde{\rho}$ *is totally odd, and the character* χ *is necessarily non-trivial. In contrast, restricting to* $L' = F'K$ *where* $K/\mathbb{Q}$ *is a quadratic imaginary extension in which* ℓ *is inert, we can find a lift* $\tilde{\rho}$ *such that* $\mathrm{Sym}^2(\tilde{\rho})$ *corresponds to* $\mathrm{BC}_{L'/F'}(\mathrm{Sym}^2(\pi))$.

Conversely, starting with a mixed-parity π, *we can produce such a* χ *and* $\tilde{\rho}$.

PROOF. In the notation of the previous Lemma, we have

$$\mathrm{Sym}^2(\pi) \otimes \frac{\det(\tilde{\rho})}{\omega_\pi} \cong \Pi \sim_{w,\infty} \mathrm{Sym}^2(\tilde{\rho})|_{\Gamma_{F'}}.$$

The character $\det(\tilde{\rho})/\omega_\pi$ is L-algebraic, hence equals $\chi|\cdot|^m$ for a finite-order character χ and an integer m. Replacing π by $\pi \otimes |\cdot|^{m/2}$, which is still mixed-parity, we are done except for the last part.

Assume instead that we can take $\chi = 1$. In particular, $\omega_\pi \sim_{w,\infty} \det(\tilde{\rho})$, and thus $\det(\tilde{\rho})(c_v) = \omega_{\pi_v}(-1)$. By the description of discrete series representations of $\mathrm{GL}_2(\mathbb{R})$ and purity of ω_π, we know that the Langlands parameter $\phi_v\colon W_\mathbb{R} \to \mathrm{GL}_2(\mathbb{C})$ takes the form

$$z \mapsto \begin{pmatrix} (z/\bar{z})^{\frac{k_v-1}{2}} & 0 \\ 0 & (z/\bar{z})^{\frac{1-k_v}{2}} \end{pmatrix} |z|_\mathbb{C}^{-w/2}$$
$$j \mapsto \begin{pmatrix} 0 & (-1)^{k_v-1} \\ 1 & 0 \end{pmatrix}.$$

Hence $\omega_{\pi_v}(-1) = \det(\phi_v)(j) = (-1)^{k_v}$. Since $\mathrm{Ad}^0(\tilde{\rho})|_{\Gamma_{F'}} \sim_{w,\infty} \mathrm{Ad}^0(\pi)$ is regular algebraic self-dual cuspidal, Proposition *A* of [**Tay12**] shows that $\mathrm{Ad}^0(\tilde{\rho})$ is odd, and thus for all $v|\infty$, the eigenvalues of $\mathrm{Ad}^0(\tilde{\rho})(c_v)$ are $\{-1, 1, -1\}$. We conclude that $\det(\tilde{\rho})(c_v) = -1$ for all $v|\infty$, proving the oddness claim for $\tilde{\rho}$, and contradicting, when we assume $\chi = 1$, the fact that the k_v are both odd and even.

The remaining claims, constructing from π such a $\tilde{\rho}$ and χ, were established in Proposition 2.6.8. □

This concludes the proof of Theorem 2.7.2.

2.8. Spin examples

To address Conrad's question in general, we will have to re-cast the arguments of §2.7 in terms of root data. In the meantime we give another pleasantly concrete example, but one that relies on root-theoretic manipulation, in the hopes of easing the transition to the general case; moreover, specializing to discrete series/regular examples, we can also still exploit known results about automorphic Galois representations. The previous example (lifting across the surjection

$\mathrm{GL}_2(\overline{\mathbb{Q}}_\ell) \to \mathrm{PGL}_2(\overline{\mathbb{Q}}_\ell)$) is really just the first case of a family of spin examples,

$$\begin{array}{ccc} & & \mathrm{GSpin}_{2n+1}(\overline{\mathbb{Q}}_\ell) \\ & \overset{\tilde{\rho}}{\nearrow} & \downarrow \\ \Gamma_F & \xrightarrow[\rho]{} & \mathrm{SO}_{2n+1}(\overline{\mathbb{Q}}_\ell), \end{array}$$

where ρ is geometric but has no geometric lift $\tilde{\rho}$. F will be a totally real field, and we will again see that over CM fields, geometric lifts do in fact exist. The case $n = 1$ will amount to the contents of §2.7.

We first recall the basic setup for (odd) Spin groups. We will also make heavy use of this notation, and its obvious analogues for even Spin groups, later on in some of our examples of 'motivic lifting' (see §4.2 and §4.3).

$$\begin{CD}
@. 1 @. 1 \\
@. @VVV @VVV \\
1 @>>> \{\pm 1\} @>>> \mathrm{Spin}_{2n+1} @>>> \mathrm{SO}_{2n+1} @>>> 1\\
@. @VVV @VVV @| \\
1 @>>> \mathbb{G}_m @>>> \mathrm{GSpin}_{2n+1} @>>> \mathrm{SO}_{2n+1} @>>> 1\\
@. @V{z\mapsto z^2}VV @V{\nu}VV \\
@. \mathbb{G}_m @= \mathbb{G}_m \\
@. @VVV @VVV \\
@. 1 @. 1
\end{CD}$$

where ν is the Clifford norm. We can then identify

$$\begin{CD}
\mathrm{GSpin}_{2n+1} @<{\sim}<< \frac{\mathbb{G}_m \times \mathrm{Spin}_{2n+1}}{\{1,(-1,c)\}} \\
@V{\nu}VV @VV{(x,g)\mapsto x^2}V \\
\mathbb{G}_m @= \mathbb{G}_m
\end{CD}$$

where c is the non-trivial central element of Spin_{2n+1}. This yields an identification of Lie algebras

$$\mathfrak{gspin}_{2n+1} \xleftarrow{\sim} \mathfrak{gl}_1 \times \mathfrak{so}_{2n+1},$$

where $\mathfrak{gl}_1$ is identified with the center of $\mathfrak{gspin}_{2n+1}$. Alternatively, GSpin_{2n+1} is the dual group to GSp_{2n}, and it will be convenient to keep both normalizations of (based) root data for GSpin_{2n+1}–one from the spin description, one from the dual description–in mind. Let $(X, \Delta, X^\vee, \Delta^\vee)$ be the based root datum for GSp_{2n} (say defined with respect to $J_{2n} = \left(\begin{smallmatrix} 0 & 1_n \\ -1_n & 0 \end{smallmatrix}\right)$), with its diagonal maximal torus T and the Borel $B \supset T$ for which $e_i - e_{i+1}$ and $2e_n - e_0$ form a set of positive simple roots, with

$$e_i \colon diag(t_1, \dots, t_n, \nu t_1^{-1}, \dots, \nu t_n^{-1}) \mapsto t_i$$
$$e_0 \colon diag(t_1, \dots, t_n, \nu t_1^{-1}, \dots, \nu t_n^{-1}) \mapsto \nu.$$

Then for GSpin_{2n+1}, we have the based root datum (with $X^\vee$ the character group)

$$X^\vee = \bigoplus_{i=0}^{n} \mathbb{Z}e_i^*$$
$$\Delta^\vee = \{\alpha_i^\vee = e_i^* - e_{i+1}^*\}_{i=1}^{n-1} \cup \{\alpha_n^\vee = e_n^*\}$$
$$X = \bigoplus_{i=0}^{n} \mathbb{Z}e_i$$
$$\Delta = \{\alpha_i = e_i - e_{i+1}\}_{i=1}^{n-1} \cup \{\alpha_n = 2e_n - e_0\},$$

with X and $X^\vee$ in the duality $\langle e_i, e_j^* \rangle = \delta_{ij}$.

Alternatively, since Spin_{2n+1} is the connected, simple, simply-connected group with Lie algebra $\mathfrak{so}_{2n+1}$, we can write its character group as the weight lattice of $\mathfrak{so}_{2n+1}$, i.e., as the submodule of $\oplus_{i=1}^{n} \mathbb{Q}\chi_i$ spanned by $\chi_1, \ldots, \chi_n, \frac{\chi_1 + \ldots + \chi_n}{2}$, and its co-character lattice as those $\sum_{i=1}^{n} c_i \lambda_i$ such that $c_i \in \mathbb{Z}$ and $\sum c_i \in 2\mathbb{Z}$. The duality is $\langle \chi_i, \lambda_j \rangle = \delta_{ij}$. Letting χ_0 and λ_0 generate $X^\bullet(\mathrm{GL}_1)$ and $X_\bullet(\mathrm{GL}_1)$, respectively, the description of GSpin_{2n+1} as $\frac{\mathrm{GL}_1 \times \mathrm{Spin}_{2n+1}}{\{1,(-1,c)\}}$ then leads to another description of the root datum as

$$Y^\bullet = \bigoplus_{i=1}^{n} \mathbb{Z}\chi_i \oplus \mathbb{Z}(\chi_0 + \frac{\chi_1 + \ldots + \chi_n}{2}) \subset \bigoplus_{i=0}^{n} \mathbb{Q}\chi_i$$
$$\Delta^\bullet = \{\chi_i - \chi_{i+1}\}_{i=1}^{n-1} \cup \{\chi_n\}$$
$$Y_\bullet = \bigoplus_{i=1}^{n} \mathbb{Z}(\lambda_i + \frac{\lambda_0}{2}) \oplus \mathbb{Z}\lambda_0$$
$$\Delta_\bullet = \{\lambda_i - \lambda_{i+1}\}_{i=1}^{n-1} \cup \{2\lambda_n\}.$$

We summarize by comparing the two descriptions:

LEMMA 2.8.1. *There is an isomorphism of based root data*

$$(X^\vee, \Delta^\vee, X, \Delta) \cong (Y^\bullet, \Delta^\bullet, Y_\bullet, \Delta_\bullet)$$

given by

$$e_i^* \mapsto \chi_i \quad \text{for } i = 1, \ldots, n;$$
$$e_0^* \mapsto \chi_0 - \frac{\chi_1 + \ldots + \chi_n}{2};$$
$$e_i \mapsto \lambda_i + \frac{\lambda_0}{2} \quad \text{for } i = 1, \ldots, n-1;$$
$$e_0 \mapsto \lambda_0.$$

For example, the center of GSpin_{2n+1} is generated by the co-character $e_0 \leftrightarrow \lambda_0$. The Clifford norm is given by the character $2\chi_0 \leftrightarrow 2e_0^* + \sum_1^n e_i^*$.

Returning to our representation ρ, recall that to each $v|\ell$ and $\iota \colon \overline{\mathbb{Q}}_\ell \hookrightarrow \mathbb{C}_{F_v}$ we can associate the Sen operator

$$\Theta_{\rho,\iota} := \Theta_{\rho|_{\Gamma_{F_v}},\iota} \in \mathfrak{so}_{2n+1} \otimes_{\overline{\mathbb{Q}}_\ell,\iota} \mathbb{C}_{F_v}.$$

(The place v is implicit in ι.) Since ρ is Hodge-Tate, we can identify $\Theta_{\rho,\iota}$ up to conjugation with a diagonal element

$$\begin{pmatrix} m_1 & & & & & & \\ & \ddots & & & & & \\ & & m_n & & & & \\ & & & 0 & & & \\ & & & & -m_n & & \\ & & & & & \ddots & \\ & & & & & & -m_1 \end{pmatrix},$$

where $m_j = m_j(\iota)$ are all integers.[27]

PROPOSITION 2.8.2. *Let $\rho\colon \Gamma_F \to \mathrm{SO}_{2n+1}(\overline{\mathbb{Q}}_\ell)$ be a geometric representation as above, with F totally real. Then ρ lifts to a geometric, GSpin_{2n+1}-valued representation if and only if the parity of $\sum_j m_j(\iota)$ is independent of $v|\ell$ and $\iota\colon \overline{\mathbb{Q}}_\ell \hookrightarrow \mathbb{C}_{F_v}$.*

REMARK 2.8.3. By itself, this result is formal. Later we will see how such ρ arise, at least when ρ is regular; but note that this Proposition makes no such assumption.

PROOF. By Proposition 2.1.4, some lift $\tilde{\rho}$, necessarily unramified almost everywhere, exists. All possible lifts differ from this $\tilde{\rho}$ by $e_0 \circ \chi$ for some $\chi\colon \Gamma_F \to \overline{\mathbb{Q}}_\ell^\times$. As before, we can write the multiplier character $\nu(\tilde{\rho})\colon \Gamma_F \to \overline{\mathbb{Q}}_\ell^\times$ as $\chi^2\chi_0$, where χ_0 has finite order, and then replace $\tilde{\rho}$ by a twist having $\nu(\tilde{\rho})$ of finite-order (Lemma 3.2.2 will explain this procedure in general). Then, since finite-order characters have Hodge-Tate weights zero, functoriality of the Sen operator implies that for all $\iota\colon \overline{\mathbb{Q}}_\ell \hookrightarrow \mathbb{C}_{F_v}$, $\Theta_{\tilde{\rho},\iota}$ maps to $(0, \Theta_{\rho,\iota})$ in $(\mathfrak{gl}_1)_\iota \oplus (\mathfrak{so}_{2n+1})_\iota$.[28]

Now consider the spin representation $\mathrm{GSpin}_{2n+1} \xrightarrow{r_{\mathrm{spin}}} \mathrm{GL}_{2^n}$. In the above root datum notation this corresponds to the highest weight $-e_0^* \leftrightarrow -\chi_0 + \frac{\chi_1 + \ldots + \chi_n}{2}$. The image of $\Theta_{\tilde{\rho},\iota}$ is then a semi-simple element with one eigenvalue $\frac{m_1+\ldots+m_n}{2}$, and all eigenvalues congruent to this (half-integer) modulo $\mathbb{Z}$. In particular, $r_{\mathrm{spin}} \circ \tilde{\rho}|_{\Gamma_{F_v}}$ is Hodge-Tate, and thus de Rham, if and only if for all $\iota\colon \overline{\mathbb{Q}}_\ell \hookrightarrow \mathbb{C}_{F_v}$, $m_1(\iota)+\ldots+m_n(\iota)$ is even. If all (for all $v|\ell$ and all ι) of these sums are odd, then we twist by a character χ with all Hodge-Tate-Sen weights $1/2$ (see Lemma 2.3.17) to get a new $\tilde{\rho}$, now geometric, lifting ρ. On the other hand, Lemma 2.3.17 shows that for F totally real we cannot twist $\tilde{\rho}$ in a similar way if some $\sum_j m_j(\iota)$ are even and others are odd. □

COROLLARY 2.8.4. *Let $\rho\colon \Gamma_F \to \mathrm{SO}_{2n+1}(\overline{\mathbb{Q}}_\ell)$ be as in Proposition 2.8.2. Let L/F be a CM extension. Then $\rho|_{\Gamma_L}$ has a geometric lift.*

PROOF. With the framework of the previous proof, this follows by the same argument as Lemma 2.7.5. □

If we make additional assumptions so that potential automorphy theorems apply, we can of course say more. The next two lemmas merely exploit some very deep recent results.

LEMMA 2.8.5. *Assume that $\rho\colon \Gamma_F \to \mathrm{SO}_{2n+1}(\overline{\mathbb{Q}}_\ell) \subset \mathrm{GL}_{2n+1}(\overline{\mathbb{Q}}_\ell)$ as in the previous proposition moreover satisfies:*

---

[27] We choose the 'anti-diagonal' orthogonal pairing, so that SO_{2n+1} has a maximal torus consisting of diagonal matrices.

[28] Writing $\mathfrak{g}_\iota$ as a short-hand for $\mathfrak{g} \otimes_{\overline{\mathbb{Q}}_\ell,\iota} \mathbb{C}_{F_v}$.

- *For all* $v|\ell$, $\rho|_{\Gamma_{F_v}}$ *is regular and potentially diagonalizable;*
- $\bar{\rho}|_{\Gamma_{F(\zeta_\ell)}}$ *is irreducible.*
- $\ell \geq 2(2n+2)$.

Then after some totally real base change F'/F, there exists a regular L-algebraic self-dual cuspidal automorphic representation π of $\mathrm{GL}_{2n+1}(\mathbf{A}_{F'})$ such that $\pi \sim_{w,\infty} \rho$.

PROOF. This is immediate from Theorem C (=Corollary 4.5.2) of [**BLGGT14**], since ρ is automatically totally odd self-dual.[29] □

LEMMA 2.8.6. *Let $\rho \sim_{w,\infty} \pi$ be as in Lemma 2.8.5. Then π descends to a cuspidal automorphic representation on $\mathrm{Sp}_{2n}(\mathbf{A}_{F'})$, in a way compatible with unramified and archimedean local L-parameters.*

PROOF. This follows from Arthur's classification of automorphic representations of classical groups ([**Art13**]). Namely, $\rho^\vee \cong \rho$ implies that $\pi^\vee \cong \pi$, and $\det(\rho) = 1$ implies that $\omega_\pi = 1$. π is cuspidal, so the associated formal A-parameter ϕ (see §1.4 of [**Art13**]) is simple generic and so comes from a unique simple twisted endoscopic datum G_ϕ as in Theorem 1.4.1 of [**Art13**]; since $2n+1$ is odd, the parameter ϕ therefore factors through either $\mathrm{SO}_{2n+1}(\mathbb{C})$ or $\mathrm{O}_{2n+1}(\mathbb{C})$, but the latter case is ruled out since $\omega_\pi = 1$. It follows that $G_\phi = \mathrm{Sp}_{2n}/F$.[30] The local statement follows from [**Art13**, Theorem 1.4.2]. □

PROPOSITION 2.8.7. *Continuing with the assumptions (and conclusions) of the previous two lemmas, $\rho|_{\Gamma_{F'}}$ has a geometric lift to $\mathrm{GSpin}_{2n+1}(\overline{\mathbb{Q}}_\ell)$ if and only if π admits an L-algebraic extension to $\mathrm{GSp}_{2n}(\mathbf{A}_F)$. In the other direction, we have, as in Lemma 2.6.10, an automorphic construction of infinitely many such ρ, whose local behavior[31] we can additionally specify at any finite number of places, that do not admit a geometric lift.*

PROOF. (Some details of this argument are omitted, deferring to more general arguments in the next section.) At each $v|\infty$, we can write the L-parameter of π_v as

$$\phi_v \colon W_{F_v} \to \mathrm{SO}_{2n+1}(\mathbb{C}) \times W_{F_v}$$

$$\phi_v|_{W_{\overline{F}_v}} : z \mapsto \begin{pmatrix} z^{m_1}\bar{z}^{l_1} & & & & & & \\ & \ddots & & & & & \\ & & z^{m_n}\bar{z}^{l_n} & & & & \\ & & & 1 & & & \\ & & & & z^{-m_n}\bar{z}^{-l_n} & & \\ & & & & & \ddots & \\ & & & & & & z^{-m_1}\bar{z}^{-l_1}. \end{pmatrix},$$

with all $m_i, l_i \in \mathbb{Z}$. Since the transfer of π to $\mathrm{GL}_{2n+1}(\mathbf{A}_F)$ is cuspidal, Clozel's archimedean purity theorem (see Lemme 4.9 of [**Clo90**]) implies that $m_i + l_i = 0$

[29]Note that those authors always work with C-algebraic automorphic representations, so the statements of their theorems always have an extra, but easily unraveled, twist.

[30]One would like to say that since ρ lands in $\mathrm{SO}_{2n+1}(\overline{\mathbb{Q}}_\ell)$, $\mathrm{Sym}^2(\rho)$ contains the trivial representation and therefore $L(s, \pi, \mathrm{Sym}^2)$ has a pole at $s = 1$; this would allow us to descend π to $\mathrm{Sp}_{2n}(\mathbf{A}_F)$ using [**CKPSS04**]. Unfortunately, nothing is known a priori about $L(s, V)$ where $\mathrm{Sym}^2(\rho) = \mathbf{1} \oplus V$; in particular, we can't say immediately it is non-vanishing at $s = 1$, so this argument doesn't seem to work. The argument we have given allows us to deduce that $L(s, \pi, \mathrm{Sym}^2)$ has a simple pole at $s = 1$, and therefore $L(s, V)$ is non-vanishing at $s = 1$.

[31]i.e. inertial type

for all i (and all v). To extend π in an L or W-algebraic fashion, the key point is to construct an appropriate extension of the central character. Any lift $\tilde{\phi}_v$ to $\mathrm{GSpin}_{2n+1}(\mathbb{C})$ of the parameter ϕ_v must, on $W_{\overline{F}_v}$, take the form $z \mapsto z^{\tilde{\mu}_v} \bar{z}^{\tilde{\nu}_v}$, where

$$\tilde{\mu}_v = \sum_{i=1}^n m_{v,i}(\lambda_i + \frac{\lambda_0}{2}) + \mu_{v,0}\lambda_0$$

$$\tilde{\nu}_v = -\sum_{i=1}^n m_{v,i}(\lambda_i + \frac{\lambda_0}{2}) + \nu_{v,0}\lambda_0.$$

The central character of this extension is (by pairing with $2\chi_0$, the Clifford norm; this procedure for computing central characters is explained in general in [**Lan89**])

$$\omega_v \colon z \mapsto z^{\sum_{i=1}^n m_{v,i} + 2\mu_{v,0}} \bar{z}^{-\sum_{i=1}^n m_{v,i} + 2\nu_{v,0}}.$$

If we can choose an extension of π to an automorphic representation $\tilde{\pi}$ of $\mathrm{GSp}_{2n}(\mathbf{A}_{F'})$ with finite-order central character, then this calculation shows that $\tilde{\phi}_v$ (the local L-parameter for $\tilde{\pi}_v$) is always 'W-algebraic,'[32] and it is 'L-algebraic' if and only if $\sum_{i=1}^n m_{v,i}$ is even. The result would follow easily, twisting our given $\tilde{\pi}$ by $|\cdot|^{1/2}$ if all $\sum_i m_{v,i}$ are odd (just as in the Galois case). That we can find such an extension $\tilde{\pi}$ with finite-order central character will be proven in much more generality in Proposition 3.1.14.

The construction of geometric $\rho \colon \Gamma_F \to \mathrm{SO}_{2n+1}(\overline{\mathbb{Q}}_\ell)$ having no geometric lift to $\mathrm{GSpin}_{2n+1}(\overline{\mathbb{Q}}_\ell)$ (and with specified local behavior) follows as in Lemma 2.6.10, applying Clozel's limit multiplicity formula to $G = \mathrm{Sp}_{2n}$ over F, transferring these forms to GL_{2n+1} (via [**Art13**]), and then invoking the Paris Book Project, or, more precisely, Remark 7.6 of [**Shi11**].

□

We have therefore generalized some of the results of §2.7 (the case $n = 1$).

[32]Using this term abusively for the obvious local analogue.

CHAPTER 3

Galois and automorphic lifting

This chapter begins (§3.1) by addressing the natural automorphic analogue of Conrad's lifting question. It is much easier to see what should be true in this setting, and the proofs are simpler as well. Equipped with the intuition coming from the automorphic case, we address the original Galois-theoretic question in §3.2. In §3.3, we combine the results of §3.1 and 3.2 to compare, *modulo the Fontaine-Mazur-Langlands conjecture*, descent problems for certain automorphic representations and Galois representations. The closing section, §3.4, is of a rather different nature, assembling a few results about the images of compatible systems of ℓ-adic Galois representations. The results of that section continue the attempt to compare aspects of the automorphic and Galois-theoretic formalisms.

3.1. Lifting W-algebraic representations

In this section, we make the simplifying assumption that G is a connected semi-simple split group over F. This semi-simplicity assumption is not essential, and in §3.2 we will work more generally. Let $\widetilde{Z}$ be an F-split *torus* containing Z_G, and set $\widetilde{G} = (\widetilde{Z} \times G)/Z_G$, as before, with maximal torus $\widetilde{T} = (\widetilde{Z} \times T)/Z_G$ and center $\widetilde{Z}$. In each case Z_G is embedded anti-diagonally. We will study lifting problems for automorphic representations from G to $\widetilde{G}$, and also from G to intermediate (connected reductive) groups $G \subset \mathbf{G} \subset \widetilde{G}$; here the reader should think of $G \subset \mathbf{G}$ as the general case, but any such $\mathbf{G}$ embeds in a $\widetilde{G}$ whose center $\widetilde{Z}$ is a torus, simply by embedding the center $\mathbf{Z}$ of $\mathbf{G}$ into a torus $\widetilde{Z}$. We focus on the case $G \subset \widetilde{G}$ because the general case reduces to this one.[1] Note that the cases of greatest interest are when G is simply-connected (so $G^\vee$ is adjoint), such as SL_n or Sp_{2n}, but we do not assume this.

Now let π be a (unitary) cuspidal automorphic representation of $G(\mathbf{A}_F)$ that is W-algebraic. We are interested in the problem of lifting π to a W-algebraic automorphic representation $\tilde{\pi}$ of $\widetilde{G}(\mathbf{A}_F)$; when π is moreover L-algebraic, we similarly ask whether an L-algebraic lift exists. By 'lift,' we simply mean the most naive thing: the restriction $\tilde{\pi}|_{G(\mathbf{A}_F)}$ contains π. Corollary 3.1.6 justifies this convention, showing that under the L-morphism ${}^L\widetilde{G} \to {}^LG$, the L-packet of π is a functorial transfer of the L-packet of $\tilde{\pi}$.

3.1.1. Notation and central character calculation. In order to do computations in terms of root data, we choose coordinates (using invariant factors)

[1] To distinguish the two cases, consider for instance $G = \mathrm{Spin}_{2n} \subset \mathbf{G} = \mathrm{GSpin}_{2n}$, which dualizes to a surjection $\mathrm{GSO}_{2n} \twoheadrightarrow \mathrm{PSO}_{2n}$. The center of $\mathbf{G}$ is not connected.

such that

$$X^\bullet(\widetilde{Z}) = \bigoplus_{i=1}^{r} \mathbb{Z}w_i \oplus \bigoplus_{j=1}^{s} \mathbb{Z}w'_j,$$

and the kernel of $X^\bullet(\widetilde{Z}) \to X^\bullet(Z_G)$ is

$$\bigoplus_{i=1}^{r} d_i\mathbb{Z}w_i \oplus \bigoplus_{j=1}^{s} \mathbb{Z}w'_j.$$

This is $X^\bullet(S)$ for the torus $S = \widetilde{Z}/Z_G$. We then write $X^\bullet(Z_G) = \oplus_1^r(\mathbb{Z}/d_i\mathbb{Z})\bar{w}_i$. To relate parameters for $\widetilde{G}$ to those for G, we use the Cartesian diagram

$$\begin{array}{ccc} X^\bullet(\widetilde{T}) & \longrightarrow & X^\bullet(\widetilde{Z}) \\ \downarrow & & \downarrow \\ X^\bullet(T) & \longrightarrow & X^\bullet(Z_G), \end{array}$$

representing elements of $X^\bullet(\widetilde{T})$ as pairs $(\chi_T, \chi_{\widetilde{Z}})$ that are congruent in $X^\bullet(Z_G)$. Extending scalars to $\mathbb{Q}$ (or any characteristic zero field), $X^\bullet(\widetilde{T})_\mathbb{Q} \xrightarrow{\sim} X^\bullet(T)_\mathbb{Q} \oplus X^\bullet(\widetilde{Z})_\mathbb{Q}$. We will usually (out of sad necessity) take F to be either CM or totally real. In the former case (or whenever an archimedean place is complex) we compute local central characters (see page 21 ff. of [**Lan89**]) as follows:

Suppose v is complex, so there is an isomorphism $\iota_v\colon F_v \xrightarrow{\sim} \mathbb{C}$. Let

$$\mathrm{rec}_v(\pi_v)\colon z \mapsto \iota_v(z)^{\mu_v}\bar{\iota}(z)^{\nu_v} \in T^\vee(\mathbb{C})$$[2]

be the associated Langlands parameter, with $\mu_v, \nu_v \in X^\bullet(T)_\mathbb{C}$ with $\mu_v - \nu_v \in X^\bullet(T)$. Write $\sum_{i=1}^r [\mu_v - \nu_v]_i \bar{w}_i$ for the projection of $\mu_v - \nu_v$ to $X^\bullet(Z_G)$. Choose any $\mu_{v,i}, \nu_{v,i} \in \mathbb{C}$ with $\mu_{v,i} - \nu_{v,i}$ an integer projecting to $[\mu_v - \nu_v]_i \in \mathbb{Z}/d_i\mathbb{Z}$. We can then regard $\tilde{\mu}_v = (\mu_v, \sum_1^r \mu_{v,i}w_i)$ and $\tilde{\nu}_v = (\nu_v, \sum_1^r \nu_{v,i}w_i)$ as elements of $X^\bullet(\widetilde{T})_\mathbb{C}$ parametrizing an extension of the local L-parameter of π_v to a parameter for $\widetilde{G}(F_v)$. Identifying, via our chosen basis, $\widetilde{Z}(F_v)$ with $(\mathbb{C}^\times)^{r+s}$, the central character $\omega_{\tilde{\pi}_v}$ of this lift is simply

$$\omega_{\tilde{\pi}_v}\colon (z_1, \ldots, z_r, z'_1, \ldots, z'_s) \mapsto \prod_{i=1}^{r} z_i^{\mu_{v,i} - \nu_{v,i}} |z_i|_\mathbb{C}^{\nu_{v,i}}.$$

Restricting to $Z_G(F_v) \cong \mu_{d_1} \times \cdots \times \mu_{d_r} \subset (\mathbb{C}^\times)^r$, we find that the central character ω_{π_v} is given by

$$\omega_{\pi_v}\colon (\zeta_1, \ldots, \zeta_r) \mapsto \prod_{i=1}^{r} \zeta_i^{\mu_{v,i} - \nu_{v,i}},$$

where each ζ_i is a d_i^{th} root of unity. Clearly this is independent of the choice of lifts $\tilde{\mu}_v, \tilde{\nu}_v$.

DEFINITION 3.1.1. If F is a CM field, and π is a (unitary) automorphic representation of $G(\mathbf{A}_F)$ with archimedean parameters μ_v, ν_v as above at each $v|\infty$,

---

[2] From now on, we write $z, \bar{z}$ in place of $\iota_v(z), \bar{\iota}_v(z)$.

then we define $\tilde{\omega} = \tilde{\omega}(\pi)$ to be any choice of Hecke character of $\widetilde{Z}$ lifting ω_π and such that

$$\tilde{\omega}_v(z_1, \dots, z'_s) = \prod_{i=1}^{r} (z_i/|z_i|)^{\mu_{v,i}-\nu_{v,i}}.$$

This is a unitary, type A Hecke character whose existence is assured by Lemma 2.3.6.

When v is real, the lifting process is more complicated (see [**Lan89**]), and the central character computation depends very much on where the L-parameter of π_v sends an element of $W_{\mathbb{R}} - W_{\mathbb{C}}$. We can avoid this, however:

DEFINITION 3.1.2. Let π be as in the previous definition, but now suppose F is totally real. Then we define $\tilde{\omega}$ to be any choice of finite-order Hecke character of $\widetilde{Z}$ extending ω_π (existence again by Lemma 2.3.6).

3.1.2. Generalities on lifting from $G(\mathbf{A}_F)$ to $\widetilde{G}(\mathbf{A}_F)$. We now try to find an automorphic representation $\tilde{\pi} \subset L^2_{\text{cusp}}(\widetilde{G}(F)\backslash\widetilde{G}(\mathbf{A}_F), \tilde{\omega})$[3] lifting π, where $\tilde{\omega}$ is any (unitary) lift of the central character ω_π. We say that we are in the Grunwald-Wang special case if one of the pairs (F, d_i) is in the usual Grunwald-Wang special case. Let $H_{\mathbf{A}_F} = G(\mathbf{A}_F)\widetilde{Z}(\mathbf{A}_F)$ (a normal subgroup of $\widetilde{G}(\mathbf{A}_F)$), and let $H_F = H_{\mathbf{A}_F} \cap \widetilde{G}(F)$.

LEMMA 3.1.3. *We are in the Grunwald-Wang special case if and only if H_F strictly contains $G(F)\widetilde{Z}(F)$.*

PROOF. Recall the characters $d_i w_i \in X^\bullet(S) = X^\bullet(\widetilde{G})$ (since G is semi-simple). These induce an isomorphism

$$H_F/G(F)\widetilde{Z}(F) \xrightarrow{\prod d_i w_i} \prod_{i=1}^{r} \left(F^\times \cap (\mathbf{A}_F^\times)^{d_i}\right)/(F^\times)^{d_i}.$$

$\square$

There are various ways to show forms on $G(\mathbf{A}_F)$ extend to $\widetilde{G}(\mathbf{A}_F)$; the argument we use here is modeled on one of Flicker for the case $(\widetilde{G}, G) = (\mathrm{GSp}_{2n}, \mathrm{Sp}_{2n})$ (see Proposition 2.4.3 of [**Fli06**]).

PROPOSITION 3.1.4. *Let π be a cuspidal automorphic representation of $G(\mathbf{A}_F)$, with central character ω_π. If we are not in the Grunwald-Wang special case, then for any extension $\tilde{\omega}$ of ω_π to a Hecke character of $\widetilde{Z}$, there exists a cuspidal automorphic representation $\tilde{\pi}$ of $\widetilde{G}(\mathbf{A}_F)$ lifting π, and having central character $\tilde{\omega}$. If we are in the Grunwald-Wang special case, then for at least one extension of ω_π to*

$$\prod_{i=1}^{r} C_F[d_i] \supset \prod_{i=1}^{r} \mu_{d_i}(F)\backslash\mu_{d_i}(\mathbf{A}_F),$$

and for any extension of this character to a Hecke character $\tilde{\omega}$ of $\widetilde{Z}$, there exists a cuspidal $\tilde{\pi}$ lifting π with central character $\tilde{\omega}$.

[3]That is, the space of measurable functions on $\widetilde{G}(F)\backslash\widetilde{G}(\mathbf{A}_F)$ with (unitary) central character $\tilde{\omega}$, and square-integrable modulo $\widetilde{Z}(F_\infty)^0$–or, equivalently, modulo $A_{\widetilde{G}}(\mathbb{R})^0$, where $A_{\widetilde{G}}$ is the maximal $\mathbb{Q}$-split central torus in $\mathrm{Res}_{F/\mathbb{Q}}(\widetilde{G})$.

For any intermediate connected reductive group $G \subset \mathbf{G} \subset \widetilde{G}$, any cuspidal automorphic representation of G extends to a cuspidal automorphic representation of $\mathbf{G}$.

REMARK 3.1.5. In particular, in all cases, for all Hecke characters $\tilde{\omega}$ extending ω_π, there exists a Hecke character $\tilde{\omega}'$ having the same infinity-type, and a cuspidal representation $\tilde{\pi}$ of $\widetilde{G}(\mathbf{A}_F)$ lifting π, with central character $\tilde{\omega}'$.

PROOF. Choose an extension $\tilde{\omega}\colon \widetilde{Z}(F)\backslash\widetilde{Z}(\mathbf{A}_F) \to \mathrm{S}^1$ of ω_π. By extending functions along $\widetilde{Z}(\mathbf{A}_F)$ via $\tilde{\omega}$, we embed the space V_π of π into the space of cusp-forms on $G(F)\widetilde{Z}(F)\backslash H_{\mathbf{A}_F}$, and thereby obtain an extension of π to a representation $\pi_{\tilde{\omega}}$ of $H_{\mathbf{A}_F}$. We construct an intertwining map

$$\mathrm{Ind}_{H_{\mathbf{A}_F}}^{\widetilde{G}(\mathbf{A}_F)}(\pi_{\tilde{\omega}}) \xrightarrow{U} L^2_{\mathrm{cusp}}(\widetilde{G}(F)\backslash\widetilde{G}(\mathbf{A}_F), \tilde{\omega}),$$

where the induction consists of functions $F\colon \widetilde{G}(\mathbf{A}_F) \to V_\pi$ with the usual left-$H_{\mathbf{A}_F}$-equivariance, and the requirement of compact support modulo $H_{\mathbf{A}_F}$. Write $\delta_1\colon V_\pi \to \mathbb{C}$ for evaluation at 1 of the cusp forms in V_π, and set

$$L(F) = \sum_{G(F)\widetilde{Z}(F)\backslash\widetilde{G}(F)\ni u} \delta_1(F_u),$$

where we write F_u for the value at u of $F \in \mathrm{Ind}(\pi_{\tilde{\omega}})$. This sum is in fact finite: $\widetilde{G}(F)$ is discrete in $\widetilde{G}(\mathbf{A}_F)$, the fibers of $G(F)\widetilde{Z}(F)\backslash\widetilde{G}(F) \to H_{\mathbf{A}_F}\backslash\widetilde{G}(\mathbf{A}_F)$ are finite, and we have assumed F has compact support modulo $H_{\mathbf{A}_F}$. It is also well-defined because for $\gamma \in G(F)\widetilde{Z}(F)$,

$$\delta_1(F_{\gamma u}) = \delta_1(\gamma \cdot F_u) = F_u(\gamma) = F_u(1) = \delta_1(F_u).$$

The map U is then given by

$$U(F)\colon g \mapsto L(I(g)F),$$

where $I(\cdot)$ denotes the $\widetilde{G}(\mathbf{A}_F)$-action on the induction. U is clearly $\widetilde{G}(\mathbf{A}_F)$-equivariant and has output $U(F)$ which is left-$\widetilde{G}(F)$-equivariant (by its construction as an average) and has central character $\tilde{\omega}$; $U(F)$ is a cusp form by a simple calculation using the fact that the unipotent radical of any parabolic of $\widetilde{G}$ is in fact contained in G.

To see that $U \neq 0$, take a non-zero form $f \in V_\pi$; we may assume (by translating) $f(1) \neq 0$. Then define $F \in \mathrm{Ind}(\pi_{\tilde{\omega}})$ by

$$F_h\colon h' \mapsto f(h'h)$$

if $h \in H_{\mathbf{A}_F}$, and $F_g = 0$ for $g \notin H_{\mathbf{A}_F}$. Then

$$U(F)(1) = \sum_u \delta_1(F_u) = \sum_u \begin{cases} f(u) & \text{if } u \in H_{\mathbf{A}_F}; \\ 0 & \text{otherwise.} \end{cases}$$

So, if we are not in the special case, we just get $U(F)(1) = f(1) \neq 0$. If we are in the special case, then for each pair (F, d_i) in the special case there is an element $\alpha_i \in F^\times$ which is everywhere locally a d_i^{th}-power–say $\alpha_i = \beta_i^{d_i}$, but not globally a d_i^{th}-power. By abuse of notation, we also write α_i for an element of $\widetilde{G}(F)$ such that $(d_i w_i)(\alpha_i)$ is this element of $F^\times$ and the other characters $d_j w_j$ $(j \neq i)$ and

w'_j of $\widetilde{G}$ are trivial on α_i.[4] Regarding β_i as an element of the ith componenet of $\widetilde{Z}(\mathbf{A}_F)$, we can therefore write $\alpha_i = \beta_i \cdot \gamma_i$, where each γ_i lies in $G(\mathbf{A}_F)$. If only one pair (F, d_i) is in the special case, then the above expression for $U(F)(1)$ becomes $f(1) + f(\alpha_i) = f(1) + \tilde{\omega}(\beta_i) f(\gamma_i)$. This time we normalize f so that $f(\gamma_i) \neq 0$ (rather than $f(1) \neq 0$), and so necessarily either $f(1) + \tilde{\omega}(\beta_i) f(\gamma_i) \neq 0$, or this expression is non-zero for any $\tilde{\omega}$ extending the *other* extension of ω_π to a character of $C_F[d_i]$. Thus we can choose at least one initial extension to $C_F[d_i]$, and thereafter the argument proceeds as in the non-exceptional case. If multiple (F, d_i) are in the special case, the same argument applies: arrange some $f(\gamma_i) \neq 0$, and if $U(F)(1) = 0$, change the extension of ω_π to $C_F[d_i]$ just in this i^{th} component.

Finally, given that $U \neq 0$, we take $\tilde{\pi}$ to be the image of any irreducible constituent of $\mathrm{Ind}_{H_{\mathbf{A}_F}}^{\widetilde{G}(\mathbf{A}_F)}(\pi_{\tilde{\omega}})$ that survives under U, and we claim this is the desired extension of $\pi_{\tilde{\omega}}$. The isomorphism classes of $H_{\mathbf{A}_F}$-representations appearing in the restriction to $H_{\mathbf{A}_F}$ of the full induction are precisely the conjugates $\pi_{\tilde{\omega}}^g$, for $g \in \widetilde{G}(\mathbf{A}_F)$, and $\widetilde{G}(\mathbf{A}_F)$-stability of $\tilde{\pi}$ implies that all of these in fact appear in $\tilde{\pi}|_{H_{\mathbf{A}_F}}$. In particular, this latter restriction contains $\pi_{\tilde{\omega}}$.

For the final claim, we extend π from G to $\widetilde{G}$, and then restrict to $\mathbf{G}$; any constituent of the restriction can be taken as the desired extension. □

We need to supplement this by understanding to what extent π is a 'functorial transfer' of $\tilde{\pi}$, with respect to the L-homomorphism ${}^L\widetilde{G} \twoheadrightarrow {}^LG$. Somewhat more generally, let $G \xrightarrow{\eta} \widetilde{G}$ be a morphism of connected reductive groups over F_v, with abelian kernel and cokernel; here we either take v to be archimedean, or finite such that G and $\widetilde{G}$ are unramified over F_v. By assumption, there is a dual L-homomorphism ${}^L\eta \colon {}^L\widetilde{G} \to {}^LG$. In both the archimedean and unramified cases, the L-parameters ϕ_v and $\tilde{\phi}_v$, as well as the corresponding L-packets Π_{ϕ_v} and $\Pi_{\tilde{\phi}_v}$, of π_v and $\tilde{\pi}_v$ have been defined (as will be explained in Corollary 3.1.6).

COROLLARY 3.1.6. *Continue in the setting of the proposition. Then π is a weak transfer of $\tilde{\pi}$, which is also a strong transfer at archimedean places. More precisely, for all places v of F that are either archimedean or such that G is unramified at v, the restriction $\tilde{\pi}_v|_{G(F_v)}$ is a finite direct sum of elements of Π_{ϕ_v}, and $\tilde{\phi}_v$ reduces to ϕ_v (i.e. ${}^L\eta \circ \tilde{\phi}_v = \phi_v$ up to $G^\vee(\mathbb{C})$-conjugacy).*

PROOF. For infinite places v, the assertion follows from desideratum (iv) on page 30 of [**Lan89**]. Since the construction of the correspondence is inductive, the verification necessarily stretches out through §3 of that paper; for the first (and most important) case of discrete series, see page 43.

Now we treat the unramified case. Suppose $\tilde{\pi}_v$ is unramified, i.e. there exists a hyperspecial maximal compact subgroup $\widetilde{K}_v$ of $\widetilde{G}(F_v)$ such that $\tilde{\pi}_v^{\widetilde{K}_v} \neq 0$. Then for some unramified character $\tilde{\chi}_v$ of $\widetilde{T}(F_v)$, $\tilde{\pi}_v$ is a sub-quotient of $I_{\widetilde{B}(F_v)}^{\widetilde{G}(F_v)} \tilde{\chi}_v$. The natural map $G(F_v)/B(F_v) \to \widetilde{G}(F_v)/\widetilde{B}(F_v)$ is an isomorphism, so restriction of functions gives an isomorphism of $G(F_v)$ representations

$$\left(I_{\widetilde{B}(F_v)}^{\widetilde{G}(F_v)} \tilde{\chi}_v\right)|_{G(F_v)} \xrightarrow{\sim} I_{B(F_v)}^{G(F_v)}(\tilde{\chi}_v|_{B(F_v)}).$$

[4]We use the assumption that G is split over F.

Write χ_v for $\tilde{\chi}_v|_{B(F_v)}$. It is an unramified character of $T(F_v)$, and under the identification of unramified characters of $T(F_v)$ with $\mathrm{Hom}(X_\bullet(T)^{\Gamma_{F_v}}, \mathbb{C}^\times)$ (and the analogue for $\widetilde{T}$), and thus with the space of unramified L-parameters, χ_v corresponds to the parameter $\phi_v = {}^L\eta \circ \tilde{\phi}_v$. By definition of unramified L-packets (see [**Bor79**, §10.4]), to see that any constituent π_v of $\tilde{\pi}_v|_{G(F_v)}$ lies in the packet Π_{ϕ_v}, we need only check that π_v has invariants under some hyperspecial maximal compact subgroup of $G(F_v)$. To see this, let u be a non-zero vector in $\tilde{\pi}_v^{\widetilde{K}_v}$. Decomposing $\tilde{\pi}_v|_{G(F_v)} = \oplus_1^M \pi_i^{\oplus m_i}$ (with the π_i distinct isomorphism classes; in fact, the multiplicities m_i are all equal), we see that u lies in one of the isotypic componenets $\pi_i^{\oplus m_i}$, since the induction $I(\chi_v)$ can only have one $K_v := G(F_v) \cap \widetilde{K}_v$- invariant line. Fixing a decomposition $u = \sum_{j=1}^{m_i} u_j$ with u_j in the jth copy of π_i, we find that the u_j are themselves also K_v-invariant, hence π_i contains a K_v-invariant vector (this in turn implies the multiplicities m_i must equal 1). Since $\widetilde{G}(F_v)$ acts transitively on the isomorphism classes π_i, each π_i, and in particular our given π_v, contains some vector of the form $\tilde{\pi}_v(g)u$, which is gK_vg^{-1}-invariant; gK_vg^{-1} is also a hyperspecial maximal compact subgroup, so we're done. □

REMARK 3.1.7. According to expected properties of local L-packets, π should be a strong transfer of $\tilde{\pi}$ (see §10.3 of [**Bor79**]), but of course this statement is meaningless until the local Langlands correspondence is known for G and $\widetilde{G}$.

Eventually, we will also want to understand the ambiguity in the choice of $\tilde{\pi}$. Strictly speaking, we only need this in one direction of Proposition 3.1.14 below, but as a general problem, its importance is basic. We will need to assume something from local representation theory, which was originally conjectured always to hold, more generally for quasi-split G and $\widetilde{G}$, by Adler and Prasad ([**AP06**]); in fact, in certain quasi-split, but not split, cases, Adler and Prasad have found counterexamples to their conjecture (these and related questions will be taken up in [**APP16**]).

HYPOTHESIS 3.1.8. *Let v be any place F, and let G and $\widetilde{G}$ be as above. Then for any irreducible smooth representation $\tilde{\pi}_v$ of $\widetilde{G}(F_v)$, the restriction $\tilde{\pi}_v|_{G(F_v)}$ decomposes with multiplicity one.*

The motivation for this hypothesis is the uniqueness of Whittaker models for quasi-split groups, and indeed it always holds for generic $\tilde{\pi}_v$. The analogous statement for v archimedean is straightforward in many cases: for instance, for a simple G, it is easy because $\mathbb{R}^\times/(\mathbb{R}^\times)^n$ is cyclic for all n. Moreover, the structure of archimedean L-packets may be well-enough understood to resolve the question in general for archimedean v, but we do not pursue this here. For v non-archimedean, we have verified it in the unramified case (see the proof of Lemma 3.1.6), and in general the hypothesis is known for the following pairs $(\widetilde{G}, G)$: $(\mathrm{GL}_n, \mathrm{SL}_n)$ (from the theory of, possibly degenerate, Whittaker models; see [**LL79**] for the $n = 2$ case), and any pair $(GU(V), U(V))$ where $(V, \langle , \rangle)$ is a vector space over F_v equipped with a non-degenerate symmetric or skew-symmetric form $\langle , \rangle$, and $GU(V)$, respectively $U(V)$, denote the similitude and isometry groups of the pairing (see Theorem 1.4 of [**AP06**]). So, whatever the status of the general hypothesis, the discussion below applies to some interesting cases.

LEMMA 3.1.9. *Assume Hypothesis 3.1.8. Suppose $\tilde{\pi}$ and $\tilde{\pi}'$ are two lifts of π (i.e. their restrictions contain π) with the same central character $\tilde{\omega}$.*

- *There exist continuous idele characters $\alpha_i \colon \mathbf{A}_F^\times \to \mathbb{C}^\times$, for $i = 1, \dots, r$, such that, in the notation of §3.1.1,*

$$\tilde{\pi} \cong \tilde{\pi}' \cdot \left(\prod_{i=1}^{r} \alpha_i \circ d_i w_i \right),$$

  and each $\alpha_i^{d_i}$ factors as a genuine Hecke character $\alpha_i^{d_i} \colon \mathbf{A}_F^\times / F^\times \to \mathbb{C}^\times$.
- *Since Hecke characters satisfy purity, so do the characters α_i.*

REMARK 3.1.10. The point of the Lemma is that the idele characters α_i *need not be Hecke characters*, i.e. need not factor through C_F. We are particularly interested in the constraint on the infinity type of α_i, which by the lemma must be just as rigid as the constraints for Hecke characters. This will be applied in Proposition 3.1.14.

PROOF. As before, write $H_{\mathbf{A}_F} = G(\mathbf{A}_F)\widetilde{Z}(\mathbf{A}_F)$, and write $H_v = G(F_v)\widetilde{Z}(F_v)$ for its local analogue. By hypothesis, both restrictions $\tilde{\pi}|_{H_{\mathbf{A}_F}}$ and $\tilde{\pi}'|_{H_{\mathbf{A}_F}}$ contain $\pi \boxtimes \tilde{\omega}$, so this holds everywhere locally as well. Lemma 2.4 of [**GK82**] (applying Hypothesis 3.1.8) implies that for all finite v, there exists a smooth character $\alpha_v \colon \widetilde{G}(F_v)/H_v \to \mathbb{C}^\times$ such that $\tilde{\pi}_v \cong \tilde{\pi}'_v \cdot \alpha_v$. By duality (for finite abelian groups), we can extend this to a character $\bar{\alpha}_v$:

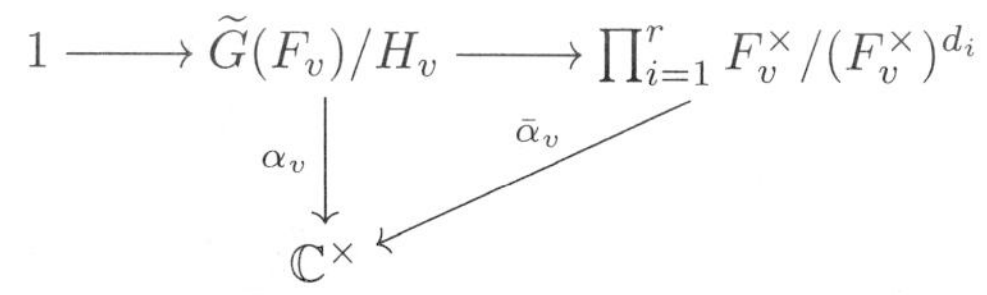

We express α_v in coordinates as

$$\alpha_v = (\alpha_{i,v} \circ (d_i w_i))_{i=1}^r$$

for smooth characters $\alpha_{i,v}$ of $F_v^\times$ trivial on $(F_v^\times)^{d_i}$. Globally, we then have

$$\tilde{\pi} \cong \tilde{\pi}' \cdot \left(\prod_{i=1}^{r} (\otimes'_v \alpha_{i,v}) \circ d_i w_i \right).$$

Here the $\alpha_i := \otimes'_v \alpha_{i,v}$ are continuous[5] characters $\mathbf{A}_F^\times \to \mathbb{C}^\times$, but they need not be Hecke characters. Taking central characters, however, we see that each $\alpha_i^{d_i}$ *is* in fact a Hecke character. □

REMARK 3.1.11. Under Hypothesis 3.1.8, one should be able to refine the arguments of this section to produce a multiplicity formula for cuspidal automorphic representations of $G(\mathbf{A}_F)$ in terms of those of $\widetilde{G}(\mathbf{A}_F)$, as in Lemma 6.2 of [**LL79**]). One can also axiomatize the passage between local Langlands conjectures for G and for $\widetilde{G}$; this would include verifying compatibility of the conjectural formulae for the sizes of L-packets (a template, in the case of (GSp, Sp), is given in the paper of [**GT10**]; their arguments will clearly apply much more generally).

[5]Almost all $\alpha_{i,v}$ are unramified, since the same holds for $\tilde{\pi}_v$ and $\tilde{\pi}'_v$.

3.1.3. Algebraicity of lifts: The ideal case. Now we return to questions of algebraicity, handling the CM and totally real cases in turn. In each case, we carefully choose an extension of the central character of π (as described in Definitions 3.1.1 and 3.1.2), and then find a $\tilde{\pi}$ as in Proposition 3.1.4 with that central character, and whose restriction to $G(\mathbf{A}_F)$ contains π. This will yield enough information about the archimedean L-parameters of $\tilde{\pi}$ to deduce the desired algebraicity statements. To give a clean argument with broad conceptual scope, we will assume that π is tempered at infinity[6]. As remarked before, this is not a serious assumption for algebraic representations: it is satisfied for forms having cuspidal transfer to some GL_N. For non-tempered forms, analogous results for the discrete spectrum can be deduced from Arthur's conjectures.

PROPOSITION 3.1.12. *Let F be CM, with $\pi, \tilde{\pi}$ as above, and with $\tilde{\omega}$ the (type A) extension of ω_π described in Definition* 3.1.1. *Assume π_∞ is tempered.*

(1) *If π is L-algebraic, then $\tilde{\pi}$ is L-algebraic. In particular, there exists an L-algebraic lift of π.*
(2) *If π is W-algebraic, then $\tilde{\pi}$ is W-algebraic.*

Similarly, any L-algebraic (respectively, W-algebraic) π on G extends to an L-algebraic (respectively, W-algebraic) automorphic representation of $\mathbf{G}(\mathbf{A}_F)$.

PROOF. We use the notation of §3.1.1. We may of course take all $\mu_{v,i}$ and $\nu_{v,i}$ to be integers. By the choice of central character $\tilde{\omega}$, the archimedean L-parameter for $\tilde{\pi}_v$ corresponds to

$$\tilde{\mu}_v = (\mu_v, \sum \frac{\mu_{v,i} - \nu_{v,i}}{2} w_i) \in X^\bullet(T)_\mathbb{C} \oplus X^\bullet(\widetilde{Z})_\mathbb{C},$$

and

$$\tilde{\nu}_v = (\nu_v, \sum \frac{\nu_{v,i} - \mu_{v,i}}{2} w_i) \in X^\bullet(T)_\mathbb{C} \oplus X^\bullet(\widetilde{Z})_\mathbb{C}.$$

We write this parameter as an obviously integral term plus a defect:

$$\tilde{\mu}_v = (\mu_v, \sum \mu_{v,i} w_i) + (0, -\sum \frac{\mu_{v,i} + \nu_{v,i}}{2} w_i),$$

and likewise for $\tilde{\nu}_v$. Note that this lies in $X(\widetilde{T})$ if and only if for all $i = 1, \dots, r$, we have $\frac{\mu_{v,i}+\nu_{v,i}}{2} \in d_i\mathbb{Z}$. The discussion is so far general; if we now assume π_v is tempered, then for all $\lambda \in X_\bullet(T)$, the character

$$z \mapsto z^{\langle \mu_v, \lambda \rangle} \bar{z}^{\langle \nu_v, \lambda \rangle}$$

is unitary, i.e.

$$Re(\langle \mu_v + \nu_v, X_\bullet(T) \rangle) = 0.$$

W-algebraic representations of course have real infinitesimal character, so $\mu_v = -\nu_v$, and therefore in the initial choice of lift we may assume $\mu_{v,i} = -\nu_{v,i}$ for all i. Then obviously $\tilde{\mu}_v$ is L-algebraic.

If π is only W-algebraic, then we have to check that $2\tilde{\mu}_v = (2\mu_v, \sum(\mu_{v,i} - \nu_{v,i})w_i)$ lies in $X^\bullet(\widetilde{T})$. This element of $X^\bullet(T) \oplus X^\bullet(\widetilde{Z})$ represents an element of $X^\bullet(\widetilde{T})$ if and only if $2\mu_v$ maps to $\sum_i [\mu_v - \nu_v]_i \bar{w}_i$ in $X^\bullet(Z_G)$. The latter is also the image of $\mu_v - \nu_v$, so it is equivalent to ask that $\mu_v + \nu_v \in X^\bullet(T)$ map to zero in $X^\bullet(Z_G)$. As above, temperedness of π_v guarantees this, so we are done.

[6]No assumption at finite places.

For the final claim, we again apply the result to the inclusion $G \subset \widetilde{G}$, and then restrict to $\mathbf{G}$; restriction of course preserves L-algebraicity and W-algebraicity. $\square$

REMARK 3.1.13. This result would immediately extend to totally imaginary fields if we knew that ω_π extended to a type A_0 Hecke character of $\widetilde{Z}(\mathbf{A}_F)$. As noted in Remark 2.4.10, this would follow from the ('CM descent') conjectures of §2.4.

We now turn to the case of totally real F. I am grateful to Brian Conrad for urging a coordinate-free formulation of the obstruction in part 2 of the proposition.

PROPOSITION 3.1.14. *Now suppose F is totally real, with $\pi, \tilde{\omega}, \tilde{\pi}$ as before ($\tilde{\omega}$ finite-order as in Definition 3.1.2). Alternatively, let F be arbitrary, but assume that ω_π admits a finite-order extension $\tilde{\omega}$. Continue to assume π_∞ is tempered.*

1. *If π is L-algebraic, then it admits a W-algebraic lift $\tilde{\pi}$.*
2. *Assume Hypothesis 3.1.8 for the 'only if' direction of this statement. Assume F is totally real, and consider this W-algebraic lift $\tilde{\pi}$. Then the images of μ_v and ν_v under $X^\bullet(T) \to X^\bullet(Z_G)$ lie in $X^\bullet(Z_G)[2]$, and π admits an L-algebraic lift if and only if these images are independent of $v|\infty$.*

Similarly, any L-algebraic π on G admits a W-algebraic lift to $\mathbf{G}$, and it admits an L-algebraic lift if (and, under Hypothesis 3.1.8, only if) the images of the μ_v in $\mathrm{coker}\,(X^\bullet(\mathbf{Z})_{\mathrm{tor}} \to X^\bullet(Z_G)) = X^\bullet(Z_G \cap \mathbf{Z}^0)$ *are independent of $v|\infty$.*

PROOF. We continue with the parameter notation of the previous proof. Let π be L-algebraic. First we check that since $\tilde{\omega}$ can be chosen finite-order (automatic in the totally real case, but an additional assumption at complex places) $\mu_v - \nu_v$ maps to zero in $X^\bullet(Z_G)$. If v is imaginary, then $\tilde{\omega}$ cannot be finite order unless ω_{π_v} is trivial, hence μ_v and ν_v themselves are trivial in $X^\bullet(Z_G)$. If v is real, then in $T^\vee \subset G^\vee$ we have the relation

$$\phi(j) z^{\mu_v} \bar{z}^{\nu_v} \phi(j)^{-1} = z^{\nu_v} \bar{z}^{\mu_v},$$

writing ϕ for the L-parameter and $j \in W_{F_v} - W_{\overline{F}_v}$. Now, $\phi(j) \in N_{G^\vee}(T^\vee)$ represents an element w of the Weyl group of $G^\vee$, and this yields the relations $w\mu_v = \nu_v$ and $w\nu_v = \mu_v$. Thus $\mu_v - \nu_v = \mu_v - w\mu_v$, which lies in the root lattice Q of G. [This holds for $w\chi - \chi$ for any $\chi \in X^\bullet(T)$: reduce to the case of simple reflections, where it is clear from the defining formula.] Restricting characters to Z_G factors through a perfect duality $Z_G \times X^\bullet(T)/Q \to \mathbb{Q}/\mathbb{Z}$, so $\mu_v - w\mu_v \in Q$ has trivial image in $X^\bullet(Z_G)$.

Any lift with finite-order central character has parameters $\tilde{\mu}_v = (\mu_v, 0)$ and $\tilde{\nu}_v = (\nu_v, 0)$; this pair yields a well-defined representation of $W_{\overline{F}_v}$, i.e. $\tilde{\mu}_v - \tilde{\nu}_v \in X^\bullet(\widetilde{T})$, and its projections to T and $\widetilde{Z}$ are what they have to be, so they are the only possible parameters; that this extends (possibly non-uniquely) to an L-parameter on the whole of W_{F_v} follows from general theory (Langlands' Lemma), but we do not need the details of this extension. Now we use the assumption that π_v is tempered: as in the previous proposition, we see that $\mu_v = -\nu_v$. Hence $2\mu_v$ maps to zero in $X^\bullet(Z_G)$, and so $2(\mu_v, 0) \in X^\bullet(\widetilde{T})$, i.e. $(\mu_v, 0)$ is W-algebraic.

For F totally real, we now show the second part of the proposition. By Lemma 3.1.9, if a second lift $\tilde{\pi}'$ is L-algebraic, then

$$\tilde{\pi}' \cong \tilde{\pi} \cdot \left(\prod_{i=1}^{r} \alpha_i \circ d_i w_i \right),$$

where each $\alpha_i^{d_i}$ is a Hecke character of our totally real field, and therefore (since $\tilde{\pi}$ and $\tilde{\pi}'$ both have rational infinitesimal character) takes the form $\chi_i \cdot |\cdot|^{d_i y_i}$ for a finite-order character χ_i and a rational number y_i. The infinitesimal character of $\tilde{\pi}'_v$ (for $v|\infty$) then corresponds to $(\mu_v, \sum y_i d_i w_i) \in X^\bullet(\widetilde{T})_{\mathbb{Q}}$, which is integral if and only if $y_i \in \frac{1}{d_i}\mathbb{Z}$ and $y_i d_i$ is the image of μ_v in $\mathbb{Z}/d_i\mathbb{Z}$ for each i. The claim follows easily.

It follows immediately that an L-algebraic π on G extends to a W-algebraic automorphic representation on $\mathbf{G}$. For the last claim, note that

$$\operatorname{coker}\left(X^\bullet(\mathbf{Z})_{\mathrm{tor}} \to X^\bullet(Z_G)\right)$$

is canonically isomorphic to $X^\bullet(Z_G \cap \mathbf{Z}^0)$. Our previous arguments apply to the inclusion (enlarging a *subgroup* of Z_G to a torus) $G \subset (G \times \mathbf{Z}^0)/(Z_G \cap \mathbf{Z}^0) \xrightarrow{\sim} \mathbf{G}$ (this map is an isomorphism since $\mathbf{G}$ is connected reductive with derived group G), showing the existence of an L-algebraic extension of π precisely when the image of μ_v in $X^\bullet(Z_G \cap \mathbf{Z}^0)$ is independent of $v|\infty$. □

COROLLARY 3.1.15. *Let F be totally real and π be L-algebraic on the F-group G.*

(1) *Suppose π_∞ is tempered. If all d_i are odd, then $\tilde{\pi}$ is L-algebraic. In particular, if the simple factors of G are all type A_{2n} (i.e. SL_{2n+1}), or of type E_6, E_8, F_4, G_2,[7] then any L-algebraic π has an L-algebraic lift.*
(2) *Assume $F \neq \mathbb{Q}$ (still totally real). For F-groups G that are simple simply-connected (and split) of type B_n, C_n, D_{2n}, and E_7, there exist, assuming Hypothesis 3.1.8 for the pair $(G, \widetilde{G})$, L-algebraic π on G, tempered at infinity, that admit no L-algebraic lift to $\widetilde{G}$.*

PROOF. The first part is immediate from the proof of the proposition: for d_i odd, if the image of $2\mu_v$ in $\mathbb{Z}/d_i\mathbb{Z}$ is trivial, then so of course is the image of μ_v. For the second part, we use existence results for automorphic forms that are discrete series at infinity as in Lemma 2.6.10 and Proposition 2.8.7. For a semi-simple, simply-connected group, ρ lies in the weight lattice, so discrete series representations will be L-algebraic. Choose a quotient $G \twoheadrightarrow G' \twoheadrightarrow G_{\mathrm{ad}}$, with T' the induced maximal torus of G', such that $X^\bullet(T')/Q \cong X^\bullet(Z_G)[2]$ under the isomorphism $X^\bullet(T)/Q \xrightarrow{\sim} X^\bullet(Z_G)$.Then we can find a discrete series representation of $G(F_\infty)$ such that at two infinite places v_1 and v_2, the parameters have the form (on $\mathbb{C}^\times$)

$$z \mapsto z^{\rho + \mu_{v_i}} \bar{z}^{-\rho - \mu_{v_i}},$$

where the μ_{v_i} are distinct in $X^\bullet(T')/Q$.[8] This ensures that the $\rho + \mu_{v_i}$ have distinct images in $X^\bullet(Z_G)[2]$. It remains therefore to check that the split real

[7] A_{2n} and E_6 are the interesting examples here, since in the other cases the adjoint group is simply-connected. The index $[X^\bullet(T) : Q]$ for the simply-connected simple group of type E_6 is $\mathbb{Z}/3\mathbb{Z}$.

[8] We have shifted what was previously denoted μ_v by ρ in this case because of the description of discrete series L-parameters as arising from elements of $\rho + X^\bullet(T)$. Note that ρ mapst to $X^\bullet(Z_G)[2]$ since 2ρ is in the root lattice Q.

forms of these groups, except for A_{2n-1}, actually admit discrete series. Type C_n was treated previously. For type B_n, the (real) split group is $\mathrm{Spin}(n, n+1)$, which always admits discrete series (real orthogonal groups $\mathrm{SO}(p,q)$ have discrete series whenever pq is even). Likewise, for type D_n, the split form $\mathrm{Spin}(n,n)$ has discrete series if and only if n is even. The split real form of E_7 also admits discrete series: its maximal compact subgroup is $\mathrm{SU}(8)$, which contains a compact Cartan isomorphic to $(\mathrm{S}^1)^7$ (see the table in Appendix 4 of [**Kna02**]). □

REMARK 3.1.16.

- Arthur constructs in [**Art02**] a conjectural candidate for the automorphic Langlands group $\mathcal{L}_F$, also giving an 'automorphic' construction of a candidate for the motivic Galois group $\mathcal{G}_F$. (In §4.1 we will discuss motivic Galois groups in detail, explaining both conjectural and unconditional constructions.) The ability to extend L-algebraic representations of G to L-algebraic representations of $\widetilde{G}$ is essential to Arthur's conjectural construction (see [**Art02**, §6]) of a morphism $\mathcal{L}_F \to \mathcal{G}_F$. Roughly, the factor of the motivic Galois group $\mathcal{G}_F$ associated to π (or rather its almost-everywhere system of Hecke eigenvalues; this construction is restricted to suitably 'primitive' π) is defined as a sub-group $\mathcal{G}_\pi$ of $\widetilde{G}^\vee \times \mathcal{T}_F$, with $\mathcal{T}_F$ the Taniyama group (see §4.1.6 for background on $\mathcal{T}_F$); then $\mathcal{L}_F$ and $\mathcal{G}_F$ are constructed as suitable fiber products over varying pairs (G,π).[9] The natural motivic Galois representation corresponding to 'the motive of π' is the projection to $G^\vee$, and implicit in the construction is the fact that this lifts to $\widetilde{G}^\vee$. Corollary 3.1.15 then gives concrete examples of automorphic representations (including ones that ought to be 'primitive,' arranging the local behavior suitably) for which the Tannakian formalism cannot behave in this way; that is, while Arthur's construction is consistent with our discussion in the CM case (Proposition 3.1.12), it must be modified for totally real fields F.
- If our lifting results were generalized to quasi-split groups, we could presumably include D_n for n odd in the second part of the Proposition, since the non-split quasi-split form has signature $(n+1, n-1)$, for which the associated orthogonal group admits discrete series. Similarly, we could include type A_{2n-1} by using the quasi-split unitary group $\mathrm{SU}(n,n)$ instead of SL_{2n}.

3.2. Galois lifting: The general case

Now we discuss a general framework for Conrad's lifting problem. We consider lifting problems of the form

$$\begin{array}{ccc} & & \widetilde{H}(\overline{\mathbb{Q}}_\ell) \\ & \overset{\tilde{\rho}}{\nearrow} & \downarrow \\ \Gamma_F & \xrightarrow[\rho]{} & H(\overline{\mathbb{Q}}_\ell), \end{array}$$

[9]For non-split G, one works with the semi-direct product $\widetilde{G}^\vee \rtimes \mathcal{T}_F$, with $\mathcal{T}_F$ acting via its projection to the global Weil group W_F.

for any surjection $\widetilde{H} \twoheadrightarrow H$, with central torus kernel, of linear algebraic groups over $\overline{\mathbb{Q}}_\ell$. In this picture, we want to understand when a geometric ρ does or does not admit a geometric lift $\tilde{\rho}$. We will see that the general case reduces to the case where $\widetilde{H}$ and H are connected reductive, and so to reap the psychological benefits of the work of §3.1, we will begin by considering the case $\widetilde{H} = \widetilde{G}^\vee$, $H = G^\vee$, where $G \subset \widetilde{G}$ is an inclusion of connected reductive[10] split F-groups, constructed by extending the center Z_G of G to a central torus $\widetilde{Z}$; that is

$$\widetilde{G} = (G \times \widetilde{Z})/Z_G.$$

The quotient $\widetilde{Z}/Z_G$ is a torus S, and we get an exact sequence

$$1 \to S^\vee \to \widetilde{G}^\vee \to G^\vee \to 1,$$

and a canonical isomorphism of Lie algebras $\tilde{\mathfrak{g}}^\vee \cong \mathfrak{g}^\vee \oplus \mathfrak{s}^\vee$. We again caution the reader that, even if $\widetilde{H}$ and H are assumed connected reductive, this is *not* the most general situation: namely, the center $\widetilde{Z}$ need not be connected in order to have $\widetilde{G}^\vee \twoheadrightarrow G^\vee$ a surjection with central torus kernel. Nevertheless, at least for F totally imaginary, we will reduce the general analysis to this case. To start, we elaborate on some of the associated group theory, slightly recasting the notation of the previous section to take into account the fact that G may have positive-dimensional center. The exact sequence

$$1 \to Z_G \to \widetilde{Z} \to S \to 1$$

gives rise, by applying $X^\bullet(\cdot)$ and then $\mathrm{Hom}(\cdot, \mathbb{Z})$, to an exact sequence

$$1 \to \mathrm{Hom}(X^\bullet(Z_G), \mathbb{Z}) \to X_\bullet(\widetilde{Z}) \to X_\bullet(S) \to \mathrm{Ext}^1(X^\bullet(Z_G), \mathbb{Z}) \to 1.$$

Using invariant factors, it is convenient to fix a basis of $X^\bullet(\widetilde{Z})$ such that the inclusion $X^\bullet(S) \subset X^\bullet(\widetilde{Z})$ is in coordinates

$$\begin{array}{ccccc}
X^\bullet(S) & \longrightarrow & X^\bullet(\widetilde{Z}) & \longrightarrow & X^\bullet(Z_G) \\
\| & & \| & & \| \\
\bigoplus_{i=1}^r \mathbb{Z} d_i w_i & \longrightarrow & \bigoplus_{i=1}^{r+s} \mathbb{Z} w_i & \longrightarrow & \bigoplus_{i=1}^r \mathbb{Z}/d_i\mathbb{Z}\bar{w}_i \oplus \bigoplus_{j=1}^s \mathbb{Z}\bar{w}_{r+j}
\end{array}$$

Let w_i^* denote the dual basis for $X^\bullet(\widetilde{Z}^\vee)$, and let ν_i^* denote the basis of $X^\bullet(S^\vee)$ dual to $d_i w_i$; in particular, w_i^* maps to $d_i \nu_i^*$ under $X^\bullet(\widetilde{Z}^\vee) \to X^\bullet(S^\vee)$.

We work with the maximal torus $\widetilde{T} = (T \times \widetilde{Z})/Z_G$ of $\widetilde{G}$, and deduce from its definition an exact sequence

$$1 \to X^\bullet(\widetilde{T}) \to X^\bullet(T) \oplus X^\bullet(\widetilde{Z}) \to X^\bullet(Z_G) \to 1,$$

and thus a (crucial) exact sequence

(3.1)
$$1 \to \mathrm{Hom}(X^\bullet(Z_G), \mathbb{Z}) \to X^\bullet(T^\vee) \oplus X^\bullet(\widetilde{Z}^\vee) \to X^\bullet(\widetilde{T}^\vee) \to \mathrm{Ext}^1(X^\bullet(Z_G), \mathbb{Z}) \to 1.$$

---

[10] Unlike in §3.1, whose results were largely intended to motivate the results of this section, we no longer require G to be semi-simple.

Note also that there is a canonical isomorphism $X^\bullet(\widetilde{Z}^\vee) \xrightarrow{\sim} X^\bullet(\widetilde{G}^\vee)$, which follows from exactness of the row in the diagram:

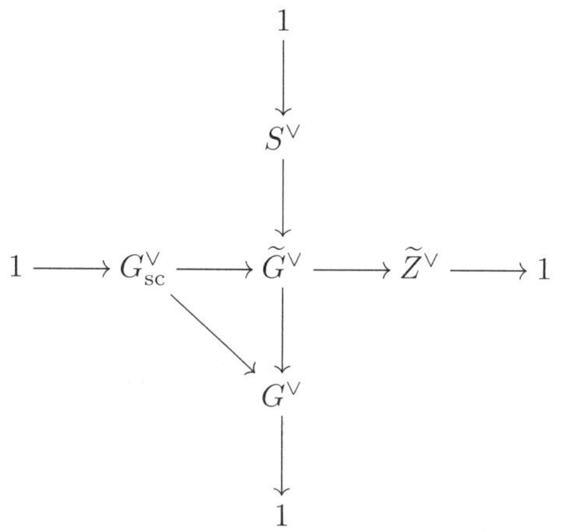

where $G^\vee_{\mathrm{sc}}$ denotes the simply-connected cover of the derived group of $G^\vee$. Recall that we index Sen operators $\Theta_{\rho,\iota}$ by embeddings $\iota\colon \overline{\mathbb{Q}}_\ell \hookrightarrow \mathbb{C}_{F_v}$, where v is a place above ℓ. Given a geometric $\rho\colon \Gamma_F \to G^\vee(\overline{\mathbb{Q}}_\ell)$, Conrad's result (see Theorem 2.1.6, and Remark 2.1.5) implies that a lift $\tilde{\rho}$ is geometric if and only if it is Hodge-Tate at all places above ℓ; we will use this from now on without comment. That is, we need to arrange $\tilde{\rho}$ such that each Sen operator $\Theta_{\tilde{\rho},\iota} \in \mathrm{Lie}(\widetilde{G}^\vee)_\iota$ is conjugate to an element of $\mathrm{Lie}(\widetilde{T}^\vee)_\iota$ that pairs integrally with all of $X^\bullet(\widetilde{T}^\vee)$[11] under the natural map

$$X^\bullet(\widetilde{T}^\vee) \xrightarrow{\mathrm{Lie}} \mathrm{Hom}(\mathrm{Lie}(\widetilde{T}^\vee), \overline{\mathbb{Q}}_\ell).$$

Here is our starting-point:

LEMMA 3.2.1. *There exists some lift $\tilde{\rho}$ of ρ. Any other lift is of the form*

$$\tilde{\rho}(\sum_{i=1}^{r} \nu_i \circ \chi_i)\colon g \mapsto \tilde{\rho}(g) \cdot \prod_{i=1}^{r} (\nu_i \circ \chi_i)(g),$$

where the $\nu_i = d_i w_i$ range over the above basis of $X_\bullet(S^\vee)$, and each $\chi_i\colon \Gamma_F \to \overline{\mathbb{Q}}_\ell^\times$ is a continuous character.

PROOF. A lift exists by Proposition 2.1.4. Although continuous cohomology does not in general have good δ-functorial properties, short exact sequences do give (not very) long exact sequences on H^0 and H^1, so any lift has the form $\tilde{\rho}(\chi)$ for some $\chi\colon \Gamma_F \to S^\vee(\overline{\mathbb{Q}}_\ell)$. We compose with the dual characters ν_i^* in $X^\bullet(S^\vee)$ to put χ in the promised form. □

The following lemma is the general substitute for choosing lifts with finite-order Clifford norm in the spin examples; this result is also implicit in the proof of 2.1.4, but a little warm-up with our notation is perhaps helpful.

---

[11]This condition is independent of the way $\Theta_{\tilde{\rho},\iota}$ is conjugated into $\mathrm{Lie}(\widetilde{T}^\vee)_\iota$, since: (1) we already know that $\Theta_{\tilde{\rho},\iota}$ pairs integrally with the roots, which all lie in $X^\bullet(T^\vee)$; (2) the ambiguity in conjugating into $\mathrm{Lie}(\widetilde{T}^\vee)_\iota$ is an element of the Weyl group; (3) for any weight $\lambda \in X^\bullet(\widetilde{T}^\vee)$ and w in the Weyl group, $w\lambda - \lambda$ lies in the root lattice. Compare the proof of Proposition 3.1.14.

LEMMA 3.2.2. *Let $\rho\colon \Gamma_F \to G^\vee(\overline{\mathbb{Q}}_\ell)$ be a geometric representation. Then there exists a lift $\tilde{\rho}\colon \Gamma_F \to \widetilde{G}^\vee(\overline{\mathbb{Q}}_\ell)$ such that, for all $v|\ell$ and all $\iota\colon \overline{\mathbb{Q}}_\ell \hookrightarrow \mathbb{C}_{F_v}$, the Sen operator $\Theta_{\tilde{\rho},\iota}$ pairs integrally with all of $X^\bullet(\widetilde{Z}^\vee) \cong X^\bullet(\widetilde{G}^\vee)$.*

PROOF. It suffices to find a lift whose composition with all elements of $X^\bullet(\widetilde{Z}^\vee)$ is Hodge-Tate. We use the bases of the various character groups specified above. In particular, composing an initial lift $\tilde{\rho}$ with the various $w_i^* \in X^\bullet(\widetilde{G}^\vee)$, $i = 1, \dots, r$, we can write

$$w_i^* \circ \tilde{\rho} = \chi_i^{d_i}\chi_{i,0}\colon \Gamma_F \to \overline{\mathbb{Q}}_\ell^\times,$$

where the χ_i and $\chi_{i,0}$ are Galois characters with $\chi_{i,0}$ finite-order. Then we consider the new lift

$$\tilde{\rho}' = \tilde{\rho}(\sum_{i=1}^{r}(d_i w_i) \circ \chi_i^{-1}),$$

which has the advantage that $w_i^* \circ \tilde{\rho}' = (w_i^* \circ \tilde{\rho}) \cdot \chi_i^{-d_i}$ is finite-order for all $i = 1, \dots, r$. Moreover, for the characters w_{r+j}^*, $j = 1, \dots, s$, namely, the sub-module $\mathrm{Hom}(X^\bullet(Z_G), \mathbb{Z}) \subset X^\bullet(\widetilde{Z}^\vee)$, the compositions $w_{r+j}^* \circ \tilde{\rho}'$ are all geometric, since ρ is. Therefore $\alpha \circ \tilde{\rho}'$ is geometric for all $\alpha \in X^\bullet(\widetilde{Z}^\vee)$, as desired. □

Returning to equation (3.1) on page 76, we see that the obstruction to geometric lifts then comes from $\mathrm{Ext}^1(X^\bullet(Z_G)_{\mathrm{tor}}, \mathbb{Z})$. For any weight $\lambda \in X^\bullet(\widetilde{T}^\vee)$, there is a positive integer $d \in \mathbb{Z}$ such that $d\lambda \in X^\bullet(T^\vee) \oplus X^\bullet(\widetilde{Z}^\vee)$, so with a $\tilde{\rho}$ as produced by Lemma 3.2.2, $\Theta_{\tilde{\rho},\iota}$ pairs integrally with $d\lambda$ for all ι. We obtain a well-defined class, independent of the choice of $\tilde{\rho}$ as constructed in the proof of Lemma 3.2.2 (namely, with the compositions $w_i^* \circ \tilde{\rho}$ finite-order for $i = 1, \dots, r$),

$$\langle \lambda, \Theta_{\tilde{\rho},\iota} \rangle \in \mathbb{Q}/\mathbb{Z},$$

which can also be interpreted as the common value modulo $\mathbb{Z}$ of the eigenvalues of $\Theta_{r_\lambda \circ \tilde{\rho},\iota} = \mathrm{Lie}(r_\lambda) \circ \Theta_{\tilde{\rho},\iota}$; here, and in what follows, we denote by r_λ the irreducible representation of $\widetilde{G}^\vee$ associated to the (dominant) weight $\lambda \in X^\bullet(\widetilde{T}^\vee)$. This pairing factors through a map

$$\mathrm{Ext}^1(X^\bullet(Z_G), \mathbb{Z}) \to \mathbb{Q}/\mathbb{Z},$$

and since for any finite abelian group A, the long exact sequence associated to $0 \to \mathbb{Z} \to \mathbb{Q} \to \mathbb{Q}/\mathbb{Z} \to 0$ yields an isomorphism

$$\mathrm{Hom}(A, \mathbb{Q}/\mathbb{Z}) \xrightarrow{\sim} \mathrm{Ext}^1(A, \mathbb{Z}),$$

we can make the following definition:

DEFINITION 3.2.3. Let $\theta_{\rho,\iota}$ be the element of $X^\bullet(Z_G)_{\mathrm{tor}}$ canonically corresponding to the above map $\mathrm{Ext}^1(X^\bullet(Z_G), \mathbb{Z}) \to \mathbb{Q}/\mathbb{Z}$.

To make further progress, we need to assume that ρ satisfies certain Hodge-Tate weight symmetries.

HYPOTHESIS 3.2.4. *Let H be a linear algebraic group and $\rho\colon \Gamma_F \to H(\overline{\mathbb{Q}}_\ell)$ a geometric Galois representation with connected reductive algebraic monodromy group $H_\rho = \overline{(\rho(\Gamma_F))}^{\mathrm{Zar}}$. We formulate the following Hodge-Tate symmetry hypothesis for such a ρ:*

- *Let r be any irreducible algebraic representation $r\colon H_\rho \to \mathrm{GL}(V_r)$. Then:*
  (1) *For $\tau\colon F \hookrightarrow \overline{\mathbb{Q}}_\ell$, the set $\mathrm{HT}_\tau(r \circ \rho)$ depends only on $\tau_0 = \tau|_{F_{\mathrm{cm}}}$.*

(2) *Writing* HT_{τ_0} *for this set common to all* τ *above* τ_0, *there exists an integer* w *such that*[12]

$$\mathrm{HT}_{\tau_0 \circ c}(r \circ \rho) = w - \mathrm{HT}_{\tau_0}(r \circ \rho),$$

for c *the unique complex conjugation on* F_{cm}.

For ρ *whose algebraic monodromy group is reductive but not necessarily connected, the corresponding hypothesis is simply that some finite restriction (with connected monodromy group) of* ρ *satisfies the above.*

In particular, we note for later use that the lowest weight in $\mathrm{HT}_{\tau_0 \circ c}(r \circ \rho)$ is w minus the highest weight in $\mathrm{HT}_{\tau_0}(r \circ \rho)$. We will see that to establish lifting results for geometric representations $\rho \colon \Gamma_F \to G^\vee(\overline{\mathbb{Q}}_\ell)$, we will only need to know Hypothesis 3.2.4 for some easily identifiable finite collection of compositions $r_\lambda \circ \rho$, $\lambda \in X^\bullet(T^\vee)$, but we do not make that explicit here. Most important, Hypothesis 3.2.4 should in fact be no additional restriction on ρ, because of the following conjecture, which would follow from various versions of the conjectural Fontaine-Mazur-Langlands correspondence and Tate conjecture (Conjecture 1.2.1):

CONJECTURE 3.2.5. *Let* F *be a number field, and let* $\rho \colon \Gamma_F \to \mathrm{GL}_N(\overline{\mathbb{Q}}_\ell)$ *be an irreducible geometric Galois representation. Then:*

(1) *For* $\tau \colon F \hookrightarrow \overline{\mathbb{Q}}_\ell$, *the set* $\mathrm{HT}_\tau(\rho)$ *depends only on* $\tau_0 = \tau|_{F_{\mathrm{cm}}}$. *(This will still hold if* ρ *is geometric but reducible.)*
(2) *Writing* HT_{τ_0} *for this set common to all* τ *above* τ_0, *there exists an integer* w *such that*

$$\mathrm{HT}_{\tau_0 \circ c}(\rho) = w - \mathrm{HT}_{\tau_0}(\rho),$$

for c *the unique complex conjugation on* F_{cm}.

Unfortunately, for an abstract Galois representation, this conjecture will be extremely difficult to establish. The next lemma explains it in the automorphic case; for a motivic variant, see Corollary 4.1.26.[13]

LEMMA 3.2.6. *Suppose* $\rho \colon \Gamma_F \to \mathrm{GL}_N(\overline{\mathbb{Q}}_\ell)$ *is an irreducible geometric representation. If* ρ *is automorphic in the sense of Conjecture* 1.2.1, *corresponding to a cuspidal automorphic representation* π *of* $\mathrm{GL}_N(\mathbf{A}_F)$, *and if we assume that Proposition* 2.4.8 *is unconditional for* π *(i.e., admit Hypothesis* 2.4.6*), then Conjecture* 3.2.5 *holds for* ρ.

PROOF. This is immediate from the passage between infinity-types and Hodge-Tate weights, Conjecture 2.4.8, and Clozel's archimedean purity lemma (which was proven as part of Proposition 2.5.8). □

We can now understand when geometric lifts ought to exist; the proof proceeds by reduction to the following key case:

[12] Interpreted in the obvious way.

[13] But note for now that in the most basic motivic cases, where the Galois representation is given by $H^j(X_{\overline{F}}, \overline{\mathbb{Q}}_\ell)$ for some smooth projective variety X/F, the Hodge-Tate symmetries are immediate (even when this representation is reducible) from the ℓ-adic comparison isomorphism of [**Fal89**]; if j is even, the symmetries similarly hold for primitive cohomology. What is not obvious is that if this Galois representation decomposes, that the irreducible factors all satisfy the conjecture.

PROPOSITION 3.2.7. *Let F be a totally imaginary field, and let $\rho\colon \Gamma_F \to G^\vee(\overline{\mathbb{Q}}_\ell)$ be a geometric representation with algebraic monodromy group equal to the whole of $G^\vee$. Assume ρ satisfies Hypothesis* 3.2.4. *Then ρ admits a geometric lift $\tilde{\rho}\colon \Gamma_F \to \widetilde{G}^\vee(\overline{\mathbb{Q}}_\ell)$.*

PROOF. Choose a lift $\tilde{\rho}$ as supplied by Lemma 3.2.2. Recall that our weight-bookkeeping is done within the diagram

$$\begin{array}{ccccccc} & & 0 & & & & \\ & & \uparrow & & & & \\ X^\bullet(\widetilde{Z}^\vee) & \longrightarrow & X^\bullet(S^\vee) & \longrightarrow & \mathrm{Ext}^1(X^\bullet(Z_G),\mathbb{Z}) & \longrightarrow & 0 \\ & & \uparrow & & & & \\ & & X^\bullet(\widetilde{T}^\vee) & & & & \\ & & \uparrow & & & & \\ & & X^\bullet(T^\vee) & & & & \\ & & \uparrow & & & & \\ & & 0 & & & & \end{array}$$

We have the elements $\nu_i^* \in X^\bullet(S^\vee)$ dual to $d_i w_i \in X^\bullet(S)$. Their images in $\mathrm{Ext}^1(X^\bullet(Z_G),\mathbb{Z})$ form a basis. Let λ_i be a (dominant weight) lift to $X^\bullet(\widetilde{T}^\vee)$ that also satisfies $\langle \lambda_i, d_j w_j\rangle = \delta_{ij}$. Note that the value $\langle \Theta_{\tilde{\rho},\iota}, \lambda_i\rangle \in \mathbb{Q}/\mathbb{Z}$ does not depend on the choice of lift of ν_i^*, and it clearly lies in $\frac{1}{d_i}\mathbb{Z}/\mathbb{Z}$, so we write it in the form $\frac{k_{\iota,i}}{d_i} + \mathbb{Z}$ for an integer $k_{\iota,i}$. By considering the geometric representation $r_{d_i\lambda_i} \circ \tilde{\rho}$, we find that $\Theta_{r_{d_i\lambda_i}\circ\tilde{\rho},\iota}$ has eigenvalues that depend only on τ_ι (by Lemma 2.2.7; see that lemma for the notation τ_ι as well), and thus we can write $k_{\tau,i}$ in place of $k_{\iota,i}$. (Equivalently, we can work with the elements $\theta_{\rho,\tau} \in X^\bullet(Z_G)_{\mathrm{tor}}$.) These classes $\frac{k_{\tau,i}}{d_i}$ mod $\mathbb{Z}$ serve as both highest and lowest τ-labeled Hodge-Tate weights (modulo $\mathbb{Z}$) for $r_{\lambda_i} \circ \tilde{\rho}$; we deduce that, modulo $d_i\mathbb{Z}$, the highest and lowest τ-labeled weights of $r_{d_i\lambda_i} \circ \tilde{\rho}$ are both congruent to $k_{\tau,i}$[14].

Now, the geometric Galois representations $(r_{d_i\lambda_i}) \circ \tilde{\rho}$ for $i = 1, \ldots, r$ are irreducible, because we have assumed that $G^\vee$ is the monodromy group of ρ. Applying Hypothesis 3.2.4, we deduce that k_τ depends only on $\tau_0 = \tau|_{F_{\mathrm{cm}}}$, along with the symmetry relation

$$k_{\tau_0,i} + k_{\tau_0 \circ c,i} \equiv w_i \mod d_i$$

for some integer w_i (for all $\tau_0\colon F_{\mathrm{cm}} \hookrightarrow \overline{\mathbb{Q}}_\ell$).

This relation allows us, by Lemma 2.3.16, to find Galois characters $\hat{\psi}_i\colon \Gamma_F \to \overline{\mathbb{Q}}_\ell^\times$ with $\mathrm{HT}_\tau(\hat{\psi}_i) \in \frac{k_{\tau,i}}{d_i} + \mathbb{Z}$ for all τ. We then form the twist $\tilde{\rho}' := \tilde{\rho}(\sum_i (d_i w_i) \circ \hat{\psi}_i^{-1})$; recall that $\langle d_i w_i, \lambda_j\rangle = \delta_{ij}$. This new lift is then geometric:

$$\langle \Theta_{\tilde{\rho}',\iota}, \lambda_i\rangle = \langle \Theta_{\tilde{\rho},\iota}, \lambda_i\rangle + \langle \Theta_{\sum(d_j w_j)\circ\hat{\psi}_j^{-1},\iota}, \lambda_i\rangle \equiv \frac{k_{\tau_\iota,i}}{d_i} - \frac{k_{\tau_\iota,i}}{d_i} \equiv 0 \mod \mathbb{Z}.$$

[14]Note that all eigenvalues of $\Theta_{r_{\lambda_i}\circ\tilde{\rho}}$ are congruent modulo $\mathbb{Z}$; this does not imply that all elements of $\mathrm{HT}_\tau(r_{d_i\lambda_i}\circ\tilde{\rho})$ are congruent modulo $d_i\mathbb{Z}$, but this congruence does hold for the highest and lowest weights.

(Recall it suffices to check $\tilde{\rho}'$ is Hodge-Tate, by Conrad's result, quoted here as Theorem 2.1.6.) □

COROLLARY 3.2.8. *Let $\rho\colon \Gamma_F \to G^\vee(\overline{\mathbb{Q}}_\ell)$ be geometric with algebraic monodromy group $G^\vee$; maintain the notation of the previous proof, but now suppose F is totally real. Then ρ has a geometric lift if and only if for varying ι, the elements $\theta_{\rho,\iota}$ (see Definition 3.2.3) in $X^\bullet(Z_G)_{\mathrm{tor}}$ are independent of ι.*

REMARK 3.2.9. More concretely, to determine whether ρ has a geometric lift, apply the following criterion:

(1) Ignore any i for which d_i is odd; these do not obstruct geometric lifting;
(2) Then for fixed i the integers $k_{\tau,i} \mod d_i$ are all $\frac{w_i}{2}$ translated by a two-torsion class in $\mathbb{Z}/d_i\mathbb{Z}$;
(3) ρ has a geometric lift if and only if each of these classes $k_{\tau,i}$ (or, equivalently, the associated two-torsion class) is independent of τ.

PROOF. By the previous proof and Lemma 2.3.17, ρ has a geometric lift if and only if the classes $k_{\tau,i} \mod d_i$ (fixed i, varying τ) are independent of τ. The weight-symmetry relation becomes $2k_{\tau,i} \equiv w_i \mod d_i$, and the claim follows easily. □

Over totally imaginary fields, we can now reduce the general lifting problem to the special case of full monodromy:

THEOREM 3.2.10. *Let F be totally imaginary, and let $\pi\colon \widetilde{H} \twoheadrightarrow H$ be any surjection of linear algebraic groups with central torus kernel. Suppose $\rho\colon \Gamma_F \to H(\overline{\mathbb{Q}}_\ell)$ is a geometric representation, with arbitrary image, satisfying Hypothesis 3.2.4. Then ρ admits a geometric lift $\tilde{\rho}\colon \Gamma_F \to \widetilde{H}(\overline{\mathbb{Q}}_\ell)$.*

PROOF. We may assume $\widetilde{H}$ and H are reductive since π induces an isomorphism on unipotent parts (we take this observation from Conrad, who has exploited this reduction in the arguments of [**Con11**]).

We next show that the theorem holds if $H_\rho := \overline{\rho(\Gamma_F)}^{\mathrm{Zar}} \subset H$ is connected. In that case, let $\widetilde{H}_\rho$ be the preimage in $\widetilde{H}$ of H_ρ. Then $\widetilde{H}_\rho \to H_\rho$ is a surjection of connected reductive groups with central torus kernel, and we may write $H_\rho = G^\vee$, $\widetilde{H}_\rho = \widetilde{G}^\vee$ where $G \subset \widetilde{G}$ is an inclusion of connected reductive groups of the form $\widetilde{G} = (G \times \widetilde{Z})/Z_G$ for some inclusion of Z_G into a multiplicative group $\widetilde{Z}$. If $\widetilde{Z}$ is not connected, Proposition 3.2.7 does not immediately apply, so we embed $\widetilde{Z}$ into a torus $\widetilde{\mathbf{Z}}$, with corresponding inclusions $G \subset \widetilde{G} \subset \widetilde{\mathbf{G}}$. Dually, $\widetilde{\mathbf{G}}^\vee \twoheadrightarrow G^\vee$ is a quotient to which we can apply Proposition 3.2.7, and then projecting from $\widetilde{\mathbf{G}}^\vee$ to $\widetilde{G}^\vee$, we obtain a geometric lift $\tilde{\rho}\colon \Gamma_F \to \widetilde{H}_\rho(\overline{\mathbb{Q}}_\ell) \subset \widetilde{H}(\overline{\mathbb{Q}}_\ell)$ of ρ.

For arbitrary ρ (i.e. H_ρ not necessarily connected), let F'/F be a finite extension such that $\overline{\rho(\Gamma_{F'})}^{\mathrm{Zar}} = H_\rho^0$ is connected. Over F', we can therefore find a geometric lift $\tilde{\rho}_{F'}\colon \Gamma_{F'} \to \widetilde{H}_\rho(\overline{\mathbb{Q}}_\ell) \subset \widetilde{H}(\overline{\mathbb{Q}}_\ell)$, letting $\widetilde{H}_\rho$ as before denote the preimage of H_ρ in $\widetilde{H}$. Thus, for *any* lift $\tilde{\rho}_0\colon \Gamma_{F'} \to \widetilde{H}(\overline{\mathbb{Q}}_\ell)$ of $\rho|_{\Gamma_{F'}}$, there exists a character $\hat{\psi}\colon \Gamma_{F'} \to S^\vee(\overline{\mathbb{Q}}_\ell)$ such that $\tilde{\rho}_0 \cdot \hat{\psi}$ is geometric. In particular, letting $\tilde{\rho}\colon \Gamma_F \to \widetilde{H}(\overline{\mathbb{Q}}_\ell)$ be any lift over F itself (with rational Hodge-Tate-Sen weights), there is a $\hat{\psi}_{F'}\colon \Gamma_{F'} \to S^\vee(\overline{\mathbb{Q}}_\ell)$ such that $\tilde{\rho}|_{\Gamma_{F'}} \cdot \hat{\psi}_{F'}$ is geometric. But by Corollary 2.3.16, there is a Galois character $\hat{\psi}\colon \Gamma_F \to S^\vee(\overline{\mathbb{Q}}_\ell)$ whose labeled Hodge-Tate-Sen weights descend those of $\hat{\psi}_{F'}$. It follows immediately that $\tilde{\rho} \cdot \hat{\psi} \colon \Gamma_F \to \widetilde{H}(\overline{\mathbb{Q}}_\ell)$ is a geometric lift of ρ. □

REMARK 3.2.11. (1) The theorem also lets us make explicit precisely which sets of labeled Hodge-Tate weights can be achieved in a geometric lift $\tilde{\rho}$. We will exploit this in §4.2.

(2) We will not treat the general totally real case here, since a somewhat different approach seems more convenient in that case. For a succinct, coordinate-free treatment of the totally real case, and another perspective on the arguments of this section, see [**Pat14b**], where the following general result is obtained: for any $\widetilde{H} \twoheadrightarrow H$ as in our lifting setup, we write $H^0 = G^\vee$ and $(\widetilde{H})^0 = \widetilde{G}^\vee$, where $\widetilde{G} = (G \times \widetilde{Z})/Z_G$; here $\widetilde{Z}$ is not necessarily connected. We can define in this generality elements

$$\theta_{\rho,\tau} \in \operatorname{coker}\left(X^\bullet(\widetilde{Z})_{\mathrm{tor}} \to X^\bullet(Z_G)_{\mathrm{tor}}\right),$$

and ρ admits a geometric lift to $\widetilde{H}$ if and only if the $\theta_{\rho,\tau}$ are independent of τ. The argument described in [**Pat14b**] has the disadvantage of not making explicit the parity obstruction (assuming 'Hodge symmetry') found in Proposition 3.2.8; for this reason we have retained the two different expositions of the totally real case.

The method of proof of Proposition 3.2.7 and Theorem 3.2.10 also implies the following local result, which emerged from a conversation with Brian Conrad. Here local algebraicity of Hodge-Tate representations replaces the appeal to the theory of algebraic Hecke characters. In fact, this connection holds more generally when K is the fraction field of a complete discrete valuation ring with perfect residue field: see [**CCO14**, §3.9].

COROLLARY 3.2.12. *Let $\widetilde{H} \twoheadrightarrow H$ be a central torus quotient, and let $\rho \colon \Gamma_{F_v} \to H(\overline{\mathbb{Q}}_\ell)$ be a Hodge-Tate representation of Γ_{F_v}, for $F_v/\mathbb{Q}_\ell$ finite. Then there exists a Hodge-Tate lift $\tilde{\rho} \colon \Gamma_{F_v} \to \widetilde{H}(\overline{\mathbb{Q}}_\ell)$.*

PROOF. We sketch a proof, replacing appeal to Hypothesis 3.2.4 and the existence of certain Hecke characters by the simpler observation that local class field theory lets us find the necessary twisting characters $\hat{\psi} \colon \Gamma_{F_v} \to \overline{\mathbb{Q}}_\ell^\times$ by hand. That is, $\mathcal{O}_{F_v}$ sits in $I_{F_v}^{\mathrm{ab}}$ in finite index, and on the former we can define the character

$$x \mapsto \prod_{\tau \colon K \hookrightarrow \overline{\mathbb{Q}}_\ell} \tau(x)^{k_\tau}$$

for any integers k_τ; then up to a finite-order character we take a dth root for an integer d to build (having extended to all of Γ_{F_v}) characters with any prescribed set of rational τ-labeled Hodge-Tate weights (τ running over all $F_v \hookrightarrow \overline{\mathbb{Q}}_\ell$). This observation suffices (invoking Lemma 2.2.7 in the full generality of Hodge-Tate, rather than merely de Rham, representations) for the previous arguments to carry through. □

This Corollary combined with [**Con11**, Proposition 6.5] implies the following stronger result:

COROLLARY 3.2.13. *Let $\widetilde{H} \twoheadrightarrow H$ be a central torus quotient, and let $\rho\colon \Gamma_{F_v} \to H(\overline{\mathbb{Q}}_\ell)$ be a representation of Γ_{F_v}, for $F_v/\mathbb{Q}_\ell$ finite, satisfying a basic p-adic Hodge theory property*[15] $\mathbf{P}$. *Then there exists a lift $\tilde{\rho}\colon \Gamma_{F_v} \to \widetilde{H}(\overline{\mathbb{Q}}_\ell)$ also satisfying* $\mathbf{P}$.

3.3. Applications: Comparing the automorphic and Galois formalisms

In §3.2, we took the Fontaine-Mazur-Langlands conjecture (or rather its weakened form Hypothesis 3.2.4) relating geometric Galois representations $\Gamma_F \to \mathrm{GL}_n(\overline{\mathbb{Q}}_\ell)$ to L-algebraic automorphic representations of $\mathrm{GL}_n(\mathbf{A}_F)$ as input to establish some of our lifting results. Now we want to apply these lifting results to give some evidence for the relationship between automorphic forms and Galois representations on groups other than GL_n. We will touch on the Buzzard-Gee conjecture, certain cases of the converse problem, and some general thoughts about comparing descent problems on the (ℓ-adic) Galois and automorphic sides. We first digress to discuss what 'automorphy' of an ${}^LG(\overline{\mathbb{Q}}_\ell)$-valued representation even means.

3.3.1. Notions of automorphy. As usual we have fixed $\iota_\infty\colon \overline{\mathbb{Q}} \hookrightarrow \mathbb{C}$ and $\iota_\ell\colon \overline{\mathbb{Q}} \hookrightarrow \overline{\mathbb{Q}}_\ell$. It is sometimes more convenient simply to fix an isomorphism $\iota_{\ell,\infty}\colon \mathbb{C} \to \overline{\mathbb{Q}}_\ell$, and to regard $\overline{\mathbb{Q}}$ as the subfield of algebraic numbers in $\mathbb{C}$. G is a connected reductive F-group, and we take a $\overline{\mathbb{Q}}$-form of the L-group ${}^LG$. For an automorphic representation π of $G(\mathbf{A}_F)$ and $\rho\colon \Gamma_F \to {}^LG(\overline{\mathbb{Q}}_\ell)$, always assumed continuous and composing with Γ_F-projection to the identity, there are (at least) four relations between such ρ and π that might be helpful, of which only the first two can be stated unconditionally. First we describe analogues that restrict to either the automorphic or Galois side, borrowing some ideas from [**Lap99**] (and, inevitably, [**LL79**]). For automorphic representations π and π' of $G(\mathbf{A}_F)$, three notions of similarity are:

- $\pi \sim_w \pi'$ if almost everywhere locally, the (unramified, say) L-parameters are $G^\vee(\mathbb{C})$-conjugate.
- $\pi \sim_{w,\infty} \pi'$ if almost everywhere locally, and at infinity, the L-parameters are conjugate. This also makes sense unconditionally, since archimedean local Langlands is known.
- $\pi \sim_{ew} \pi'$ if everywhere locally, the L-parameters are $G^\vee(\mathbb{C})$-conjugate; this only makes sense if one knows the local Langlands conjecture for G.
- $\pi \sim_s \pi'$ is the most fanciful condition: if the conjectural automorphic Langlands group $\mathcal{L}_F$ exists, so π and π' give rise to representations $\mathcal{L}_F \to {}^LG(\mathbb{C})$, then this condition requires these representations to be globally $G^\vee(\mathbb{C})$-conjugate.

A fifth notion would compare L-parameters in a particular finite-dimensional representation of $G^\vee$.

We can unconditionally make the same sort of comparisons between ℓ-adic Galois representations, writing $\rho \sim_w \rho'$, $\rho \sim_{w,\infty} \rho'$, $\rho \sim_{ew} \rho'$, and $\rho \sim_s \rho'$. By equivalence 'at infinity' here, we mean that at real places the actions of complex conjugation are conjugate, and at places above ℓ, the associated Sen operators (i.e. labeled Hodge-Tate data) are conjugate. Since it is conjectured, but totally out of reach, that frobenius elements act semi-simply in a geometric Galois representation, we should only compare 'frobenius semi-simplifications' in these definitions of local

[15] Namely: crystalline, semi-stable, de Rham, or Hodge-Tate.

equivalence. For some nice examples, Lapid's paper ([**Lap99**]) studies the difference between $\sim_w$, $\sim_{ew}$, and $\sim_s$ for certain Artin representations (and, when possible, the corresponding comparison on the automorphic side).

We then have corresponding ways to relate an ℓ-adic ρ and an automorphic π:

- $\rho \sim_w \pi$
- $\rho \sim_{w,\infty} \pi$
- $\rho \sim_{ew} \pi$
- $\rho \sim_s \pi$

First, write $\rho \sim_w \pi$ if for almost all unramified v (for ρ and π), $\rho|^{\mathrm{ss}}_{W_{F_v}}$ is $G^\vee(\overline{\mathbb{Q}}_\ell)$-conjugate to $\mathrm{rec}_v(\pi_v)\colon W_{F_v} \to G^\vee(\mathbb{C}) \rtimes \Gamma_F$; implicit is the assumption that the local parameter lands in $G^\vee(\overline{\mathbb{Q}}) \rtimes \Gamma_F$, so that we can apply $\iota_\ell \circ \iota_\infty^{-1}$. Note that this definition does not distinguish between π and other elements of its (conjectural) global L-packet $L(\pi)$. The other relations are straightforward modifications (for compatibility with complex conjugation, we take the condition in Conjecture 3.2.1 of [**BG11**]), except for $\rho \sim_s \pi$. Writing $\mathcal{G}_{F,E}(\sigma)$ for the motivic Galois group for motives[16]over F with E-coefficients, using a Betti realization via $\sigma\colon F \hookrightarrow \mathbb{C}$, one might hope that after fixing $E \hookrightarrow \mathbb{C}$ as well, one would obtain a map of pro-reductive groups over $\mathbb{C}$, $\mathcal{L}_F \to (\mathcal{G}_{F,E})(\sigma) \otimes_E \mathbb{C}$ (compare Remark 3.1.16). If π corresponds to a representation $\mathrm{rec}(\pi)$ of $\mathcal{L}_F$, and ρ arises from (completing at some finite place of E) a representation ρ_E of $\mathcal{G}_{F,E}(\sigma)$, it makes sense to ask whether $\mathrm{rec}(\pi)$ factors through $\mathcal{G}_{F,E}(\sigma)(\mathbb{C})$, and whether the resulting representation is globally $G^\vee(\mathbb{C})$-conjugate to (the complexification via $E \hookrightarrow \mathbb{C}$ of) ρ_E. Of course, any discussion of the automorphic Langlands group and its relation with the motivic Galois group is pure speculation; but these heuristics do provide context for the basic problems raised in Question 1.1.9 and Conjecture 1.1.10, as well as the work of §4.2.

Although we don't actually require it, it is helpful to keep in mind a basic lemma of Steinberg, which implies that $\sim_w$ can be checked by checking in all finite-dimensional representations:

LEMMA 3.3.1. *Let x and y be two semi-simple elements of a connected reductive group $G^\vee$ over an algebraically closed field of characteristic zero. If x and y are conjugate in every (irreducible) representation of $G^\vee$, or even merely have the same trace, then they are in fact conjugate in $G^\vee$.*

PROOF. For semi-simple groups, this is Corollary 3 (to Theorem 2) in Chapter 3 of [**Ste74**]; the proof extends to the reductive case (and even more generally, see Proposition 6.7 of [**Bor79**]). The key point is that the characters of finite-dimensional representations of $G^\vee$ restrict to a basis of the ring of Weyl-invariant regular functions on a maximal torus; these in turn separate conjugacy classes in the torus. □

3.3.2. Automorphy of projective representations.

Throughout this section, *we assume the Fontaine-Mazur-Langlands conjecture on automorphy of geometric* (GL_N-valued) *Galois representations*. It suffices to

[16]Either for absolutely Hodge cycles, for motivated cycles, or, assuming the standard conjectures, for homological cycles. Again, we will deal more precisely with motivic Galois groups in §4.1.

take a version that matches unramified (almost everywhere) and Hodge-theoretic parameters; to be precise, assume Part 3 of Conjecture 1.2.1, and note that this includes the requirement that cuspidality is equivalent to irreducibility under the automorphic-Galois correspondence. We will show how our lifting results–both automorphic and Galois-theoretic–give rise to a 'Fontaine-Mazur-Langlands'-type correspondence between algebraic automorphic representations of $\mathrm{SL}_N(\mathbf{A}_F)$ and $\mathrm{PGL}_N(\overline{\mathbb{Q}}_\ell)$-valued geometric Γ_F-representations.[17] The starting point is the following consequence of the results of §3.2:

COROLLARY 3.3.2. *Let F be totally imaginary. Then any geometric $\rho\colon \Gamma_F \to \mathrm{PGL}_n(\overline{\mathbb{Q}}_\ell)$ is weakly automorphic, i.e. there exists an L-algebraic automorphic representation π of $\mathrm{SL}_n(\mathbf{A}_F)$ such that $\rho \sim_{w,\infty} \pi$ (or $\rho \sim_{ew} \pi$, if we assume a form of Fontaine-Langlands-Mazur that matches local factors everywhere).*

PROOF. We have seen that ρ lifts to a geometric $\tilde{\rho}\colon \Gamma_F \to \mathrm{GL}_n(\overline{\mathbb{Q}}_\ell)$, which by assumption is automorphic, corresponding to some $\tilde{\pi}$ on GL_n/F. The irreducible constituents of $\tilde{\pi}|_{\mathrm{SL}_n(\mathbf{A}_F)}$ form a global L-packet whose local unramified parameters correspond to those of ρ. □

We now give descent arguments that extend this automorphy result to F totally real. First we need a couple of elementary lemmas.

LEMMA 3.3.3. *Let L/F be a cyclic, degree d, extension of number fields, with σ a generator of $\mathrm{Gal}(L/F)$. Let χ be a Hecke character of L, and let δ be any Hecke character whose restriction to $C_F \subset C_L$ is $\delta = \delta_{L/F}$, a fixed order d character that cuts out the extension L/F. Assume that $\chi^{1+\sigma+\dots+\sigma^{d-1}} = 1$. Then for a unique integer $i = 0, \dots, d-1$, $\chi\delta^i$ is of the form $\psi^{\sigma-1}$ for a Hecke character ψ of L.*

PROOF. We may assume χ is unitary. Write C^D as usual for the Pontryagin dual of a locally compact abelian group C. We have the following exact sequences:

$$\begin{array}{ccccccc}
 & & 1 & & & & \\
 & & \uparrow & & & & \\
 & & \mathrm{Gal}(L/F) & & & & \\
 & & \uparrow & & & & \\
1 & \longrightarrow & C_F & \longrightarrow & C_L & \xrightarrow{\sigma-1} & C_L \\
 & & {\scriptstyle N_{L/F}}\uparrow & & & & \\
 & & C_L & & & &
\end{array},$$

[17] For some unconditional results in this direction, see [**Pat14b**, §4].

dualizing to

$$\begin{array}{ccccccc}
 & & 1 & & & & \\
 & & \downarrow & & & & \\
 & & \mathrm{Gal}(L/F)^D & & & & \\
 & & \downarrow & & & & \\
1 & \longleftrightarrow & C_F^D & \xleftarrow{\ \mathrm{res}\ } & C_L^D & \xleftarrow{\sigma-1} & C_L^D \\
 & & \downarrow{\scriptstyle N_{L/F}} & & & & \\
 & & C_L^D & & & & .
\end{array}$$

By assumption, $N_{L/F} \circ \mathrm{res}(\chi) = 1$, so $\mathrm{res}(\chi) = \delta^{-i} \in \mathrm{Gal}(L/F)^D$ for some integer i, unique modulo d. Then $\mathrm{res}(\chi\delta^i) = 1$, and we are done by exactness of the horizontal diagram. □

We need a special case of an ℓ-adic analogue of the remark after Statement A of [**LR98**];[18] that remark is in turn the (much easier) analogue, for complex representations of the Weil group, of the main result of their paper. We first record the simple case that we need, and then out of independent interest we prove a general ℓ-adic analogue of the Lapid-Rogawski result.

LEMMA 3.3.4. *Let L/F be a quadratic CM extension of a totally real field F, with $\sigma \in \Gamma_F$ generating* Gal(L/F) *Suppose $\hat{\psi} \colon \Gamma_L \to \overline{\mathbb{Q}}_\ell^\times$ is a Galois character such that $\hat{\psi}^{1-\sigma}$ is geometric. Then there exists a unitary, type A Hecke character ψ of L such that $\psi^{1-\sigma}$ is the type A_0 Hecke character associated to $\hat{\psi}^{1-\sigma}$.*

PROOF. Write $(\hat{\psi}^{1-\sigma})_{\mathbf{A}}$ for the Hecke character associated to $\hat{\psi}^{1-\sigma}$. By the previous lemma, it suffices to check that $(\hat{\psi}^{1-\sigma})_{\mathbf{A}}$ is trivial on $C_F \subset C_L$. If a finite place v of L is split over a place v_F of F, and unramified for $\hat{\psi}$, then for a uniformizer ϖ_v of F_{v_F} (embedded into the L_v and $L_{\sigma v}$ components of $\mathbf{A}_L^\times$),

$$(\hat{\psi}^{1-\sigma})_{\mathbf{A}}(\varpi_v, \varpi_v) = \hat{\psi}^{1-\sigma}(\mathrm{fr}_v)\hat{\psi}^{1-\sigma}(\mathrm{fr}_{\sigma v}) = \frac{\hat{\psi}(\mathrm{fr}_v)}{\hat{\psi}(\mathrm{fr}_{\sigma v})} \cdot \frac{\hat{\psi}(\mathrm{fr}_{\sigma v})}{\hat{\psi}(\mathrm{fr}_v)} = 1.$$

Similarly for v inert, $(\hat{\psi}^{1-\sigma})_{\mathbf{A}}(\varpi_v) = \frac{\hat{\psi}(\mathrm{fr}_v)}{\hat{\psi}(\sigma \mathrm{fr}_v \sigma^{-1})} = 1$. The Hecke character $(\hat{\psi}^{1-\sigma})_{\mathbf{A}}|_{C_F}$ is therefore trivial. To see that we may choose ψ to be unitary and type A, we invoke Corollary 2.3.9: decomposing ψ as in that result, both the $|\cdot|^w$ and 'Maass' components descend to the totally real subfield F, so dividing out by them yields a new ψ, now unitary type A, and with $\psi^{1-\sigma}$ unchanged. □

LEMMA 3.3.5. *Let L/F be cyclic of degree d, with $\sigma \in \Gamma_F$ restricting to a generator of* Gal(L/F). *Suppose $\rho \colon \Gamma_L \to \mathrm{GL}_n(\overline{\mathbb{Q}}_\ell)$ is an irreducible continuous representation satisfying $\rho^\sigma \cong \rho \cdot \chi$ for some character $\chi \colon \Gamma_L \to \overline{\mathbb{Q}}_\ell^\times$. Suppose further that χ is geometric (in particular, this holds if ρ is geometric), necessarily of weight zero, so we may regard it as a character $\chi_{\mathbf{A}} = \prod_{w \in |L|} \chi_w \colon C_L \to \mathbb{C}^\times$.*

[18] In the proof of Corollary 3.3.6 below, we could replace appeal to this lemma by simply citing Statement B of [**LR98**].

Then the (finite-order) restriction of $\chi_{\mathbf{A}}$ *to* $C_F \subset C_L$ *cannot factor through a non-trivial character of* $\mathrm{Gal}(L/F)$.

PROOF. Iterating the relation $\rho^\sigma \cong \rho\cdot\chi$, we obtain $\rho \cong \rho\cdot\chi^{1+\sigma+\ldots+\sigma^{d-1}}$, so that $\chi^{1+\sigma+\ldots+\sigma^{d-1}}$ is finite-order. Each of the characters χ^{σ^i} has some common weight w, so $d\cdot w = 0$, and thus $w = 0$. Moreover, writing as usual p_{ι_w} for the algebraic parameter of χ_w (with respect to a choice $\iota_w \colon L \hookrightarrow \mathbb{C}$ representing the place w), we have $\sum p_{\iota_w} = 0$ as ι_w ranges over a $\mathrm{Gal}(L/F)$-orbit of such embeddings. If L has a real embedding, then χ has finite-order, and the passage from $\chi\colon \Gamma_L^{\mathrm{ab}} \to \overline{\mathbb{Q}}_\ell^\times$ to $\chi_{\mathbf{A}}\colon C_L \to \mathbb{C}^\times$ is simply via the reciprocity map $C_L \twoheadrightarrow \Gamma_L^{\mathrm{ab}}$.[19] In particular, $\chi(x) = \chi_{\mathbf{A}}(x')$ for any representative x' in C_L of the image of x in Γ_L^{ab}. If on the other hand L is totally imaginary, then continuing to write $x' = (x'_w)_{w\in|L|}$, we have

$$\chi(x) = \prod_{w\nmid\ell\infty}\chi_w(x'_w)\prod_{w|\ell}\left(\chi_w(x'_w)\prod_{\tau\colon L_w\hookrightarrow\overline{\mathbb{Q}}_\ell}\tau(x'_w)^{p_{\iota^*_{\infty,\ell}(\tau)}}\right).$$

If we further assume that the representative x' can be chosen in $C_F \subset C_L$, with elements $x'_v \in F_v^\times$ giving rise to all x'_w for $w|v$, then we can rewrite

$$\prod_{w|\ell}\prod_{\tau\colon L_w\hookrightarrow\overline{\mathbb{Q}}_\ell}\tau(x'_w)^{p_{\iota^*_{\infty,\ell}(\tau)}}$$

as

$$\prod_{v|\ell}\prod_{\tau\colon F_v\hookrightarrow\overline{\mathbb{Q}}_\ell}\prod_{w|v,\tilde{\tau}|\tau}\tau(x'_v) = \prod_{v|\ell}\prod_{\tau\colon F_v\hookrightarrow\overline{\mathbb{Q}}_\ell}\tau(x'_v)^{\sum_{w|v,\tilde{\tau}|\tau}p_{\iota^*_{\infty,\ell}(\tilde{\tau})}} = 1.$$

The last equality follows since $\chi^{1+\sigma+\ldots+\sigma^{d-1}}$ is finite-order, and we are summing $p_{\iota^*_{\infty,\ell}(\tilde{\tau})}$ over a full $\mathrm{Gal}(L/F)$-orbit. A similar argument shows that

$$\prod_{w|\infty}\chi_w(x'_w) = 1.$$

We conclude that in this case ($x' \in C_F$), $\chi(x)$ can be computed simply as $\chi_{\mathbf{A}}(x')$.

Let V be the space on which ρ acts. Then the isomorphism $\rho^\sigma \cong \rho\cdot\chi$ yields an operator $A \in \mathrm{Aut}(V)$ satisfying $A\rho^\sigma = \rho\cdot\chi A$. Fix $g \in \Gamma_L$, and for any $x \in \Gamma_L$ we compute

$$\begin{aligned}\mathrm{tr}(\rho(g)A) &= \mathrm{tr}(\rho^\sigma(x)\rho(g)A\rho^\sigma(x)^{-1})\\ &= \mathrm{tr}(\rho^\sigma(x)\rho(g)\rho(x^{-1})\chi(x^{-1})A)\\ &= \chi(x^{-1})\,\mathrm{tr}(\rho({}^\sigma x g x^{-1})A).\end{aligned}$$

(Here ${}^\sigma x = \sigma x\sigma^{-1}$.) So, if we can find an $x \in \Gamma_L$ such that ${}^\sigma x g x^{-1} = g$ and $\chi(x^{-1}) \neq 1$, then we will have $\mathrm{tr}(\rho(g)A) = 0$. Doing this for all $g \in \Gamma_L$, we see that by Schur's Lemma ρ cannot be irreducible, else $A = 0$. Now, $y = g^{-1}\sigma \in \Gamma_F$ satisfies ${}^\sigma y g y^{-1} = g$, so $x = y^d \in \Gamma_L$ does as well. It suffices to show that if $\chi|_{C_F}$ cuts out the extension L/F, then $\chi(x) \neq 1$. In fact, the image of x in Γ_L^{ab} lies in the image of the transfer $\mathrm{Ver}\colon \Gamma_F^{\mathrm{ab}} \to \Gamma_L^{\mathrm{ab}}$. Explicitly,

$$\mathrm{Ver}(y) = \prod_{i=0}^{d-1}\sigma^i(g^{-1}\sigma)\phi(\sigma^i g^{-1}\sigma)^{-1},$$

[19]Throughout this argument we implicitly use our fixed embeddings ι_ℓ, ι_∞, but omit any reference to them for notational simplicity.

where $\phi\colon \Gamma_F \to \{\sigma^i\}_{i=0,\dots,d-1}$ records the representative of the Γ_L-coset of an element of Γ_F. It is then easily seen[20] that

$$\mathrm{Ver}(y) = (g^{-1}\sigma)^d = x,$$

so by class field theory $x \in \Gamma_L^{\mathrm{ab}}$ is represented by an element x' of $C_F \subset C_L$ under the reciprocity map rec_L. This element is a generator of $C_F/N_{L/F}C_L$, since y lifts a generator of $\mathrm{Gal}(L/F)$, and thus $\chi(x) = \chi_{\mathbf{A}}(x') \neq 1$ if $\chi_{\mathbf{A}}|_{C_F}$ factors through a non-trivial character of $\mathrm{Gal}(L/F)$. □

Finally we can (conditionally) prove automorphy of geometric projective representations over totally real fields.

COROLLARY 3.3.6. *Let F be totally real. Continue to assume Part 3 of Conjecture 1.2.1 (Fontaine-Mazur-Langlands). Then for any geometric $\rho\colon \Gamma_F \to \mathrm{PGL}_n(\overline{\mathbb{Q}}_\ell)$, there exists an L-algebraic π on SL_n/F such that $\rho \sim_w \pi$.*

PROOF. We will first treat the case of ρ having irreducible lifts to $\mathrm{GL}_n(\overline{\mathbb{Q}}_\ell)$. Choose a lift $\tilde{\rho}$ with finite-order determinant, a CM quadratic extension L/F, and, by Theorem 3.2.7, a Galois character $\hat{\psi}\colon \Gamma_L \to \overline{\mathbb{Q}}_\ell^\times$ such that $\tilde{\rho}|_{\Gamma_L}\cdot\hat{\psi}^{-1}$ is geometric. Let $\tilde{\pi}$ be the cuspidal automorphic representation of $\mathrm{GL}_n(\mathbf{A}_L)$ corresponding to this geometric twist. Write σ for the nontrivial element of $\mathrm{Gal}(L/F)$, so that $\tilde{\pi}^\sigma \cong \tilde{\pi}\cdot\chi$ where χ is the Hecke character corresponding to the geometric Galois character $\hat{\psi}^{1-\sigma}$. Appealing either to Lemma 3.3.4 or Lemma 3.3.5, we can write $\chi = \psi^{1-\sigma}$ for a unitary type A Hecke character ψ. Then $\tilde{\pi}\cdot\psi$ is σ-invariant, and by cyclic (prime degree) descent, there is a cuspidal representation π of $\mathrm{GL}_n(\mathbf{A}_F)$ whose base-change is $\tilde{\pi}\cdot\psi$. We want to compare the projectivization of the unramified parameters of π with the unramified restrictions $\rho|_{\Gamma_{F_v}}$.

To do so, we repeat the argument but instead with infinitely many (disjoint) quadratic CM extensions L_i/F, showing that in all cases the descent π to $\mathrm{GL}_n(\mathbf{A}_F)$ gives an L-packet for $\mathrm{SL}_n(\mathbf{A}_F)$ that is independent of the field L_i. $\tilde{\rho}$ is still a fixed lift with finite-order determinant, and we can write, for each $\tau\colon F \hookrightarrow \overline{\mathbb{Q}}_\ell$, $\mathrm{HT}_\tau(\tilde{\rho}) \in \frac{k_\tau}{n} + \mathbb{Z}$ for some integer k_τ, which we fix (rather than just its congruence class mod n). For some integer w, we have the purity relation $2k_\tau \equiv w \mod n$, as follows, for instance, from (geometric) liftability after a CM base-change;[21] as with k_τ, we fix an actual integer w, not just the congruence class. Now, for each such τ, let $\iota\colon F \hookrightarrow \mathbb{C}$ be the archimedean embedding associated via ι_∞, ι_ℓ (elsewhere denoted $\iota_{\infty,\ell}^*(\tau)$). For each L_i, fix an embedding $\tau(i)\colon L_i \hookrightarrow \overline{\mathbb{Q}}_\ell$ extending τ, so that the other extension is $\tau(i)\circ c$. Likewise write $\iota(i)$ and $\iota(i)\circ c = \overline{\iota(i)}$ for the corresponding complex embeddings. We can then construct Galois characters $\hat{\psi}_i\colon \Gamma_{L_i} \to \overline{\mathbb{Q}}_\ell^\times$ such that

$$\mathrm{HT}_{\tau(i)}(\hat{\psi}_i) = \frac{k_\tau}{n}$$
$$\mathrm{HT}_{\tau(i)\circ c}(\hat{\psi}_i) = \frac{w-k_\tau}{n},$$

[20] Note that $\phi(\sigma^{d-1}g^{-1}\sigma) = 1$, while otherwise $\phi(\sigma^i g^{-1}\sigma) = \sigma^{i+1}$.

[21] After such a base-change L/F, the character $\hat{\psi}$ twisting $\tilde{\rho}$ to a geometric representation will have Hodge-Tate-Sen weights congruent to $\frac{k_\tau}{n} \in \mathbb{Q}/\mathbb{Z}$ at both embeddings $L \hookrightarrow \overline{\mathbb{Q}}_\ell$ above τ; the integer w is then the weight of the Hecke character associated to $\hat{\psi}^n$.

such that $\tilde{\rho}|_{\Gamma_{L_i}} \cdot \hat{\psi}_i^{-1}$ is geometric, corresponding to an L-algebraic cuspidal $\tilde{\pi}_i$ on GL_n/L_i. As before, we find a Hecke character ψ_i of L_i such that $\psi_i^{1-\sigma_i}$ is the type A_0 Hecke character associated to $\hat{\psi}_i^{1-\sigma_i}$; here we write σ_i for the non-trivial element of $\mathrm{Gal}(L_i/F)$, but of course all the σ_i are just induced by complex conjugation. Again, for all i we find cuspidal automorphic representations π_i of $\mathrm{GL}_n(\mathbf{A}_F)$ such that $\mathrm{BC}_{L_i/F}(\pi_i) = \tilde{\pi}_i \cdot \psi_i$. Restricting to composites L_iL_j, we have the comparison

$$\tilde{\rho}|_{L_iL_j}\hat{\psi}_i^{-1}|_{L_iL_j} \cdot \left(\frac{\hat{\psi}_i}{\hat{\psi}_j}|_{L_iL_j}\right) = \tilde{\rho}|_{L_iL_j}\hat{\psi}_j^{-1}|_{L_iL_j},$$

and thus

$$\mathrm{BC}_{L_iL_j/L_i}(\tilde{\pi}_i\psi_i) \cdot \mathrm{BC}_{L_iL_j}\left(\frac{\psi_j}{\psi_i}\right) \cdot \mathrm{BC}_{L_iL_j}\left(\frac{\hat{\psi}_i}{\hat{\psi}_j}\right) = \mathrm{BC}_{L_iL_j/L_j}(\tilde{\pi}_j\psi_j),^{22}$$

so finally

$$\mathrm{BC}_{L_iL_j}(\pi_i) \cdot \mathrm{BC}_{L_iL_j}\left(\frac{\psi_j}{\psi_i} \cdot \frac{\hat{\psi}_i}{\hat{\psi}_j}\right) = \mathrm{BC}_{L_iL_j}(\pi_j).$$

If the character $\frac{\psi_j}{\psi_i} \cdot \frac{\hat{\psi}_i}{\hat{\psi}_j}$ is finite-order–in the next paragraph, we check that we may assume this–it cuts out a cyclic extension L'/L_iL_j, and we have $\mathrm{BC}_{L'}(\pi_i) = \mathrm{BC}_{L'}(\pi_j)$. L'/F is solvable, however, so the characterization of the fibers of solvable base-change in [**Raj02**] implies that π_i and π_j are twist-equivalent, hence that $\pi_i|_{\mathrm{SL}_n(\mathbf{A}_F)}$ and $\pi_j|_{\mathrm{SL}_n(\mathbf{A}_F)}$ define the same L-packet of $\mathrm{SL}_n(\mathbf{A}_F)$. Let us denote by π_0 any representative of this global L-packet. Now consider places v of F that are split in a given L_i/F. The semi-simple part $\rho(\mathrm{fr}_v)^{\mathrm{ss}}$ is equal (in $\mathrm{PGL}_n(\overline{\mathbb{Q}}_\ell)$) to $(\tilde{\rho}\hat{\psi}^{-1}(\mathrm{fr}_w))^{\mathrm{ss}}$ for any $w|v$, and this is conjugate in $\mathrm{GL}_n(\overline{\mathbb{Q}}_\ell)$ to $\iota_{\ell,\infty}(\mathrm{rec}_w(\tilde{\pi}_w)(\mathrm{fr}_w))$, whose projectivization lies in the same $\mathrm{PGL}_n(\overline{\mathbb{Q}}_\ell)$-conjugacy class as $\iota_{\ell,\infty}(\mathrm{rec}_v(\pi_{0,v})(\mathrm{fr}_v))$. This verifies that for all such v, $\rho(\mathrm{fr}_v)^{\mathrm{ss}}$ is $\mathrm{PGL}_n(\overline{\mathbb{Q}}_\ell)$-conjugate to $\iota_{\ell,\infty}(\mathrm{rec}_v(\pi_{0,v})(\mathrm{fr}_v))$. Varying L_i/F, and remembering that π_0 is independent of this variation, we get the same result for all v split in any single quadratic CM extension L_i/F (we have to throw out a finite number of such L_i to ensure our representations remain cuspidal/irreducible), we conclude that $\rho \sim_w \pi$.

To finish the proof, we must check that $\frac{\psi_j}{\psi_i} \cdot \frac{\hat{\psi}_i}{\hat{\psi}_j}$ may indeed be assumed finite-order. First, recall that each ψ_i may be taken unitary and type A; in this case, the infinity-type is determined by the relation $\psi_i^{1-\sigma_i} = \hat{\psi}_i^{1-\sigma_i}$. Explicitly (using the above notation for the various embeddings), $\hat{\psi}_i^{\,n}$ corresponds to a Hecke character of L_i with infinity-type (where we abusively denote $\iota(i)(z)$ by simply z)

$$\mathrm{rec}_{\iota(i)}(\hat{\psi}_i^{\,n})\colon z \mapsto z^{k_\tau}\bar{z}^{w-k_\tau},$$

so

$$\mathrm{rec}_{\iota(i)}(\psi_i^{1-\sigma_i})\colon z \mapsto z^{\frac{2k_\tau - w}{n}}\bar{z}^{\frac{w-2k_\tau}{n}}.$$

(Recall that $2k_\tau \equiv w \mod n$.) We then have, under our assumptions,

$$\mathrm{rec}_{\iota(i)}(\psi_i)\colon z \mapsto \left(\frac{z}{|z|}\right)^{\frac{2k_\tau - w}{n}}.$$

---

[22]Here $\frac{\hat{\psi}_i}{\hat{\psi}_j}$ restricted to $\Gamma_{L_iL_j}$ is geometric, so we abusively write this for the associated Hecke character as well.

Of course, $\mathrm{rec}_{\iota(i)\circ c}$ is the same but with $\frac{w-2k_\tau}{n}$ in the exponent. To make the parameter comparison after restriction to a composite L_iL_j, we use the following notation for embeddings of L_iL_j into $\overline{\mathbb{Q}}_\ell$ and $\mathbb{C}$, lying above the given τ and ι:

$$\tau_1 \text{ extends } \tau(i) \text{ and } \tau(j),$$
$$\tau_2 \text{ extends } \tau(i) \text{ and } \tau(j)\circ c,$$
$$\iota_1 \text{ extends } \iota(i) \text{ and } \iota(j),$$
$$\iota_2 \text{ extends } \iota(i) \text{ and } \iota(j)\circ c.$$

We then have the conjugate embeddings $\tau_1 \circ c$, etc. Computing the τ_k-labeled weights of $\frac{\hat{\psi}_i}{\hat{\psi}_j}$, and translating them to the infinity-type at the place corresponding to ι_k, with ι_k as the chosen embedding $L_iL_j \hookrightarrow \mathbb{C}$, we then find

$$\mathrm{rec}_{\iota_1}\left(\frac{\hat{\psi}_i}{\hat{\psi}_j}\right): z \mapsto 1$$
$$\mathrm{rec}_{\iota_2}\left(\frac{\hat{\psi}_i}{\hat{\psi}_j}\right): z \mapsto (z/\bar{z})^{\frac{2k_\tau - w}{n}},$$

whereas

$$\mathrm{rec}_{\iota_1}\left(\frac{\psi_j}{\psi_i}\right): z \mapsto 1$$
$$\mathrm{rec}_{\iota_2}\left(\frac{\psi_j}{\psi_i}\right): z \mapsto (z/\bar{z})^{\frac{w-2k_\tau}{n}}.$$

We conclude that $\frac{\psi_j}{\psi_i} \cdot \frac{\hat{\psi}_i}{\hat{\psi}_j}$ is, with our normalization of the ψ_i, in fact finite-order, and the proposition follows.

Finally, we quickly treat the case of general ρ, having reducible lifts. If $\tilde{\rho}$ as above decomposes $\tilde{\rho} = \oplus_{i=1}^m \tilde{\rho}_i$, say with $\tilde{\rho}_i$ of dimension n_i, geometricity of ρ implies that over any CM L/F the same Galois character $\hat{\psi}$ twists $\tilde{\rho}_i$, for all i, to a geometric representation. We can therefore use, for all i, the same Hecke character ψ such that $\psi^{1-\sigma} = \hat{\psi}^{1-\sigma}$. As above, we invoke automorphy of $\tilde{\rho}_i\hat{\psi}^{-1}$, and, twisting by ψ, descend to a cuspidal automorphic representation Π_i of GL_{n_i}/F. The same local check (for v split in L/F) as above, but now crucially relying on the fact that ψ and $\hat{\psi}$ were independent of i, shows $\rho(\mathrm{fr}_v)$ is $\mathrm{PGL}_n(\overline{\mathbb{Q}}_\ell)$-conjugate to $\iota\left(\mathrm{rec}_v(\boxplus_{i=1}^m \Pi_{i,v})(\mathrm{fr}_v)\right)$. □

By a similar argument, we can 'construct' the Galois representations (assuming of course the GL_N correspondence) associated to (tempered) L-algebraic π on SL_n/F for F CM (or imaginary, assuming Conjecture 2.4.8) or totally real. By Proposition 3.1.12, we are reduced to the case of F totally real.

PROPOSITION 3.3.7. *Continue to assume Fontaine-Mazur-Langlands. Let F be a totally real field, and let π be an L-algebraic cuspidal automorphic representation of $\mathrm{SL}_n(\mathbf{A}_F)$. Assume that π_∞ is tempered. Then there exists a (not necessarily unique) projective representation $\rho\colon \Gamma_F \to \mathrm{PGL}_n(\overline{\mathbb{Q}}_\ell)$ satisfying $\rho \sim_w \pi$.*

PROOF. By Proposition 3.1.14, there exists a W-algebraic cuspidal (and tempered at ∞) $\tilde{\pi}$ on GL_n/F lifting π. For all but finitely many quadratic CM L/F, we can find a type A Hecke character ψ such that $\mathrm{BC}_{L/F}(\tilde{\pi}) \cdot \psi$ is L-algebraic and cuspidal on GL_n/L, hence corresponds to an irreducible geometric representation $\tilde{\rho}_L\colon \Gamma_L \to \mathrm{GL}_n(\overline{\mathbb{Q}}_\ell)$. Conjugating, we find $\tilde{\rho}_L^\sigma \equiv \tilde{\rho}_L \cdot \widehat{(\psi^{\sigma-1})}$, the twist being by the geometric character associated to the type A_0 Hecke character $\psi^{\sigma-1}$. We wish to write $\hat{\chi} = \widehat{(\psi^{\sigma-1})}$ in the form $\hat{\psi}^{\sigma-1}$ for some Galois character $\hat{\psi}\colon \Gamma_L \to \overline{\mathbb{Q}}_\ell^\times$ (reversing the process in Corollary 3.3.6). Once this is managed, we have $(\tilde{\rho}_L \cdot \hat{\psi}^{-1})$ is σ-invariant, hence descends to $\rho\colon \Gamma_F \to \mathrm{PGL}_n(\overline{\mathbb{Q}}_\ell)$. Arguing as in Corollary 3.3.6, we find that this projective descent is independent of L/F, and, again as in that proof, by varying L/F we obtain the compatibility $\rho \sim_w \pi$.

To construct $\hat{\psi}$, we take as first approximation a Galois character $\hat{\psi}$ such that $\hat{\psi}^2$ equals $\widehat{(\psi^2)}$ up to a finite-order character χ_0; recall that ψ^2 is type A_0, so we can attach the Galois character $\widehat{(\psi^2)}$. Then

$$(\hat{\psi}^{\sigma-1})^2 = (\hat{\psi}^2)^{\sigma-1} = (\widehat{(\psi^2)}\chi_0)^{\sigma-1} = \hat{\chi}^2\chi_0^{\sigma-1},$$

and consequently $\hat{\chi}$ agrees with $\hat{\psi}^{\sigma-1}$ up to a finite-order character. Twisting $\tilde{\rho}_L$, we may therefore assume that $\hat{\chi}$ has finite-order. It still satisfies $\hat{\chi}^{1+\sigma} = 1$, so invoking Lemma 3.3.3 and (a simple case of) Lemma 3.3.5 we find a finite-order Hecke character,[23] which may therefore be directly regarded as a Galois character, that casts $\hat{\chi}$ in the desired form. □

Here is another example comparing the Tannakian formalisms:

PROPOSITION 3.3.8. *Continue to assume Fontaine-Langlands-Mazur and, in the totally imaginary but non-CM case, Conjecture* 2.4.8. *Let* Π *be a cuspidal* L*-algebraic representation of* $\mathrm{GL}_n(\mathbf{A}_F)$, *and suppose that* $\rho_\Pi \cong \rho_1 \otimes \rho_2$, *where* $\rho_i\colon \Gamma_F \to \mathrm{GL}_{n_i}(\overline{\mathbb{Q}}_\ell)$. *Then there exist cuspidal automorphic representations* π_i *of* $\mathrm{GL}_{n_i}(\mathbf{A}_F)$ *such that* $\Pi = \pi_1 \boxtimes \pi_2$.

REMARK 3.3.9. As the examples in §2.6 show, sometimes the π_i cannot be taken L-algebraic. Nevertheless, by Proposition 2.5.8, they can always be taken W-algebraic.

PROOF. First suppose F is totally imaginary. For all $v|\ell$, the fact that $\rho_1|_{\Gamma_{F_v}} \otimes \rho_2|_{\Gamma_{F_v}}$ is de Rham implies that locally these Γ_{F_v}-representations are twists of de Rham representations. To see this, we apply Corollary 3.2.12 (to find a Hodge-Tate lift) and Theorem 2.1.6 (to find a de Rham lift, given that a Hodge-Tate lift exists) to the lifting problem (with central torus kernel) $\mathrm{GL}_{n_1} \times \mathrm{GL}_{n_2} \xrightarrow{\boxtimes} G_{n_1,n_2} \subset \mathrm{GL}_{n_1 n_2}$, where G_{n_1,n_2} denotes the image of the tensor product map. In particular, the projectivizations of the $\rho_i|_{\Gamma_{F_v}}$ are de Rham, so the global projective representations $\rho_i\colon \Gamma_F \to \mathrm{PGL}_{n_i}(\overline{\mathbb{Q}}_\ell)$ are geometric. We know that these geometric projective representations have geometric lifts, and we may therefore assume our original ρ_i were in fact geometric. They then correspond to L-algebraic π_i, and we have $\Pi = \pi_1 \boxtimes \pi_2$.

For F totally real, we perform a descent similar to previous arguments. Restricting to CM L/F, we find a Galois character $\hat{\psi}$ such that $\rho_1 \cdot \hat{\psi}^{-1}$ and $\rho_2 \cdot \hat{\psi}$ are geometric, corresponding to L-algebraic cuspidal π_i on GL_{n_i}/L. Writing $\hat{\psi}^{1-\sigma} = \psi^{1-\sigma}$ for a Hecke character ψ, we find $\pi_1 \cdot \psi$ and $\pi_2 \cdot \psi^{-1}$ are σ-invariant, so descend

[23] Writing $\hat{\chi} = \psi^{\sigma-1}$ where the infinity-components of ψ have the form $z^p\bar{z}^q$, we see $z^{p-q}\bar{z}^{q-p}$ is the corresponding component of $\hat{\chi}$, hence that $p = q$ at each infinite place. Twisting ψ by the base-change of a character of the totally real field F, we can then assume it is finite-order.

to cuspidal representations $\bar{\pi}_i$ of GL_{n_i}/F. Since $\mathrm{BC}_{L/F}(\Pi) = \mathrm{BC}_{L/F}(\bar{\pi}_1 \boxtimes \bar{\pi}_2)$, we deduce that Π and $\bar{\pi}_1 \boxtimes \bar{\pi}_2$ are twist-equivalent, from which the result follows. (The same argument applies to F that are neither totally real nor totally imaginary: just replace the restriction to CM extensions L/F with restrictions to totally imaginary quadratic extensions L/F.) □

These examples (and, for instance, Corollary 2.7.8) motivate a comparison of the images of $r\colon {}^L H \to {}^L G$ on the automorphic and Galois sides, when r is an L-morphism with central kernel. The most optimistic expectation (for H and G quasi-split) is that if $\ker(r)$ is a central torus, then the two descent problems for (L-algebraic) Π and (geometric) ρ_Π are equivalent; whereas if $\ker(r)$ is disconnected, there is an obstruction to the comparison, that nevertheless can be killed after a finite base-change. If G is not GL_n/F, then one will have to decide whether weak equivalence ($\pi \sim_w \rho$) suffices to connect the descent problems, or whether some stronger link (the mysterious $\pi \sim_s \rho$) must be postulated.

3.4. Monodromy of abstract Galois representations

In this section we discuss some general results about monodromy of ℓ-adic Galois representations. Much of the richness of this subject comes from its blending of two kinds of representation theories, that of finite groups, and that of connected reductive algebraic groups. We will see (Proposition 3.4.1) that the basic lifting result (Proposition 2.1.4) allows us to some extent to understand how these two representation theories interact. In §3.4.2 we develop more refined results in the 'Lie-multiplicity-free' case (see Definition 3.4.5); this situation encapsulates the essential difficulties of independence-of-ℓ questions, such questions being trivial for Artin representations.

3.4.1. A general decomposition. The following result is a simple variant of a result of Katz ([**Kat87**, Proposition 1]), which he proves for lisse sheaves on affine curves over finite fields. We can replace Katz's appeal to the Lefschetz affine theorem by Proposition 2.1.4. Recall that a Galois representation is *Lie irreducible* if it is irreducible after restriction to every finite-index subgroup (i.e., the connected component or Lie algebra of its algebraic monodromy group acts irreducibly).

PROPOSITION 3.4.1. *Let F be any number field, and let $\rho\colon \Gamma_F \to \mathrm{GL}_{\overline{\mathbb{Q}}_\ell}(V)$ be an irreducible representation of dimension n. Then either ρ is induced, or there exists $d|n$, a Lie irreducible representation τ of dimension n/d, and an Artin representation ω of dimension d such that $\rho \cong \tau \otimes \omega$. Consequently, any (irreducible) ρ can be written in the form*

$$\rho \cong \mathrm{Ind}_L^F(\tau \otimes \omega)$$

for some finite L/F and irreducible representations τ and ω of Γ_L, with τ Lie-irreducible and ω Artin.

PROOF. Let $\mathcal{G}$ denote the algebraic monodromy group of ρ, with $\mathcal{G}^0$ the connected component of the identity. Abusively writing ρ for the representation $\mathcal{G} \hookrightarrow \mathrm{GL}(V)$, we may assume $\rho|_{\mathcal{G}^0}$ is isotypic (else ρ is induced, and we are done). If $\rho|_{\mathcal{G}^0}$ is irreducible, then ρ itself is Lie-irreducible, so again we are done. Therefore, we may assume that $\rho|_{\mathcal{G}^0} \cong \tau_0^{\oplus d}$ for some $d \geq 2$, with τ_0 an irreducible representation of $\mathcal{G}^0$, and consequently a Lie-irreducible representation of Γ_L for any L/F sufficiently large that $\rho(\Gamma_L) \subset \mathcal{G}^0(\overline{\mathbb{Q}}_\ell)$. Since the irreducible Γ_L-representation

τ_0 is Γ_F-invariant, it extends to a projective representation of Γ_F. By the basic lifting result (Proposition 2.1.4), this projective representation lifts to an honest Γ_F-representation τ_1, so for some character $\chi\colon \Gamma_L \to \overline{\mathbb{Q}}_\ell^\times$,

$$\tau_1^{\oplus d}|_{\Gamma_L} \cong \rho|_{\Gamma_L} \otimes \chi.$$

The character $\alpha := \det(\rho)/\det(\tau_1^{\oplus d})$ of Γ_F has Γ_L-restriction equal to χ^{-n}, and over F itself we can find characters $\alpha_1, \alpha_0\colon \Gamma_F \to \overline{\mathbb{Q}}_\ell^\times$, with α_0 finite-order, such that $\alpha = \alpha_1^n\alpha_0$. Then $(\chi\alpha_1|_{\Gamma_L})^n = (\alpha^{-1}\alpha_1^n)|_{\Gamma_L} = \alpha_0^{-1}|_{\Gamma_L}$, and replacing τ_1 by $\tau_1 \otimes \alpha_1$, and L by a finite extension trivializing α_0, we find a Lie-irreducible representation τ of Γ_F and a finite extension L of F such that $\tau^{\oplus d}|_{\Gamma_L} \cong \rho|_{\Gamma_L}$. The Γ_F-representation

$$\omega := \mathrm{Hom}_{\Gamma_L}(\tau, \rho).$$

is therefore a d-dimensional Artin representation,[24] and the natural map $\tau \otimes \omega \to \rho$ (i.e. $v \otimes \phi \mapsto \phi(v)$) is an isomorphism of Γ_F-representations. □

COROLLARY 3.4.2. *Let $\rho\colon \Gamma_F \to \mathrm{GL}_{\overline{\mathbb{Q}}_\ell}(V)$ be a semi-simple representation (not necessarily irreducible), and suppose that ρ is Lie-isotypic, i.e. for all F'/F sufficiently large, $\rho|_{\Gamma_{F'}}$ is isotypic. Then there exists a Lie-irreducible representation τ and an Artin representation (possibly reducible) ω, both of Γ_F, such that $\rho \cong \tau \otimes \omega$.*

PROOF. Decompose ρ into irreducible Γ_F-representations as $\oplus_1^r \rho_i$. Each ρ_i is Lie-isotypic: there exists L/F and integers m_i such that $\rho_i|_{\Gamma_L} \cong \tau_0^{\oplus m_i}$ for all i, where τ_0 is a Lie-irreducible representation independent of i. By the argument of the previous proposition, after possibly enlarging L we find a Γ_F-representation τ whose restriction to L is isomorphic to τ_0, and then there are Artin representations ω_i of Γ_F such that $\tau \otimes \omega_i \cong \rho_i$. Consequently,

$$\rho \cong \tau \otimes (\bigoplus_1^r \omega_i).$$

□

REMARK 3.4.3.
- In general, the field L in Proposition 3.4.1 is not unique, even up to Γ_F-conjugacy. Examples of such non-uniqueness should not arise in the Lie-multiplicity free case (see §3.4.2), but in the Artin case there are easy examples arising from the representation theory of finite groups. Consider, for instance, the quaternion group $Q_8 = \{\pm 1, \pm i, \pm j, \pm k\}$. The (unique) irreducible two-dimensional representation of Q_8 can be written in the form $\mathrm{Ind}_{\langle x\rangle}^{Q_8}(\varepsilon_x)$, where $\langle x\rangle$ denotes one of the subgroups generated by i, j, or k, and ε_x is a generator of the character group of $\langle x\rangle$. None of these subgroups is conjugate to any of the others.[25] It would be interesting to achieve a more systematic understanding of these ambiguities.
- These structure theorems for Galois representations should have an analogue on the automorphic side. In fact, Tate has shown (2.2.3 of [**Tat79**]) the analogue of Proposition 3.4.1 for representations of the Weil group

[24] As Γ_L-representation, $\omega \cong \mathrm{Hom}_{\Gamma_L}(\tau|_{\Gamma_L}, \tau|_{\Gamma_L}^{\oplus d}) \cong \overline{\mathbb{Q}}_\ell^{\oplus d}$.

[25] Although these groups are not conjugate, they are related by (outer) automorphisms of Q_8, but applying outer automorphisms in this fashion will not in general preserve an irreducible induced character: consider the principal series of $\mathrm{GL}_2(\mathbb{F}_p)$ and the outer automorphism of A_1.

W_F, where it takes the particularly simple form that any irreducible, non-induced $\rho\colon W_F \to \mathrm{GL}_n(\mathbb{C})$ is isomorphic to $\omega \otimes \chi$, where ω is an Artin representation and $\chi\colon W_F \to \mathbb{C}^\times$ is a character. A basic question is whether we should expect Proposition 3.4.1 to hold for 'representations of $\mathcal{L}_F$.' If we had the formalism of $\mathcal{L}_F$, then to carry out the argument of the proposition with complex representations of $\mathcal{L}_F$ in place of ℓ-adic representations of Γ_F requires two ingredients:

- That a homomorphism $\mathcal{L}_F \to \mathrm{PGL}_n(\mathbb{C})$ lifts to a homomorphism $\mathcal{L}_F \to \mathrm{GL}_n(\mathbb{C})$; but this is 'implied' by Proposition 3.1.4. I should mention in this respect the theorem of Labesse ([**Lab85**]), which establishes the analogue for lifting homomorphisms $W_F \to {}^L G$ across surjections ${}^L\widetilde{G} \to {}^L G$ with central torus kernel.
- That the analogue of the character α in the proof of Proposition 3.4.1 can be written as a finite-order twist α_0 of the nth power of a character α_1. This is *not* automatic, as it is for ℓ-adic characters, but in this case we can exploit Lemma 2.3.10, which applies to the α of the Proposition, since there the restriction to L is (continuing with the notation of the Proposition) χ^{-n}.

This discussion motivates the following conjecture, whose formulation of course requires assuming deep cases of functoriality:

CONJECTURE 3.4.4. *Let π be a cuspidal automorphic representation of $\mathrm{GL}_n(\mathbf{A}_F)$; assume π is not automorphically induced from any non-trivial extension L/F. Then there exist cuspidal automorphic representations τ and ω of, respectively, $\mathrm{GL}_d(\mathbf{A}_F)$ and $\mathrm{GL}_{n/d}(\mathbf{A}_F)$ such that $\pi = \tau \boxtimes \omega$, with the following properties:*

- *for all finite extensions L/F, the base-change $\mathrm{BC}_{L/F}(\tau)$ remains cuspidal;*
- *for some finite extension L/F, $\mathrm{BC}_{L/F}(\omega)$ is isomorphic to the isobaric sum of n/d copies of the trivial representation.*

3.4.2. Lie-multiplicity-free representations. In this section, we focus on the cases antithetical to that of Artin representations, putting ourselves in the following situation. Let F be any number field, and recall from definition 1.2.3 the definition of a (weakly) compatible system of λ-adic representations of Γ_F, with coefficients in a number field E. Let

$$\rho_\lambda\colon \Gamma_F \to \mathrm{GL}_n(\overline{E_\lambda})$$

be such a (semi-simple, continuous) compatible system. Write V_λ for the space on which ρ_ℓ acts.

DEFINITION 3.4.5. We say that V_λ is Lie-multiplicity free if after any finite restriction L/F, $V_\lambda|_{\Gamma_L}$ is multiplicity-free. Equivalently,

$$\varinjlim_{L/F} \mathrm{End}_{\overline{E_\lambda}[\Gamma_L]}(V_\lambda)$$

is commutative. We will often abbreviate 'Lie-multiplicity-free' to 'LMF.'

Cases to keep in mind are Hodge-Tate regular (see Definition 2.4.2 and the preceding discussion of §2.4.1) V_λ, or V_λ of the form $H^1(A_{\overline{F}}, \mathbb{Q}_\ell)$ where A/F is an abelian variety with $\mathrm{End}^0(A_{\overline{F}})$ a commutative $\mathbb{Q}$-algebra (by Faltings's proof of the Tate conjecture). Elementary representation theory yields:

LEMMA 3.4.6. (1) *Suppose V_λ is irreducible. Then V_λ is LMF if and only if it can be written*

$$V_\lambda \cong \mathrm{Ind}_{L(\lambda)}^F(W_\lambda),$$

where we write $L(\lambda)$ to show the a priori dependence on λ if V_λ belongs to a compatible system, and where W_λ is a Lie-irreducible $\overline{E_\lambda}$-representation of $\Gamma_{L(\lambda)}$, all of whose Γ_F-conjugates remain distinct after any finite restriction.

(2) *Let W_λ be an irreducible representation of Γ_L, and assume that $V_\lambda = \mathrm{Ind}_L^F(W_\lambda)$ is LMF. Then V_λ is irreducible.*

PROOF. For (1), restrict to a finite-index subgroup of Γ_F over which V_λ decomposes into a direct sum of Lie-irreducible representations; take one such factor, and consider its stabilizer in Γ_F–V_λ is then induced from this subgroup. For (2), Mackey theory implies we need to check that $W_\lambda|_{g\Gamma_L g^{-1}\cap\Gamma_L}$ and $(gW_\lambda)|_{g\Gamma_L g^{-1}\cap\Gamma_L}$ are disjoint for all $g \in \Gamma_F - \Gamma_L$. These two representations occur as distinct factors in the $V_\lambda|_{g\Gamma_L g^{-1}\cap\Gamma_L}$, so they are disjoint since V_λ is LMF. □

For general (possibly reducible) LMF representations, there is a decomposition into a sum of terms as in the lemma. If V_λ belongs to a compatible system, we expect that the number of such factors should be independent of λ; this is an extremely difficult problem (unlike the corresponding question for Artin representations). Let us indicate the difficulties through an example.

EXAMPLE 3.4.7. Suppose f is a holomorphic cuspidal Hecke eigenform on the upper half-plane of some weight $k \geq 2$ and level N and nebentypus ϵ. We normalize f so that its q-expansion at the cusp ∞, $f = \sum a_n(f)q^n$, has leading coefficient $a_1(f) = 1$. Then work of Eichler-Shimura-Deligne (see [**Del71**]) yields a compatible system

$$\rho_{f,\lambda} \colon \Gamma_{\mathbb{Q}} \to \mathrm{GL}_2(\overline{E_\lambda})$$

of λ-adic representations (here E is the number field generated by the $a_n(f)$) characterized by the property that for all $p \nmid N$, the characteristic polynomial of $\rho_{f,\lambda}(\mathrm{fr}_p)$ is equal to

$$X^2 - a_p(f)X + p^{k-1}\epsilon(p).$$

Moreover, the eigenvalues of $\rho_{f,\lambda}(\mathrm{fr}_p)$ are p-Weil numbers of weight $k-1$, i.e. have absolute value $p^{\frac{k-1}{2}}$ in all complex embeddings, and $\rho_{f,\lambda}$ is Hodge-Tate of weights $0, 1-k$. Since f is cuspidal, we expect these Galois representations to be irreducible. This can be proven *because the Fontaine-Mazur-Langlands conjecture is known for the group* GL_1. Suppose $\rho_{f,\lambda} \cong \chi_1 \oplus \chi_2$. Then (up to re-ordering χ_1, χ_2) Theorem 2.3.13 implies[26] that χ_1 must be a finite-order character and χ_2 must be the product of a finite-order character and the $(1-k)^{th}$ power of the cyclotomic character. This contradicts the fact that the eigenvalues of $\rho_{f,\lambda}(\mathrm{fr}_p)$ are p-Weil numbers of weight $k-1$, so we have proven the asserted irreducibility.

---

[26]This special case of Theorem 2.3.13 is notably easier than the general case. It suffices to show that a character $\chi \colon \Gamma_{\mathbb{Q}} \to \overline{\mathbb{Q}}_\ell^\times$ with all labeled Hodge-Tate weights equal to zero is finite-order; this follows from global class field theory and the corresponding statement for characters $\Gamma_{\mathbb{Q}_\ell} \to \overline{\mathbb{Q}}_\ell^\times$. This latter statement is a slight improvement of a fundamental theorem of Tate: see [**Ser98**, §III.A-3].

Therefore in this section we pursue a much more modest goal: restricting to the case of irreducible compatible systems, we will be able to say something about independence of λ of the fields $L(\lambda)$ of Lemma 3.4.6.

Our basic strategy is that the places v for which $\mathrm{tr}(\rho_\lambda(\mathrm{fr}_v)$ equals zero should detect the field $L(\lambda)$. This is in marked contrast to the case of Artin representations: any irreducible (non-trivial) representation of a finite group has elements acting with trace zero. Our main tool will be the following (slight weakening of a) theorem of Rajan:

THEOREM 3.4.8 (Theorem 3 of [**Raj98**]). *Let E be a mixed characteristic non-archimedean local field, and let H/E be an algebraic group. Let X be a subscheme of H (over E), stable under the adjoint action of H. Suppose $\rho\colon \Gamma_F \to H(E)$ is a Galois representation, unramified almost everywhere, and let $C = X(E) \cap \rho(\Gamma_F)$. Denote by $H_\rho \subset H$ the algebraic monodromy group $\overline{\rho(\Gamma_F)}^{\mathrm{Zar}}$, and let $\Phi = H_\rho/H_\rho^0$ denote its group of connected components. For $\phi \in \Phi$, we write H^ϕ for the corresponding component, and we set $\Psi = \{\phi \in \Phi | H^\phi \subset X\}$. Then the density of the set of places v of F with $\rho(\mathrm{fr}_v) \in C$ is precisely $|\Psi|/|\Phi|$.*

Rajan applies this to prove[27] (Theorem 4 of [**Raj98**]) that an irreducible, but *Lie reducible*, representation necessarily has a positive density of frobenii acting with trace zero; note that this also follows immediately from Čebotarev and Proposition 3.4.1, which is a more robust version of Rajan's result (basically combining his argument with Proposition 2.1.4). Our next two results establish a converse, also extending Corollaire 2 to Proposition 15 of [**Ser81**] to its natural level of generality: that result handles the case of connected monodromy groups.

PROPOSITION 3.4.9. *Let $\rho_\lambda\colon \Gamma_F \to \mathrm{GL}_n(\overline{E_\lambda})$ be a continuous, semi-simple, LMF representation. Decompose V_λ as above, so*

$$V_\lambda \cong \bigoplus_{i=1}^{r_\lambda} \mathrm{Ind}_{L(\lambda)_i}^F(W_{\lambda,i})$$

for Lie-irreducible representations $W_{\lambda,i}$ of $\Gamma_{L(\lambda)_i}$. Then:

(1) *Up to a density zero set of places,*

$$\{v \in |F| : \mathrm{tr}(\rho_\lambda(\mathrm{fr}_v)) = 0\} = \{v : \mathrm{fr}_v \notin \bigcup_i \bigcup_{\sigma \in S_{\lambda,i}} \sigma\Gamma_{L(\lambda)_i}\sigma^{-1}\},$$

where $S_{\lambda,i}$ is a set of representatives of $\Gamma_F/\Gamma_{L(\lambda)_i}$.

(2) *Further assume that ρ_λ belongs to a compatible system $\{\rho_\lambda\}$ of λ-adic representations of Γ_F (although in contrast to definition 1.2.3, we need not assume here that the ρ_λ are geometric). Then up to a set of density zero, the set of places of F which have a split factor in $L(\lambda)_i$ for some i is independent of λ. If we further assume that all V_λ are absolutely irreducible ($r_\lambda = 1$) and all $L(\lambda)/F$ are Galois, then $L(\lambda)$ is independent of λ.*[28]

PROOF. For the first part of the Proposition, we ignore the underlying 'coefficient' number field E and just view ρ_λ as valued in $\mathrm{GL}_n(E)$ for some sufficiently

[27] In addition to the main result of his paper, a beautiful 'strong multiplicity one' theorem for ℓ-adic representations.

[28] In the non-Galois case, see Exercise 6 in [**Cp86**]!

large finite extension E of $\mathbb{Q}_\ell$. The "$\supseteq$" direction follows from the usual formula for the trace of an induced representation. To establish the reverse inclusion, let us consider, for each non-empty subset $I \subset \coprod_i S_{\lambda,i}$, the set $\mathcal{X}_I$ of places v such that $\mathrm{tr}(\rho_\lambda(\mathrm{fr}_v)) = 0$, and $\mathrm{fr}_v \in \sigma\Gamma_{L(\lambda)_{i(\sigma)}}\sigma^{-1}$ if and only if $\sigma \in I$ [Notation: if $\sigma \in I$, then $\sigma \in S_{\lambda,i}$ for a unique $i =: i(\sigma)$]. Also set

$$\Gamma_I = \bigcap_{\sigma \in I} \sigma\Gamma_{L(\lambda)_{i(\sigma)}}\sigma^{-1},$$

and let $\mathcal{G}_I$, resp. $\mathcal{G}$, denote the algebraic monodromy group of $\rho_\lambda|_{\Gamma_I}$, resp. ρ_λ. To establish the "$\subseteq$" (up to density zero) direction, we must show that $\mathcal{X}_I$ has density zero for every non-empty I. For all I, $\mathcal{G}_I$ contains $\mathcal{G}^0$, the identity component of $\mathcal{G}$. We apply Rajan's Theorem (3.4.8 above) to $\rho_\lambda|_{\Gamma_I}$: if $\mathcal{X}_I$ has positive density, then there is a full connected component $T\mathcal{G}^0 \subset \mathcal{G}_I$ on which the trace vanishes (here T is some coset representative for the component). Representing endomorphisms of V_λ in block-matrix form corresponding to the decomposition

$$V_\lambda|_{\rho_\lambda^{-1}(\mathcal{G}^0)} = \bigoplus_{i,\sigma} \sigma W_{\lambda,i},$$

we have

$$\mathrm{tr}\left[T \cdot \begin{pmatrix} * & 0 & \cdots & 0 \\ 0 & * & \cdots & 0 \\ \vdots & \vdots & \ddots & 0 \\ 0 & 0 & \cdots & * \end{pmatrix}\right] = 0,$$

where the $*$'s represent *arbitrary* elements of the $\mathrm{End}(\sigma W_{\lambda,i})$. For this, we use the fact that these constituents are (absolutely) irreducible and distinct: either apply Wedderburn theory to the semi-simple $\overline{E_\lambda}$-algebra $\overline{E_\lambda}[[\rho_\lambda(\rho_\lambda^{-1}(\mathcal{G}^0))]]$, or apply Schur's lemma (using absolute irreducibility) and an algebra version of Goursat's lemma (using multiplicity-freeness). Then the above matrix equation implies the automorphism T, written in block-matrix form, has all zeros along the (block)-diagonal. But I is non-empty, so $T \in \mathcal{G}_I$ preserves at least one subspace $\sigma W_{\lambda,i}$, and so its block-diagonal entry corresponding to that sub-space must be non-zero (invertible). This contradiction forces all components of $\mathcal{G}_I$ to have non-zero trace (generically), for all non-empty I, and thus the "$\subseteq$" direction is established.

Part 2 now follows from independence of λ of $\mathrm{tr}(\rho_\lambda(\mathrm{fr}_v))$ and Čebotarev, noting that

$$\{v : \mathrm{fr}_v \in \bigcup_i \bigcup_{\sigma \in S_{\lambda,i}} \Gamma_{\sigma(L(\lambda)_i)}\}$$

is the set of places v of F that have at least one split factor in some $L(\lambda)_i$. □

We make a few remarks about the limitations of this method:

REMARK 3.4.10. (1) Without any information about how the various V_λ decompose into irreducible sub-representations, this result yields frustratingly little, since it is easy to find disjoint collections of number fields $\{L_i\}_{i \in I}$ and $\{L'_i\}_{i \in I'}$ such that the union of primes with a split factor (or even split) in the various L_i equals the union of those with a split factor in the various L'_i. For, the simplest example, take $L_1 = \mathbb{Q}$, $L'_1 = \mathbb{Q}(i)$, $L'_2 = \mathbb{Q}(\sqrt{2})$, $L'_3 = \mathbb{Q}(\sqrt{-2})$.

(2) Even when the V_λ are irreducible, and even assuming that one field $L(\lambda_0)$ is Galois, care must be taken when the other inducing fields $L(\lambda)$ are not (known to be) Galois. We see that $L(\lambda_0)$ is contained in $L(\lambda)$ for all λ, but equality does not follow, as the following example from finite group theory shows. We want an inclusion $H \leq K \lhd G$ of groups with K normal in G, H a proper subgroup of K, and $\cup_{g \in G} gHg^{-1} = K$. Taking $K \lhd S_4$ to be the copy of the Klein four-group given by the $(2,2)$-cycles, and H to be the subgroup generated by one of these permutations, meets the requirements. Note that such examples require K to have non-trivial outer automorphism group: if G-conjugation acts by K-inner automorphisms on K, then $\cup_G gHg^{-1} = \cup_K kH^k-1 = K$ implies $H = K$, since no finite group is the union of conjugates of a proper subgroup.

(3) The group-theoretic counterexample of the previous item should not arise in practice: if we assume that W_{λ_0} can be put in a compatible-system, say with λ-adic realization $W_{\lambda,0}$, then conjecturally $W_{\lambda,0}$ will be Lie irreducible as well, and then the isomorphism

$$\mathrm{Ind}_{L(\lambda)}^F(W_\lambda) \cong \mathrm{Ind}_{L(\lambda_0)}^F(W_{\lambda,0})$$

implies, by Mackey theory, that there is a non-zero $\Gamma_{L(\lambda_0)}$-morphism

$$\mathrm{Ind}_{s(L(\lambda))}^{L(\lambda_0)}(sW_\lambda) \twoheadrightarrow W_{\lambda,0}$$

for some $s \in \Gamma_F$. By part (2) of Lemma 3.4.6, this induction is irreducible, so this map is an isomorphism. But $W_{\lambda,0}$ is (conjecturally) Lie-irreducible, so $L(\lambda_0) = L(\lambda)$.

In any case, the following corollary is the promised converse to Rajan's result:

COROLLARY 3.4.11. *An irreducible, LMF representations* $\rho_\ell \colon \Gamma_F \to \mathrm{GL}_n(\overline{\mathbb{Q}}_\ell)$ *is Lie irreducible precisely when the set of* v *with* $\mathrm{tr}(\rho_\ell(\mathrm{fr}_v)) = 0$ *has density-zero. In particular, in a compatible system of irreducible, LMF representations, Lie-irreducibility is independent of* λ.

REMARK 3.4.12. If we know general automorphic base-change for GL_n, we can formulate the conditions 'Lie irreducible' and 'Lie multiplicity free' on the automorphic side. This result then suggests how to tell whether a 'LMF' automorphic representation is automorphically induced. Finding an intrinsic characterization of the image of automorphic induction, even conjecturally, is a mystery (in contrast to its close cousin base-change), so it may come as a surprise that there should be such a simple condition at the level of Satake parameters, for this broad class of LMF representations.

I originally developed Proposition 3.4.9 to prove that a regular compatible system of representations of Γ_F for F a CM field, if induced, is necessarily induced from a CM field. See Remark 2.4.9 for an automorphic analogue. Here is a partial result; it is another application of the 'Hodge-theory with coefficients' in §2.4.1. For the most natural formulation of the result, recall the definition of purity of a weakly compatible system (Definition 1.2.4).

LEMMA 3.4.13. *Let* F *be a CM field, and let* $\mathcal{R} = \{\rho_\lambda \colon \Gamma_F \to \mathrm{GL}_n(\overline{E_\lambda})\}$ *be a weakly compatible system of* λ*-adic representations with coefficients in a number field* E. *Assume that the* ρ_λ *are (almost all) Hodge-Tate regular, and pure. Finally,* *assume that there is a single number field L such that for λ above a set of rational primes ℓ of density one,*

$$\rho_\lambda \cong \mathrm{Ind}_L^F(r_\lambda),$$

for some $\overline{E_\lambda}$-representation r_λ of Γ_L. Then L is CM.

PROOF. We may assume $E/\mathbb{Q}$ is Galois. In yet another variant of the theme of §2.4, purity implies we may take the number field E to be CM: simply observe that for any choice c of complex conjugation in $\mathrm{Gal}(E/\mathbb{Q})$, the characteristic polynomials $Q_v(X)$ of $\rho_\lambda(\mathrm{fr}_v)$ satisfy

$$^cQ_v(X) = X^n Q_v(q_v^w/X)/Q_v(0),$$

and thus for any two complex conjugations c, c', $^{cc'}\mathcal{R} \cong \mathcal{R}$. It follows that for a density one set of v, $Q_v(X)$ has coefficients in E_{cm}. The Čebotarev density theorem then yields a well-defined weakly compatible system (in the sense of Definition 1.2.3) of λ-adic representations with coefficients in E_{cm}. (This argument is taken from [**PT15**, Lemma 1.1, 1.2].[29]). Thus we take E to be CM. By regularity, we may also assume that there is a CM extension E'/E such that for all finite-index subgroups H of Γ_F, and all primes λ of E', all sub-representations $r \subset \rho_\lambda|_H$ are actually defined over E'_λ: this is an elementary argument (see [**BLGGT14**, Lemma 5.3.1(3)], with the CM refinement of [**PT15**, Lemma 1.4]), the key idea being that regularity gives us an abundant supply of elements in the image of ρ_λ with distinct eigenvalues, and that ρ_λ can then be defined over the extension of E generated by these eigenvalues. Thus, after enlarging E, we may assume all ρ_λ and r_λ act on E_λ-vector spaces.

For simplicity enlarge E to contain L_{cm}, the maximal CM subfield of L, and take its Galois closure–the result remains CM. If $L \neq L_{\mathrm{cm}}$, then we can find a (positive-density) set of ℓ (unramified in L and for the system ρ_λ) which are split in E (with, say, $\lambda|\ell$) but not in (the Galois closure of L, hence) L. Consider a non-split prime $w|\ell$ of L, above a place v of F, and the restriction

$$r_\lambda|_{\Gamma_{L_w}} : \Gamma_{L_w} \to \mathrm{GL}_n(E_\lambda) = \mathrm{GL}_n(\mathbb{Q}_\ell);$$

By Lemma 2.2.9, $\mathrm{D}_{\mathrm{dR}}(\rho_\lambda|_{\Gamma_{F_v}})$ is the image under the forgetful functor (from filtered L_w-vector spaces to filtered F_v-vector spaces) of $\mathrm{D}_{\mathrm{dR}}(r_\lambda|_{\Gamma_{L_w}})$. Since L_w does not embed in $E_\lambda = \mathbb{Q}_\ell$, we can invoke Corollary 2.4.3 to show ρ_λ is not regular, a contradiction. Therefore $L = L_{\mathrm{cm}}$. □

Combining this with Proposition 3.4.9, since regular clearly implies LMF,[30] we deduce:

COROLLARY 3.4.14. *Let F be a CM field, and let $\mathcal{R} = \{\rho_\lambda\}_\lambda$ be an absolutely irreducible, pure, Hodge-Tate regular weakly compatible system of representations of Γ_F. Suppose that when we write*

$$\rho_\lambda \cong \mathrm{Ind}_{L(\lambda)}^F(W_\lambda),$$

where W_λ is Lie irreducible, the extensions $L(\lambda)/F$ are Galois. Then the field $L = L(\lambda)$ is independent of λ, and L is itself CM.

---

[29] Our definition of compatible system is somewhat weaker than that given in [**PT15**]; for our purposes, we can just ignore the part of the proof of [**PT15**, Lemma 1.1] that computes Hodge numbers.

[30] This is the one case in which the LMF condition is provably independent of ℓ.

REMARK 3.4.15. One way of interpreting this result is that to study regular motives, compatible systems, or algebraic automorphic representations over CM fields, we will never have to leave the comfort of CM fields; essentially all progress in the study of automorphic Galois representations is currently restricted to this context. This is in particular the case for Galois representations occurring as irreducible sub-quotients of the cohomology of a Shimura variety.

CHAPTER 4

Motivic lifting

As noted in the introduction (see Question 1.1.9), the results of Part 3 raise more questions than they resolve. In this chapter, we discuss some cases of the motivic analogue of Conrad's lifting question; this is also the natural framework for the problem of finding *compatible* lifts of a compatible system of Galois representations. Ideally, we would be able to work in Grothendieck's category of pure motives, defined using the relation of homological equivalence on algebraic cycles (see §4.1.2 for a brief review). This category can only be proven to have the desirable categorical properties–namely, equivalence to the category of representations of some pro-reductive ('motivic Galois') group–if we assume Grothendieck's Standard Conjectures on algebraic cycles, which are far out of reach. (For the basic formalism (as relevant for the theory of motives) of algebraic cycles and precise statement of the Standard Conjectures, see [**Kle68**].) We therefore need an unconditional variant of the motivic Galois formalism, and we adopt André's approach, using his theory of motivated cycles ([**And96b**]). In §4.1 we review André's theory, prove some supplementary results needed for the application to motivic lifting, and then treat the motivic lifting problem in the potentially abelian case. In §4.2 we prove an arithmetic refinement of André's work ([**And96a**]) on, roughly speaking, the motivated theory of hyperkähler varieties. This provides a motivic analogue of Theorem 3.2.10 in many non-abelian examples. Finally, in §4.3, we speculate on a generalized Kuga-Satake construction, of which the results of §4.2 are the 'classical' case; we then prove this for H^2 of an abelian variety, generalizing known results for abelian surfaces.

4.1. Motivated cycles: Generalities

4.1.1. Lifting Hodge structures. In trying to produce a motivic analogue either of Wintenberger's or of my lifting theorem, one is naturally led to try to lift all cohomological, rather than merely the ℓ-adic, realizations of a 'motive.' The easiest such lifting problem is for real Hodge-structures, which are parametrized by representations of the Deligne torus $\mathbb{S} = \mathrm{Res}_{\mathbb{C}/\mathbb{R}}(\mathbb{G}_m)$. Recall that $X^\bullet(\mathbb{S}) = \mathbb{Z}\alpha_1 \oplus \mathbb{Z}\alpha_2$, where α_1 and α_2 are the first and second projections in the isomorphism

$$\mathbb{S}(\mathbb{C}) = (\mathbb{C} \otimes_{\mathbb{R}} \mathbb{C})^\times \xrightarrow{\sim} \mathbb{C}^\times \times \mathbb{C}^\times$$
$$z_1 \otimes z_2 \mapsto (z_1 z_2, \overline{z_1} z_2).$$

Here $\mathbb{S}(\mathbb{R}) \subset \mathbb{S}(\mathbb{C})$ is $\mathbb{C}^\times = (\mathbb{C} \otimes_{\mathbb{R}} \mathbb{R})^\times$, and the $\mathrm{Gal}(\mathbb{C}/\mathbb{R})$-action, induced by $z_1 \otimes z_2 \mapsto z_1 \otimes \overline{z_2}$, is given by $c\colon (w, z) \mapsto (\bar{z}, \bar{w})$. In particular,

$$(c \cdot \alpha_1)(w, z) = c(\alpha_1(\bar{z}, \bar{w})) = z,$$

i.e. $(c \cdot \alpha_1) = \alpha_2$, and similarly $(c \cdot \alpha_2) = \alpha_1$. The group of characters over $\mathbb{R}$, denoted $X^\bullet_{\mathbb{R}}(\mathbb{S})$, is then $\mathbb{Z}(\alpha_1 + \alpha_2)$, where $\alpha_1 + \alpha_2 = N$ is the norm, satisfying $N(z) = z\bar{z}$ on $\mathbb{R}$-points.

Let $\widetilde{H}_0 \to H_0$ be a surjection of linear algebraic groups over $\mathbb{R}$ with kernel equal to a central torus Z_0. We are interested in the lifting problem for algebraic representations over $\mathbb{R}$:

$$\begin{array}{ccc} & & \widetilde{H}_0 \\ & \overset{\tilde{h}}{\nearrow} & \downarrow \pi \\ \mathbb{S} & \xrightarrow[h]{} & H_0. \end{array}$$

Any such h lands in some (typically non-split) maximal torus T_0, and any lift will land in the preimage $\pi^{-1}(T_0) =: T_0'$, which is a maximal torus of $\widetilde{H}_0$. We are therefore reduced to studying the dual diagram of free $\mathbb{Z}$-modules with $\mathrm{Gal}(\mathbb{C}/\mathbb{R}) = \Gamma_{\mathbb{R}}$-action:

$$\begin{array}{ccc} & & 0 \\ & & \uparrow \\ & & X^\bullet(Z_0) \\ & & \uparrow \\ & & X^\bullet(T_0') \\ & \swarrow & \uparrow \\ X^\bullet(\mathbb{S}) & \longleftarrow & X^\bullet(T_0) \\ & & \uparrow \\ & & 0 \end{array}$$

As short-hand, we denote the vertically-aligned character groups, from bottom to top, by Y, Y', and L, so we in fact are studying the sequence

$$0 \to \mathrm{Hom}_{\Gamma_{\mathbb{R}}}(L, X^\bullet(\mathbb{S})) \to \mathrm{Hom}_{\Gamma_{\mathbb{R}}}(Y', X^\bullet(\mathbb{S})) \\ \to \mathrm{Hom}_{\Gamma_{\mathbb{R}}}(Y, X^\bullet(\mathbb{S})) \to \mathrm{Ext}^1_{\Gamma_{\mathbb{R}}}(L, X^\bullet(\mathbb{S})) \to \dots$$

Any real torus is isomorphic to a product of copies of $\mathbb{G}_m$, $\mathbb{S}$, and $\mathbb{S}^1 = \ker(N \colon \mathbb{S} \to \mathbb{G}_m)$. The character group $X^\bullet(\mathbb{S}^1)$ is $X^\bullet(\mathbb{S})/\mathbb{Z}(\alpha_1 + \alpha_2)$. A generator is the image of $\alpha_1 - \alpha_2$, which on $\mathbb{R}$-points is simply the character $z \mapsto z/\bar{z}$ of the (analytic) unit circle S^1. Complex conjugation acts as -1 on $X^\bullet(\mathbb{S}^1)$. We can therefore completely address the lifting problem by understanding morphisms and extensions between these three basic $\mathbb{Z}[\Gamma_{\mathbb{R}}]$-modules. The case of immediate interest to us will be when Z_0 is split, so L is just some number of copies of $\mathbb{Z}$ with trivial $\Gamma_{\mathbb{R}}$-action. Now,

$$\mathrm{Ext}^1_{\Gamma_{\mathbb{R}}}(\mathbb{Z}, X^\bullet(\mathbb{S})) = H^1(\Gamma_{\mathbb{R}}, X^\bullet(\mathbb{S})) = 0,$$

since $\ker(1+c) = \operatorname{im}(c-1) = \mathbb{Z}(\alpha_1 - \alpha_2)$. Therefore any $h\colon \mathbb{S} \to H_0$ lifts, and the ambiguity in lifting is a collection of elements (of order equal to the rank of Z_0) of $\operatorname{Hom}_{\Gamma_{\mathbb{R}}}(\mathbb{Z}, X^\bullet(\mathbb{S})) \cong \mathbb{Z}(\alpha_1+\alpha_2)$.[1]

EXAMPLE 4.1.1. In §4.2.2 we will consider the following setup: $V_{\mathbb{R}}$ will be an orthogonal space with signature $(m-2,2)$[2]. Write $m = 2n$ or $m = 2n+1$. The lifting problem will be

$$\begin{array}{ccc} & & \mathrm{GSpin}(V_{\mathbb{R}}) \\ & \overset{\tilde{h}}{\nearrow} & \downarrow \pi \\ \mathbb{S} & \xrightarrow[h]{} & \mathrm{SO}(V_{\mathbb{R}}), \end{array}$$

where h lands in a maximal anisotropic torus $T_0 \cong (\mathbb{S}^1)^n$. We can write

$$X^\bullet(T_0') = \left(\bigoplus_{i=1}^n \mathbb{Z}\chi_i\right) \oplus \mathbb{Z}(\chi_0 + \frac{\sum \chi_i}{2}),$$

where conjugation acts by -1 on each χ_i, $i = 1, \ldots, n$, and trivially on χ_0. Here the characters χ_i are conjugate in $\mathrm{SO}(V_{\mathbb{C}})$ to the characters denoted χ_i in §2.8 (see page 60). The torus T_0 is built out of copies of $\mathrm{SO}(2)$ embedded in $\mathrm{SO}(m-2,2)$, and these are just the usual characters

$$\begin{pmatrix} \cos(\theta) & \sin(\theta) \\ -\sin(\theta) & \cos(\theta) \end{pmatrix} \mapsto e^{i\theta}.$$

Moreover, $\oplus_{i=1}^n \mathbb{Z}\chi_i$ is the submodule $X^\bullet(T_0)$. A morphism $\mathbb{S} \to T_0 \subset \mathrm{SO}(V_{\mathbb{R}})$ is given in coordinates by $\chi_i \mapsto m_i(\alpha_1 - \alpha_2)$, for some integers m_i. A lift to a morphism $\mathbb{S} \to \mathrm{GSpin}(V_{\mathbb{R}})$ then amounts to an extension

$$\chi_0 + \frac{\sum_{i=1}^n \chi_i}{2} \mapsto \frac{\epsilon_0}{2}(\alpha_1+\alpha_2) + \frac{\sum_{i=1}^n m_i}{2}(\alpha_1 - \alpha_2),$$

where ϵ_0 is any integer having the same parity as $\sum m_i$. The Clifford norm N is given by the character $2\chi_0$, so the composition of such a lift with the Clifford norm is $\epsilon_0(\alpha_1+\alpha_2)$. In the $K3$ (or hyperkähler) examples to be considered in the next section, $m_1 = 1$ and all other $m_i = 0$, so we find there is a unique lift

$$\begin{array}{ccc} & & \mathrm{GSpin}(V_{\mathbb{R}}) \\ & \overset{\tilde{h}}{\nearrow} & \downarrow \pi \\ \mathbb{S} & \xrightarrow[h]{} & \mathrm{SO}(V_{\mathbb{R}}), \end{array}$$

where $N \circ \tilde{h}\colon \mathbb{S} \to \mathbb{G}_m$ is any *odd* power of the usual norm $\mathbb{S} \to \mathbb{G}_m$. The Kuga-Satake theory (see §4.2.2 and following) takes the norm itself, which then gives rise to weight 1 Hodge structures.

---

[1] Note that if we dealt with representations of $\mathbb{S}^1$ we would find an obstruction in $H^1(\Gamma_{\mathbb{R}}, X^\bullet(\mathbb{S}^1)) = \mathbb{Z}/2\mathbb{Z}$.

[2] This part of the discussion applies to any signature (p,q) with at least one of p or q even, so that $\mathrm{SO}(V_{\mathbb{R}})$ has a compact maximal torus.

4.1.2. Motives for homological equivalence and the Tannakian formalism. Our aim in this sub-section is to review Grothendieck's construction of the category $\mathcal{M}_F^{\mathrm{hom}}$ of pure motives for homological equivalence over a field F. As a preliminary, we provide some background on the theory of neutral Tannakian categories. We can then describe the output of the Standard Conjectures, taking for simplicity F to be an abstract (i.e., not embedded) field of characteristic zero, small enough to be embedded in $\mathbb{C}$: $\mathcal{M}_F^{\mathrm{hom}}$ is (conjecturally) a graded, semi-simple, $\mathbb{Q}$-linear neutral Tannakian category. With the exception of §4.3.2, in the rest of this paper we will not work directly with $\mathcal{M}_F^{\mathrm{hom}}$, so much of this discussion serves only to orient the reader. For more background, the reader should consult [**And04**], especially Chapter 4, or [**Sch94**].

4.1.2.1. *Neutral Tannakian categories.* We begin with an overview of the theory of neutral Tannakian categories; [**DM11**] is a very readable and thorough introduction to which we refer the reader for details (the original source is [**SR72**]). Note that we will always work with neutral Tannakian categories, which simplifies the theory considerably; for deeper aspects in the non-neutral case, see [**Del90**]. Let E be a field. The prototypical neutral Tannakian category is the category Vec_E of finite-dimensional vector spaces over E. Vec_E is an abelian category with a notion of tensor product (namely, the usual tensor product of E-vector spaces) satisfying certain natural requirements: associativity, commutativity, and existence of a unit (E regarded as an E-vector space) for the tensor product. Moreover, every object of Vec_E has a dual (more generally, internal Hom's exist), and the endomorphisms of the unit object are just the field E itself. The main theorem of the theory in this context is the trivial observation that Vec_E is equivalent to the category of finite-dimensional representations (over E) of the trivial group.

The general definition merely formalizes the notions in the previous paragraph; we will quickly make the necessary definitions, and then give a number of examples (Example 4.1.6 below).

DEFINITION 4.1.2. Let $\mathcal{C}$ be a category, and let $\otimes\colon \mathcal{C}\times\mathcal{C}\to\mathcal{C}$ be a functor satisfying the usual axioms of a symmetric monoidal category, namely:

(1) (Associativity constraint) There is a functorial isomorphism

$$A_{X,Y,Z}\colon X\otimes(Y\otimes Z)\to(X\otimes Y)\otimes Z$$

satisfying the *pentagon axiom* ([**DM11**, 1.0.1]).

(2) (Commutativity constraint) There is a functorial isomorphism

$$C_{X,Y}\colon X\otimes Y\to Y\otimes X$$

such that $C_{Y,X}\circ C_{X,Y}=\mathrm{id}_{X\otimes Y}$, and that is compatible with the associativity constraint in the sense of the *hexagon axiom* ([**DM11**, 1.0.2]).

(3) (Unit object) There is a pair $(\mathbf{1},e)$ consisting of an object $\mathbf{1}$ and an isomorphism $e\colon \mathbf{1}\to\mathbf{1}\otimes\mathbf{1}$ such that the functor $\mathcal{C}\to\mathcal{C}$ given by $X\mapsto\mathbf{1}\otimes X$ is an equivalence of categories.

We call such a $(\mathcal{C},\otimes)$, equipped with its associativity and commutativity constraints (but omitted from the notation), a *tensor category*, for short.

If $(\mathcal{C},\otimes)$ and $(\mathcal{C}',\otimes')$ are tensor categories (whose associativity and commutativity constraints we will write as A, C and A', C', respectively; unit elements will be

$(\mathbf{1}, e)$ and $(\mathbf{1}', e')$), then a *tensor functor* from $(\mathcal{C}, \otimes)$ to $(\mathcal{C}', \otimes')$ is a pair (F, k) consisting of a functor $F\colon \mathcal{C} \to \mathcal{C}$ and a functorial isomorphism $k_{X,Y}\colon F(X) \otimes F(Y) \to F(X \otimes Y)$ satisfying the following three compatibilities:

(1) For all objects X, Y, Z of $\mathcal{C}$, the diagram

$$\begin{array}{ccccc} FX \otimes (FY \otimes FZ) & \xrightarrow{\mathrm{id} \otimes k} & FX \otimes F(Y \otimes Z) & \xrightarrow{k} & F(X \otimes (Y \otimes Z)) \\ \downarrow{\scriptstyle A'} & & & & \downarrow{\scriptstyle F(A)} \\ (FX \otimes FY) \otimes FZ & \xrightarrow{k \otimes \mathrm{id}} & F(X \otimes Y) \otimes FZ & \xrightarrow{k} & F((X \otimes Y) \otimes Z) \end{array}$$

is commutative.

(2) For all objects X, Y of $\mathcal{C}$, the diagram

$$\begin{array}{ccc} FX \otimes FY & \xrightarrow{k} & F(X \otimes Y) \\ \downarrow{\scriptstyle C'} & & \downarrow{\scriptstyle F(C)} \\ FY \otimes FX & \xrightarrow{k} & F(Y \otimes X) \end{array}$$

is commutative.

(3) If $(\mathbf{1}, e)$ is a unit object of $\mathcal{C}$, then $(F(\mathbf{1}), F(e))$ is a unit object of $\mathcal{C}'$.

We have spelled out the precise conditions for (F, k) to be a tensor functor because of its relevance for the structure of $\mathcal{M}_F^{\mathrm{hom}}$: the Künneth Standard Conjecture essentially 'corrects' the fact that the Betti realization

$$H_B\colon \mathcal{M}_F^{\mathrm{hom}} \to \mathrm{Vec}_{\mathbb{Q}}$$

with the natural Künneth isomorphism $k_{X,Y} H_B(X) \otimes H_B(Y) \xrightarrow{\sim} H_B(X \times Y)$ does *not* intertwine the (obvious) commutativity constraints on $\mathcal{M}_F^{\mathrm{hom}}$ and $\mathrm{Vec}_{\mathbb{Q}}$. See the discussion surrounding Conjecture 4.1.8 for details.

For any unit object $(\mathbf{1}, e)$, we obtain isomorphisms $l_X\colon \mathbf{1} \otimes X \xrightarrow{\sim} X$ and $r_X\colon X \otimes \mathbf{1} \xrightarrow{\sim} X$.

For any additive tensor category $(\mathcal{C}, \otimes)$ (for which we require $\otimes$ to be bi-additive), and any unit object $(\mathbf{1}, e)$, $\mathrm{End}_{\mathcal{C}}(\mathbf{1}) = R$ is a commutative ring, unique up to unique isomorphism, and $\mathcal{C}$ is naturally an R-linear category.

DEFINITION 4.1.3. Let $(\mathcal{C}, \otimes)$ be a tensor category. It is *rigid* if for every object X there is a 'dual' object $X^\vee$ along with evaluation and co-evaluation morphisms

$$X^\vee \otimes X \xrightarrow{\mathrm{ev}_X} \mathbf{1}$$

$$\mathbf{1} \xrightarrow{\mathrm{coev}_X} X \otimes X^\vee$$

such that the composites (we suppress the unit isomorphisms and the associators)

$$X \xrightarrow{\mathrm{coev}_X \otimes \mathrm{id}_X} X \otimes X^\vee \otimes X \xrightarrow{\mathrm{id}_X \otimes \mathrm{ev}_X} X$$

$$X^\vee \xrightarrow{\mathrm{id}_{X^\vee} \otimes \mathrm{coev}_X} X^\vee \otimes X \otimes X^\vee \xrightarrow{\mathrm{ev}_X \otimes \mathrm{id}_{X^\vee}} X^\vee$$

are id_X and $\mathrm{id}_{X^\vee}$, respectively.

Note that in the absence of a commutativity constraint, $X^\vee$ is what would be called a 'right dual.'

Finally, we come to the main definition:

DEFINITION 4.1.4. Let E be a field. A *neutral Tannakian category* over E is a rigid abelian tensor category $(\mathcal{C}, \otimes)$ such that $\mathrm{End}(\mathbf{1}) = E$, and for which there exists a faithful, exact, E-linear tensor functor $\omega\colon \mathcal{C} \to \mathrm{Vec}_E$. Such an ω is called a *fiber functor*.

The main theorem of the theory of neutral Tannakian categories just says that every Tannakian category is equivalent to the category of representations of some affine group scheme:

THEOREM 4.1.5 ([**DM11**, Theorem 2.11]). *Let $(\mathcal{C}, \otimes)$ be a neutral Tannakian category over E, equipped with a fiber functor $\omega\colon \mathcal{C} \to \mathrm{Vec}_E$. Then the functor of E-algebras given by tensor-automorphisms of ω (see* [**Del90**, 1.9, 1.11]*) is represented by an affine group scheme G, and the functor $\mathcal{C} \to \mathrm{Rep}_E(G)$ defined by ω is an equivalence of tensor categories.*[3]

Informally, we might call G the 'Galois group' of $(\mathcal{C}, \otimes)$, whence the later terminology 'motivic Galois group' in the (conjecturally Tannakian) case of $\mathcal{M}_F^{\mathrm{hom}}$.

At last, some examples:

EXAMPLE 4.1.6. (1) The following example is crucial for the theory of motives. The category $\mathrm{Vec}_E^{\mathbb{Z}/2}$ of $\mathbb{Z}/2\mathbb{Z}$-graded vector spaces over E, with the usual graded tensor product, and with commutativity constraint given by the Koszul sign rule, i.e. $c_{V,W}\colon V \otimes W \xrightarrow{\sim} W \otimes V$ given on homogeneous tensors by $v \otimes w \mapsto (-1)^{\deg(v)\cdot\deg(w)} w \otimes v$, is a rigid E-linear abelian tensor category. Note, however, that the forgetful functor $\mathrm{Vec}_E^{\mathbb{Z}/2\mathbb{Z}} \to \mathrm{Vec}_E$ (with the obvious isomorphisms $k_{X,Y}$) is not a tensor functor, because it fails to intertwine the commutativity constraints. Indeed, $\mathrm{Vec}_E^{\mathbb{Z}/2\mathbb{Z}}$ is *not* a Tannakian category. Any rigid tensor category has an intrinsic notion of rank for every object X, given by the element of $\mathrm{End}(\mathbf{1})$ arising as the composition

$$\mathbf{1} \xrightarrow{\mathrm{coev}_X} X \otimes X^\vee \xrightarrow{c_{X,X^\vee}} X^\vee \otimes X \xrightarrow{\mathrm{ev}_X} \mathbf{1}.$$

For a $\mathbb{Z}/2\mathbb{Z}$-graded vector space $V = V_0 \oplus V_1$, it is easy to check that the rank in $\mathrm{Vec}_E^{\mathbb{Z}/2\mathbb{Z}}$ is $\dim V_0 - \dim V_1$. But this notion of rank is preserved by any tensor functor, so an obviously necessary condition to admit a fiber functor is that all objects have non-negative rank. (In fact, a deep result of Deligne gives a converse: [**Del90**, 7.1 Théorème].)

(2) The category $\mathbb{Q}$-$\mathrm{HS}^{\mathrm{pol}}$ of pure polarizable $\mathbb{Q}$-Hodge structures is a neutral Tannakian category over $\mathbb{Q}$, with fiber functor just given by taking the underlying $\mathbb{Q}$-vector space of a Hodge structure (here we take the commutativity constraint on Hodge structures to be the same as the usual constraint on vector spaces). This is a useful 'toy model' for the theory of pure motives over $\mathbb{C}$. In particular, Theorem 4.1.5 identifies $\mathbb{Q}$-$\mathrm{HS}^{\mathrm{pol}}$ with the category of representations of some affine group scheme MT over $\mathbb{Q}$; MT is the 'universal Mumford-Tate group.' We use this example to explain how properties of a Tannakian category reflect those of its Galois group (see [**DM11**, §2 'Properties of G and $\mathrm{Rep}(G)$']). From its

[3]That is, it is a tensor functor that is also an equivalence of categories; this ensures ([**DM11**, Proposition 1.11]) that there exists an inverse of F that is also a tensor functor.

Hodge-theoretic description, MT is obviously connected. This is equivalent ([**DM11**, Corollary 2.22]) to the condition that for every non-trivial representation X of MT, the strictly full subcategory of $\mathbb{Q}$-HS^{pol} whose objects are isomorphic to sub-quotients of some $X^{\oplus n}$ is not stable under $\otimes$. But if X is not the trivial Hodge structure $\mathbb{Q}$ (i.e. if X corresponds to a non-trivial representation of MT), then either it has non-zero weight, in which case the result is clear, or some $H^{p,-p}(X)$ is non-zero, for some $p \neq 0$. Then $H^{mp,-mp}(X^{\otimes m}) \neq 0$, which for m large enough cannot be the case for any sub-quotient of any $X^{\oplus n}$. In a similar spirit, MT is pro-reductive, since the category $\mathbb{Q}$-HS^{pol} is semi-simple (by polarizability; see [**DM11**, Proposition 2.23]).

4.1.2.2. *Homological motives.* We begin by sketching the construction of Grothendieck's category of (pure) homological motives over a field F. References for more details are [**Kle68**, §1] and [**Sch94**, §1]. Recall that for simplicity we take F to be a field that can be embedded in $\mathbb{C}$ (or even a specified subfield of $\mathbb{C}$)–this does not affect the construction, but does affect whether the category of motives is (conjecturally) *neutral* Tannakian. Recall that for a fixed Weil cohomology theory H^* (see [**Kle68**, §1.2]), homological equivalence defines an *adequate equivalence relation* on algebraic cycles on smooth projective varieties; for example, since $F \subset \mathbb{C}$, we will always have at our disposal Betti cohomology, H_B^*. For a smooth projective X/F, we let $A^*_{\mathrm{hom}}(X)$ denote the $\mathbb{Q}$-algebra of ($\mathbb{Q}$-linear combinations of) algebraic cycles for homological equivalence, graded by codimension. 'Adequate' ensures that the intersection product is well-defined as a linear map $A^r(X) \otimes A^s(X) \to A^{r+s}(X)$. For any two smooth projective F-varieties X and Y, with X connected of (equi-)dimension d, we can then define the space of degree r correspondences

$$C^r_{\mathrm{hom}}(X, Y) = A^{d+r}_{\mathrm{hom}}(X \times Y).$$

We obtain a composition of correspondences

$$C^r_{\mathrm{hom}}(X, Y) \otimes C^s_{\mathrm{hom}}(Y, Z) \to C^{r+s}_{\mathrm{hom}}(X, Z)$$

given by

$$f \otimes g = g \circ f \mapsto p_{13,*}\left(p_{12}^* f \cdot p_{23}^* g\right),$$

where the p_{ij} are the projections from $X \times Y \times Z$ to the products of two factors. In particular, $C^0_{\mathrm{hom}}(X, X)$ is a $\mathbb{Q}$-algebra.

Definition 4.1.7. The category $\mathcal{M}_F^{\mathrm{hom}}$ of motives over F for homological equivalence has as objects triples (X, p, m), where X is a smooth projective variety over F, m is an integer, and p is an idempotent correspondence in $C^0_{\mathrm{hom}}(X, X)$. Morphisms in $\mathcal{M}_F^{\mathrm{hom}}$ are defined by

$$\mathrm{Hom}_{\mathcal{M}_F^{\mathrm{hom}}}((X, p, m), (Y, q, n)) = q C^{n-m}_{\mathrm{hom}}(X, Y) p.$$

$\mathcal{M}_F^{\mathrm{hom}}$ is an additive, $\mathbb{Q}$-linear, pseudo-abelian category ([**Sch94**, Theorem 1.6]). There is a bi-additive functor $\otimes \colon \mathcal{M}_F^{\mathrm{hom}} \times \mathcal{M}_F^{\mathrm{hom}} \to \mathcal{M}_F^{\mathrm{hom}}$ given on objects by

$$(X, p, m) \otimes (Y, q, n) = (X \times Y, p \otimes q, m + n)$$

(the fiber product is over F; we do not here specify $\otimes$ on morphisms). This satisfies associativity and commutativity constraints induced by the tautological isomorphisms $(X \times Y) \times Z \cong X \times (Y \times Z)$ and $C \colon X \times Y \cong Y \times X$. A unit object for

$\otimes$ is given by $(\operatorname{Spec} F, \mathrm{id}, 0)$, and we can define dual objects by (again for simplicity take X of equi-dimension d)

$$(X, p, m)^\vee = (X, {}^t p, d - m);$$

In sum, these structures make $(\mathcal{M}_F^{\mathrm{hom}}, \otimes)$ into a rigid $\mathbb{Q}$-linear tensor category with $\mathrm{End}_{\mathcal{M}_F^{\mathrm{hom}}}(\mathbf{1}) = \mathbb{Q}$. Grothendieck's Standard Conjectures anticipate that moreover $\mathcal{M}_F^{\mathrm{hom}}$ should be a semi-simple neutral Tannakian category. One must prove that

- $\mathcal{M}_F^{\mathrm{hom}}$ is abelian;
- the abelian category $\mathcal{M}_F^{\mathrm{hom}}$ is moreover semi-simple;
- $\mathcal{M}_F^{\mathrm{hom}}$ possesses a fiber functor to $\mathrm{Vec}_{\mathbb{Q}}$.

We begin by discussing the question of a fiber functor. The answer is simple: $(\mathcal{M}_F^{\mathrm{hom}}, \otimes)$, with the commutativity constraint defined above, does *not* have a fiber functor! For simplicity, let us take $F \subset \mathbb{C}$, so that there is a Betti realization

$$H_B \colon \mathcal{M}_F^{\mathrm{hom}} \to \mathrm{Vec}_{\mathbb{Q}}.$$

H_B is not a tensor functor, however: the diagram

$$\begin{array}{ccc} H_B(X) \otimes H_B(Y) & \xrightarrow[\sim]{k} & H_B(X \times Y) \\ \Big\downarrow{\scriptstyle C'} & & \Big\downarrow{\scriptstyle H_B(C)} \\ H_B(Y) \otimes H_B(X) & \xrightarrow[\sim]{k} & H_B(Y \times X), \end{array}$$

given by applying the naïve commutativity constraints and the Künneth isomorphism, only commutes up to sign, since cup-product is anti-commutative. More precisely, H_B gives a tensor functor valued in $\mathrm{Vec}_{\mathbb{Q}}^{\mathbb{Z}/2\mathbb{Z}}$, and since objects of $\mathcal{M}_F^{\mathrm{hom}}$ can then have negative rank, it cannot be neutral Tannakian with the commutativity constraint induced by $X \times Y \xrightarrow{\sim} Y \times X$. The *Künneth Standard Conjecture* would give $\mathcal{M}_F^{\mathrm{hom}}$ a grading (corresponding to cohomological degree), allowing for a modified commutativity constraint, for which $H_B \colon \mathcal{M}_F^{\mathrm{hom}} \to \mathrm{Vec}_{\mathbb{Q}}$ *is* a tensor functor. More precisely, for any Weil cohomology theory H^*, the isomorphism (again taking $d = \dim X$)

$$\begin{aligned} H^{2d}(X \times X)(d) &\cong \bigoplus_i H^{2d-i}(X) \otimes H^i(X)(d) \\ &\cong \bigoplus_i H^i(X)^\vee \otimes H^i(X) \\ &\cong \bigoplus_i \mathrm{End}(H^i(X)) \end{aligned}$$

gives a cohomology class π_X^i corresponding to the composition $H^*(X) \twoheadrightarrow H^i(X) \hookrightarrow H^*(X)$.

CONJECTURE 4.1.8. *For all $i = 0, 1, \ldots, 2d$, the cohomology class π_X^i is algebraic, i.e. lies in the image of the cycle class map $A^d_{\mathrm{hom}}(X \times X) \to H^{2d}(X \times X)(d)$.*

The modified commutativity constraint on $\mathcal{M}_F^{\mathrm{hom}}$ would then be given as follows: if the original constraint is $C \colon M \otimes N \xrightarrow{\sim} N \otimes M$, we can decompose $C = \oplus_{r,s} C^{r,s}$ with

$$C^{r,s} \colon \pi^r M \otimes \pi^s N \xrightarrow{\sim} \pi^s N \otimes \pi^r M,$$

and then the correct commutativity constraint is $C' = \oplus_{r,s} (-1)^{rs} C^{r,s}$.

The Künneth Standard Conjecture is in fact implied by a stronger conjecture, the Lefschetz Standard Conjecture, which would also show that the primitive decomposition of cohomology' makes sense in the category $\mathcal{M}_F^{\mathrm{hom}}$. To state this, let X be a smooth projective variety over F, and fix an ample line bundle η on X, giving rise to the Lefschetz operator

$$L = L_{\eta,H^*} : H^i(X)(r) \to H^{i+2}(X)(r+1),$$

and the hard Lefschetz isomorphisms

$$L^{d-i} \colon H^i(X)(r) \xrightarrow{\sim} H^{2d-i}(X)(d-i+r)$$

for all $i \leq d$. As always with taking cup-product with the class of an algebraic cycle, these isomorphisms are given by algebraic correspondences.

CONJECTURE 4.1.9 (Lefschetz Standard Conjecture). *For all $i \leq d$, the inverse of L^{d-i} is given by algebraic correspondences.*

Note that *a priori* both the Künneth and Lefschetz Standard Conjectures depend on the choice of Weil cohomology. For the classical cohomologies (Betti, étale, de Rham) that are related by comparison isomorphisms (respecting cycle class maps), these conjectures do not depend on the choice of H^*.

Now we move to the question of whether $\mathcal{M}_F^{\mathrm{hom}}$ is abelian. Here there is a marvelous theorem of Jannsen:

THEOREM 4.1.10 ([**Jan92**, Theorem 1]). *For an adequate equivalence relation $\sim$, the category of motives for $\sim$-equivalence is abelian semi-simple if and only if $\sim$ is numerical equivalence. In particular, $\mathcal{M}_F^{\mathrm{hom}}$ is abelian if and only if numerical and homological equivalence coincide.*

Indeed, long before Jannsen proved his theorem, Grothendieck conjectured:

CONJECTURE 4.1.11. *Homological and numerical equivalence coincide.*

In summary, under the Standard Conjectures, we can equip $\mathcal{M}_F^{\mathrm{hom}}$ with a commutativity constraint for which it is a semi-simple neutral Tannakian category, hence:

CONJECTURE 4.1.12. *Let F be a subfield of $\mathbb{C}$. There is a pro-reductive $\mathbb{Q}$-group $\mathcal{G}_F^{\mathrm{hom}}$, the motivic Galois group for pure motives over F, and an equivalence (induced by the Betti fiber functor $H_B^* \colon \mathcal{M}_F^{\mathrm{hom}} \to \mathrm{Vec}_{\mathbb{Q}}$)*

$$\mathcal{M}_F^{\mathrm{hom}} \cong \mathrm{Rep}_{\mathbb{Q}}(\mathcal{G}_F^{\mathrm{hom}}).$$

4.1.3. Motivated cycles. In the present Part 4 of these notes, most of our results can be stated solely in terms of abelian varieties, but both the strongest assertions and the proofs will require invoking some version of the motivic Galois formalism; in the absence of the Standard Conjectures, we use André's theory of motives for motivated cycles ([**And96b**]). In this subsection, we provide a brief review, elaborating on some points for later application. Although André's theory is developed over any field F, to simplify we will always take F to be an abstract (i.e., not embedded) field of characteristic zero, small enough to be embedded in $\mathbb{C}$, and which we will eventually specify to be a number field.

In [**And96b**], André defines a $\mathbb{Q}$-linear category of motives for 'motivated cycles' whose construction mirrors the classical construction of (Grothendieck) motives for homological equivalence, but circumvents the standard conjectures by formally enlarging the group of 'cycles' to include the Lefschetz involutions. Here is the precise definition of the space of motivated cycles:

DEFINITION 4.1.13. Fix a reference Weil cohomology $H^\bullet$. A motivated cycle on X with coefficients in E is an element of $H^\bullet(X)$ of the form

$$\mathrm{pr}^{XY}_{X,*}(\alpha \cup *_L \beta),$$

where

- Y is a smooth projective F-scheme, with polarization η_Y giving rise to a 'product' polarization $\eta_{X \times Y} = [X] \otimes \eta_Y + \eta_X \otimes [Y]$, with corresponding Lefschetz involution $*_L$, on $X \times Y$;
- α and β are algebraic cycles mod $H^\bullet$-homological equivalence with E-coefficients on $X \times Y$.

We denote by $A^\bullet_{\mathrm{mot}}(X)_E$ the E-vector space of motivated cycles on X with E-coefficients. $A^\bullet_{\mathrm{mot}}(X)$ will always mean the case $E = \mathbb{Q}$.

As in §4.1.2.2, we can then define spaces $C^\bullet_{\mathrm{mot}}(X,Y)_E$ of motivated correspondences (see also [**And96b**, Définition 2]); for $X = Y$, this construction yields a graded E-algebra containing the Lefschetz involutions and the Künneth projectors π^i_X.

The first main result of this theory asserts that the spaces of motivated cycles in a precise sense do not depend on the Weil cohomology $H^\bullet$ used to define them:

THEOREM 4.1.14 (Théorème 0.3 of [**And96b**]). *For any smooth projective F-scheme X, let $A^\bullet_{\mathrm{mot}}(X)$ be the graded $\mathbb{Q}$-algebra of 'motivated cycles,' constructed with respect to a fixed Weil cohomology $H^\bullet$. $A^\bullet_{\mathrm{mot}}(X)$ contains the classes of algebraic cycles modulo homological equivalence, and there is a $\mathbb{Q}$-linear injection*

$$cl_H \colon A^\bullet_{\mathrm{mot}}(X) \to H^{2\bullet}(X)$$

extending the cycle class map for H. $A^\bullet_{\mathrm{mot}}(X)$ has the following properties:

- *$A^\bullet_{\mathrm{mot}}(X)$ depends bifunctorially (push-forward and pull-back) on X, satisfying the usual projection formula.*
- *(See [**And96b**, §2.3]) Two Weil cohomologies related by comparison isomorphisms yield canonically and functorially isomorphic algebras $A^\bullet_{\mathrm{mot}}(X)$ of motivated cycles. In particular, all the classical cohomology theories yield the same $A^\bullet_{\mathrm{mot}}(X)$.*

From here, André defines a category $\mathcal{M}_F$ of motives for motivated cycles exactly as in Definition 4.1.7, replacing algebraic correspondences with motivated correspondences.[4] André shows the endomorphism algebras of objects of $\mathcal{M}_F$ are semi-simple (finite-dimensional) $\mathbb{Q}$-algebras, from which it follows (see [**Jan92**]) that $\mathcal{M}_F$ is an abelian semi-simple category. Since the Künneth projectors π^i_X are motivated cycles, we can define the modified commutativity constraint described in §4.1.2.2, thereby making $\mathcal{M}_F$ into a neutral Tannakian category over $\mathbb{Q}$:

[4] André defines more generally a category of motives 'modeled on' a full sub-category $\mathcal{V}$ of the category of smooth projective F-schemes, with $\mathcal{V}$ assumed stable under products, disjoint union, and passage to connected components. This amounts to restricting the auxiliary varieties Y permitted in Definition 4.1.13. We will always take $\mathcal{V}$ to be all smooth projective F-schemes.

THEOREM 4.1.15 (Théorème 0.4 of [**And96b**]). *$\mathcal{M}_F$ is a neutral Tannakian category over $\mathbb{Q}$. It is graded, semi-simple, and polarized. Every classical cohomology factors through $\mathcal{M}_F$.*

REMARK 4.1.16. We will often use the short-hand $H(X)$ for the object $(X, \mathrm{id}, 0)$, and $H^i(X)$ for $(X, \pi_X^i, 0)$ (recall the notation from Definition 4.1.7). It will be clear from context when $H(X)$ refers to an object of $\mathcal{M}_F$, and when it refers to the output of any particular Weil cohomology theory H^*.

In particular, this sets in motion the formalism of motivic Galois groups, and the theory becomes a very useful circumvention of the standard conjectures. Perhaps its most unsatisfactory feature–present also in the theory of absolute Hodge cycles–is that its 'motives' are not known to give rise to compatible systems of ℓ-adic representations.

Now we specify fiber functors and no longer regard F as an abstract field. For any $\sigma \colon F \hookrightarrow \mathbb{C}$, $\mathcal{M}_F$ is Tannakian and neutralized by the σ-Betti fiber functor (denoted H_σ), so we obtain its Tannakian group, 'the' motivic Galois group, $\mathcal{G}_F(\sigma)$. For the ℓ-adic fiber functor $X \mapsto H_{\text{ét}}(X_{\overline{F}}, \mathbb{Q}_\ell)$, we denote by $\mathcal{G}_{F,\ell}$ the corresponding motivic group (over $\mathbb{Q}_\ell$). It is more convenient to choose an embedding $\sigma \colon \overline{F} \hookrightarrow \mathbb{C}$, since this allows us, via the comparison isomorphisms

$$H_{\text{ét}}(X_{\overline{F}}, \mathbb{Q}_\ell) \xrightarrow[\sigma^*]{\sim} H_{\text{ét}}(X_{\overline{F}} \otimes_{\overline{F},\sigma} \mathbb{C}, \mathbb{Q}_\ell) \cong H_\sigma(X, \mathbb{Q}) \otimes_{\mathbb{Q}} \mathbb{Q}_\ell,$$

to deduce an isomorphism $\mathcal{G}_F(\sigma) \otimes_{\mathbb{Q}} \mathbb{Q}_\ell \cong \mathcal{G}_{F,\ell}$. Eventually, F will simply be regarded as a subfield of $\mathbb{C}$, with $\overline{F}$ its algebraic closure in $\mathbb{C}$; in that case, we will omit the σ from the notation. For now, however, we retain it.

As a variant on this formalism, to any object (or collection of objects) M of $\mathcal{M}_F$, we can associate the smallest Tannakian subcategory $\langle M \rangle^{\otimes}$ generated by M (fully faithfully embedded in $\mathcal{M}_F$), and we can then look at its Tannakian group, denoted $\mathcal{G}_F^M(\sigma)$. A particularly important example comes from taking M to range over (the objects of $\mathcal{M}_F$ associated to) all finite étale F-schemes; this defines the sub-category of *Artin motives*. The fully faithful inclusion of the subcategory $\mathcal{M}_F^{\mathrm{art}}$ of Artin motives over F induces a surjection

$$\mathcal{G}_F(\sigma) \twoheadrightarrow \mathcal{G}_F^{\mathrm{art}}(\sigma) \cong \Gamma_F,$$

and when combined with the base-change functor $\mathcal{M}_F \to \mathcal{M}_{\overline{F}}$, we obtain an exact sequence[5] of pro-algebraic groups

$$1 \to \mathcal{G}_{\overline{F}}(\sigma) \to \mathcal{G}_F(\sigma) \to \Gamma_F \to 1.$$

By Proposition 6.23*d* of [**DM11**], for all ℓ there are continuous sections (homomorphisms) $\Gamma_F \to \mathcal{G}_F(\mathbb{Q}_\ell)$ (after unwinding everything, this is simply the statement that Γ_F acts on ℓ-adic cohomology).

We can similarly compare the 'arithmetic' and 'geometric' versions of the motivic Galois group of any object M of $\mathcal{M}_F$: there is a commutative diagram, where

[5]See Proposition 6.23*a*, *c* of [**DM11**]. This is a corrected, TeXed version of the original article in [**DMOS82**]; it is available at http://www.jmilne.org/math/xnotes/index.html. Note that part *b* is a modification of the (only conjectural) statement in the original article.

the vertical morphisms are surjective:

$$\begin{array}{ccccccccc} 1 & \longrightarrow & \mathcal{G}_{\overline{F}}^{M_{\overline{F}}}(\sigma) & \longrightarrow & \mathcal{G}_F^M(\sigma) & & & & \\ & & \uparrow & & \uparrow & & & & \\ 1 & \longrightarrow & \mathcal{G}_{\overline{F}}(\sigma) & \longrightarrow & \mathcal{G}_F(\sigma) & \longrightarrow & \Gamma_F & \longrightarrow & 1. \end{array}$$

LEMMA 4.1.17. *Assume that $\mathcal{G}_{\overline{F}}^{M_{\overline{F}}}(\sigma)$ is connected. Then there exists a finite extension F'/F such that $\mathcal{G}_{F'}^{M_{F'}}(\sigma)$ is connected.*

PROOF. Fix a prime ℓ, a section $s_\ell \colon \Gamma_F \to \mathcal{G}_F(\mathbb{Q}_\ell)$, and a finite extension F'/F such that the Zariski closure of the ℓ-adic representation $\rho_{M_\ell} : \Gamma_{F'} \to \mathrm{GL}(M_\ell)$ is connected. The image of $\mathcal{G}_{F'}^{M_{F'}} \otimes_{\mathbb{Q}} \mathbb{Q}_\ell$ in $\mathrm{GL}(M_\ell)$ is then equal to the product of the two connected groups $\overline{\rho_{M_\ell}(\Gamma_{F'})}^{\mathrm{Zar}}$ and $\left(\mathcal{G}_{\overline{F}}^{M_{\overline{F}}} \otimes_{\mathbb{Q}} \mathbb{Q}_\ell\right)$, hence is itself connected. □

REMARK 4.1.18. If in $\langle M_{\overline{F}}\rangle^{\otimes}$ all Hodge cycles are motivated, $\mathcal{G}_{\overline{F}}^{M_{\overline{F}}}(\sigma)$ is connected.

The assertion that Hodge cycles are motivated is a (still fiercely difficult) weakened version of the Hodge conjecture. We now describe what is essentially the only understood case of this problem. Many of the results of [**And96a**] rest on earlier work of André (Théorème 0.6.2 of [**And96b**]) showing that on a complex abelian variety, all Hodge cycles are motivated (a variant of Deligne's result that Hodge cycles on a complex abelian variety are absolutely Hodge). One useful consequence is:

COROLLARY 4.1.19 (André). *Let $A/\mathbb{C}$ be an abelian variety, and let M be the motive $H^1(A)$ (an object of $\mathcal{M}_{\mathbb{C}}$). Then the motivic group $\mathcal{G}_{\mathbb{C}}^M$ (for the Betti realization) is equal to the Mumford-Tate group $MT(A)$, and in particular is connected.*

PROOF. $MT(A)$, recall, is the smallest $\mathbb{Q}$-sub-group of $\mathrm{GL}(H^1_B(A_{\mathbb{C}}, \mathbb{Q}))$ whose $\mathbb{R}$-points contain the image of the $\mathbb{S}$-representation corresponding to the $\mathbb{R}$-Hodge structure $H^1_B(A_{\mathbb{C}}, \mathbb{R})$; by the general theory of Mumford-Tate groups, this is equal to

- the Tannakian group for the Tannakian category of $\mathbb{Q}$-Hodge structures generated by $M_B = H^1_B(A_{\mathbb{C}}, \mathbb{Q})$; and
- the subgroup of $\mathrm{GL}(M_B)$ fixing exactly the Hodge tensors in every tensor construction $T^{m,n}(M_B) := (M_B)^{\otimes m} \otimes (M_B^\vee)^{\otimes n}$.

Similarly, the motivic group $\mathcal{G}_{\mathbb{C}}^M$ is the subgroup of $\mathrm{GL}(M_B)$ fixing exactly the motivated cycles in every tensor construction $T^{m,n}(M_B)$. Of course (on any variety) all motivated cycles are Hodge cycles, so there is a quite general inclusion $MT(A) \subset \mathcal{G}_{\mathbb{C}}^M$. Applying André's result to all powers of A, we can deduce the reverse inclusion: if $t \in T^{m,n}(M_B)$ is a Hodge cycle, then (by weight considerations) $m = n$, and viewing this tensor space (via Künneth and polarization) inside $H^{2m}(A^{2m}, \mathbb{Q})(m)$, we see that t is motivated. □

The following point is implicit in the arguments of [**And96a**], but we make it explicit in part to explain an important foundational point in the theory of motivated cycles.

COROLLARY 4.1.20. *Let F be a subfield of $\mathbb{C}$, with algebraic closure $\overline{F}$ in $\mathbb{C}$, and let A/F be an abelian variety. Let M be the object of $\mathcal{M}_{\overline{F}}$ corresponding to $H^1(A_{\overline{F}})$. Then $\mathcal{G}_{\overline{F}}^M$ (for the $\overline{F} \subset \mathbb{C}$ Betti realization) is connected, equal to $MT(A_{\mathbb{C}})$.*

PROOF. We deduce this from the previous result and the following general lemma, which is of course implicit in [**And96b**]:

LEMMA 4.1.21. *Let L/K be an extension of algebraically closed fields, and let M be an object of $\mathcal{M}_K$, with base-change $M|_L$ to L. Then via the canonical isomorphism $H_{\text{ét}}(M, \mathbb{Q}_\ell) \cong H_{\text{ét}}(M|_L, \mathbb{Q}_\ell)$, the ℓ-adic motivic groups $\mathcal{G}_{K,\ell}^M$ and $\mathcal{G}_{L,\ell}^{M|_L}$ agree.*

PROOF. This follows from the fact (2.5 *Scolie* of [**And96b**]) that the comparison isomorphism (for $M = H^*(X)$) identifies the spaces of motivated cycles $A^*(X) \xrightarrow{\sim} A^*(X_L)$. This follows from standard spreading out techniques, but we provide some details, since they are omitted from [**And96b**]. Writing $L = \varinjlim_{\lambda} K_\lambda$ as the directed limit of its finite-type K-sub-algebras allows all the data required to define a motivated cycle on X_L (a smooth projective Y/L, algebraic cycles on $X_L \times Y$, the Lefschetz involution on cohomology of $X_L \times Y$) to be descended to some $S_\lambda = \mathrm{Spec}(K_\lambda)$. The general machinery allows us to assume (enlarging λ) that we have Y_λ/S_λ smooth projective (of course $X_\lambda = X \otimes_K K_\lambda$ is smooth projective), a relatively ample invertible sheaf η_λ of 'product-type' on $X_\lambda \times_{S_\lambda} Y_\lambda$, and various closed S_λ-subschemes

$$Z \hookrightarrow \mathcal{X} := X_\lambda \times_{S_\lambda} Y_\lambda$$

that are smooth over S_λ and whose linear combinations define spread-out versions of the algebraic cycles on $X_L \times Y$. The purity theorem in this context (Théorème 3.7 of Artin's Exp. XVI of [**sga73**]) yields an isomorphism

$$H_Z^{2j}(\mathcal{X}, \mathbb{Q}_\ell)(j) \xrightarrow{\sim} H^0(Z, \mathbb{Q}_\ell),$$

hence a cycle class $[Z]$ in $H^{2j}(\mathcal{X}, \mathbb{Q}_\ell)(j)$. Regarding $\bar{x}\colon \mathrm{Spec}(L) \to S_\lambda$ as a geometric point over some scheme-theoretic point x, and letting $\bar{s}\colon \mathrm{Spec}(K) \to S_\lambda$ be a geometric point over a scheme-theoretic closed point s lying in the closure of x, the cospecialization map $H^{2j}(\mathcal{X}_{\bar{s}}, \mathbb{Q}_\ell) \xrightarrow{\sim} H^{2j}(\mathcal{X}_{\bar{x}}, \mathbb{Q}_\ell)$ is an isomorphism ($\mathcal{X}/S_\lambda$ being smooth proper), and it carries the cycle class $[Z_{\bar{s}}]$ to the cycle class $[Z_{\bar{x}}]$, since these are both restrictions of $[Z]$, and the diagram

$$\begin{array}{ccc} & \nearrow & H^{2j}(\mathcal{X}_{\bar{s}}, \mathbb{Q}_\ell) \\ H^{2j}(\mathcal{X}, \mathbb{Q}_\ell) & & \downarrow \\ & \searrow & H^{2j}(\mathcal{X}_{\bar{x}}, \mathbb{Q}_\ell) \end{array}$$

commutes. □

□

4.1.4. Motives with coefficients. We will need the flexibility of working with related categories of motives with coefficients. Let E be a field of characteristic zero. Given any E-linear abelian (or in fact just additive, pseudo-abelian) category $\mathcal{M}$, and any finite extension E'/E, we can define the category $\mathcal{M}_{E'}$ of objects with coefficients in E' as either of the following (this discussion is taken from §2.1 of [**Del79**] and 3.11, 3.12 of [**DM11**]):

(1) The category of 'E'-modules in $\mathcal{M}$,' whose objects are pairs (M, α) of an object M of $\mathcal{M}$ and an embedding $\alpha\colon E' \to \mathrm{End}_{\mathcal{M}}(M)$, and whose morphisms are those commuting with these E'-structures.

(2) The pseudo-abelian envelope of the category whose objects are formally obtained from those of $\mathcal{M}$ (writing $M_{E'}$ for the object in $\mathcal{M}_{E'}$ arising from M in $\mathcal{M}$), and whose morphisms are
$$\mathrm{Hom}_{\mathcal{M}_{E'}}(M_{E'}, N_{E'}) = \mathrm{Hom}_{\mathcal{M}}(M, N) \otimes_E E'.$$
This construction is valid for infinite-dimensional E'/E.

To pass from the first to the second description, let (M, α) be as in (1), so that $\mathrm{End}_{\mathcal{M}_{E'}}(M_{E'}) = \mathrm{End}_{\mathcal{M}}(M) \otimes_E E'$ contains, via α, $E' \otimes_E E'$. This E'-algebra (via the left factor) is isomorphic to a product of fields, and there is a unique projection $e_{id}\colon E' \otimes_E E' \twoheadrightarrow E'$ in which $x \otimes 1$ and $1 \otimes x$ both map to x. Then $e_{id}(M_{E'})$ is the object of (2) corresponding to (M, α).

There is a functor $\mathcal{M} \to \mathcal{M}_{E'}$, which in the first language is $M \mapsto (M \otimes_E E', id_M \otimes id_{E'})$. (See 2.11 of [**DM11**] for a precise description of $M \otimes_E E'$.) If $\mathcal{M}$ is semi-simple, then so is $\mathcal{M}_{E'}$. If $\mathcal{M}$ is a neutral Tannakian category over E with fiber functor ω, then we can make $\mathcal{M}_{E'}$ into a neutral Tannakian category over E'. Define an E'-valued fiber functor $\omega_{E'}: \mathcal{M} \to \mathrm{Vec}_{E'}$ by $\omega_{E'}(M) = \omega(M) \otimes_E E'$. There is a diagram commuting up to canonical isomorphism,
$$\begin{array}{ccc} \mathcal{M} & \longrightarrow & \mathcal{M}_{E'} \\ & \searrow^{\omega_{E'}} & \downarrow{\scriptstyle \omega'_{E'}} \\ & & \mathrm{Vec}_{E'}, \end{array}$$
where
$$\omega'_{E'}(M, \alpha) := \omega_{E'}(M) \underset{\alpha \otimes \mathrm{id}_{E'}, E' \otimes_E E'}{\otimes} E'.$$
This tensor product (and similar tensor products) means that $E' \otimes E'$ acts on E' by the usual bi-module structure, and acts on $\omega_{E'}(M)$ via α on the left-hand copy of E' and via E'-multiplication on the right-hand copy (these two actions coincide on E). This $\omega'_{E'}$ neutralizes $\mathcal{M}_{E'}$, and so we can define the associated Tannakian group $\mathcal{G}_{E'} = \mathrm{Aut}^{\otimes}(\omega'_{E'})$. We wish to compare $\mathcal{G}_{E'}$ with $\mathcal{G} = \mathrm{Aut}^{\otimes}(\omega)$. Note that $\mathcal{M}_{E'}$ is equivalent to $\mathrm{Rep}_{E'}(\mathcal{G})$ and $\mathcal{G} \otimes_E E' = \mathrm{Aut}^{\otimes}(\omega_{E'})$. Then the composition of functors
$$\mathcal{M}_{E'} \underset{\sim}{\xrightarrow{\omega}} \mathrm{Rep}_{E'}(\mathcal{G}) \underset{\sim}{\xrightarrow{F}} \mathrm{Rep}_{E'}(\mathcal{G} \otimes_E E')$$
is given on $\mathcal{M}_{E'}$ by (F just sends an object V of $\mathrm{Rep}_{E'}(\mathcal{G})$, corresponding to an E-homomorphism $\mathcal{G} \to \mathrm{Res}_{E'/E}(\mathrm{GL}_{E'}(V))$, to $E' \otimes_{E' \otimes_E E'} V$)
$$(M, \alpha) \mapsto (\omega(M), \omega(\alpha)) \mapsto E' \underset{E' \otimes E', \omega(\alpha) \otimes 1}{\otimes} (\omega(M) \otimes E')$$

This composite is naturally isomorphic to $\omega'_{E'}$, whence a tensor-equivalence

$$\mathrm{Rep}_{E'}(\mathcal{G}_{E'}) \xrightarrow{\sim} \mathrm{Rep}_{E'}(\mathcal{G} \otimes_E E').$$

We apply these constructions to André's category of motivated motives over F, to form the variant $\mathcal{M}_{F,E}$, motives over F with coefficients in E, for E any finite extension of $\mathbb{Q}$. We denote the corresponding motivic Galois group (for the σ-Betti realization) by $\mathcal{G}_{F,E}(\sigma)$; its E-linear (pro-algebraic) representations correspond to objects of $\mathcal{M}_{F,E}$, and it is naturally isomorphic to $\mathcal{G}_{F,\mathbb{Q}}(\sigma) \otimes_{\mathbb{Q}} E$. For any object M of $\mathcal{M}_{F,E}$, we also have the corresponding motivic group $\mathcal{G}^M_{F,E}(\sigma)$.

As a more concrete variant, we can start with the ($\mathbb{Q}$-linear, semi-simple) isogeny category AV^0_F of abelian varieties over F, and form the E-linear, semi-simple category $AV^0_{F,E}$ of isogeny abelian varieties over F with complex multiplication by E. There is a (contravariant) functor $AV^0_{F,E} \to \mathcal{M}_{F,E}$. Faltings' theorem implies the following E-linear variant: fix an embedding $E \hookrightarrow \overline{\mathbb{Q}}_\ell$, and consider an object A of $AV^0_{F,E}$; then the natural map

$$\mathrm{End}_{AV^0_{F,E}}(A) \otimes_E \overline{\mathbb{Q}}_\ell \to \mathrm{End}_{\overline{\mathbb{Q}}_\ell[\Gamma_F]}(H^1(A_{\overline{F}}, \mathbb{Q}_\ell) \otimes_E \overline{\mathbb{Q}}_\ell)$$

is an isomorphism: simply take the usual isomorphism with $\mathbb{Q}$ in place of E, restrict to those endomorphisms commuting with $E \to \mathrm{End}_{AV^0_F}(A)$, and then project to the $E \hookrightarrow \overline{\mathbb{Q}}_\ell$ component of the resulting $E \otimes_{\mathbb{Q}} \overline{\mathbb{Q}}_\ell$-module.

4.1.5. Hodge symmetry in $\mathcal{M}_{F,E}$. Our next goal is to show that objects of $\mathcal{M}_{F,E}$ satisfy (unconditionally) the Hodge-Tate weight symmetries, and the more fundamental symmetries in the rational de Rham realization, needed for our abstract Galois lifting results. I expect these results are well-known to experts, but they do not seem to have been explained in their natural degree of generality, and in a context in which they can be *proven* unconditionally[6]. So, let M be an object of $\mathcal{M}_{F,E}$, of rank r. The de Rham realization M_{dR} is a filtered $F \otimes_{\mathbb{Q}} E$-module, which is moreover free of rank r. As in §2.4.1, we define the $\tau \colon F \hookrightarrow \overline{E}$-labeled weights of M as follows: $\mathrm{HT}_\tau(M)$ is an r-tuple of integers h, with h appearing with multiplicity

$$\dim_{\overline{E}} \mathrm{gr}^h \left(M_{\mathrm{dR}} \otimes_{F \otimes_{\mathbb{Q}} E, \tau \otimes 1} \overline{E} \right).$$

In talking of τ-labeled weights, there is alway an ambient over-field, in this case $\overline{E}$, but we will at times want to change this to either $\mathbb{C}$ or $\overline{\mathbb{Q}}_\ell$; that will require *fixing* an embedding $\iota \colon \overline{E} \hookrightarrow \overline{\mathbb{Q}}_\ell$ or $\iota \colon \overline{E} \hookrightarrow \mathbb{C}$. In either case, we can then speak of $\mathrm{HT}_{\iota\tau}(M)$, with E embedded into $\mathbb{C}$ or $\overline{\mathbb{Q}}_\ell$ via ι, and there is an obvious equality $\mathrm{HT}_\tau(M) = \mathrm{HT}_{\iota\tau}(M)$.

The essential point is that, as in the automorphic analogue Corollary 2.4.7, "motives have CM coefficients"; compare the assertion in [**And96b**, §4.6(iii)] that $\mathcal{G}_F$ splits (i.e., for any algebraic quotient $\mathcal{G}_F \twoheadrightarrow H$ over $\mathbb{Q}$, the connected component H^0 has a maximal torus that splits) over $\mathbb{Q}^{\mathrm{cm}}$. For lack of reference for the proof, we give some details:

[6]Our discussion is an analogue of the standard discussion in the theory of motives of CM type–see §7 of [**Ser94**]. The Galois symmetries themselves have also been postulated in §5 of [**BLGGT14**], but since that paper is only concerned with Galois representations arising from essentially conjugate self-dual automorphic representations, it has no need to deal with this symmetry as a general principle.

LEMMA 4.1.22. *Let N be an object of $\mathcal{M}_{F,E}$. Then there exists an object N_0 of $\mathcal{M}_{F,E_{\mathrm{cm}}}$ such that $N \cong N_0 \otimes_{E_{\mathrm{cm}}} E$. Consequently, $\mathrm{HT}_\tau(N)$ depends only on the restriction of τ to F_{cm}.*

PROOF. By the formalism of Lemma 2.4.1, the second claim follows from the first. By Proposition 3.3 (the analogue of the Hodge index theorem) of [**And96b**], $C^0_{\mathrm{mot}}(X,X)$ is endowed, via a choice of ample line bundle on X, with a positive-definite, $\mathbb{Q}$-valued symmetric form, which we call $\langle \cdot, \cdot \rangle$. For any sub-object $M \subset H(X)$ in $\mathcal{M}_F$, there follows a decomposition $H(X) = M \oplus M^\perp$, by positivity and the fact that $\mathcal{M}$ is a semi-simple abelian category. $\langle \cdot, \cdot \rangle$ therefore restricts to a positive form on M itself, and in particular every simple object of $\mathcal{M}_F$ carries a positive definite form. Again by semi-simplicity, any such simple object M has $\mathrm{End}(M)$ isomorphic to a division algebra D, on which we now have an involution $'$ (transpose with respect to $\langle \cdot, \cdot \rangle$) and a trace form $\mathrm{tr}_M \colon D \to \mathbb{Q}$ (given by $\phi \mapsto \mathrm{tr}(\phi|_{H_B(M)})$ such that $\mathrm{tr}_M(\phi\phi') > 0$ for all non-zero $\phi \in D$. As in the proof of the Albert classification (see [**Mum70**, §21, Theorem 2]), this implies that the center of D is either a totally real or a CM field. Thus, $\mathrm{End}(M)$ splits over a CM field; in particular, for any number field E and any factor N of $M \otimes_{\mathbb{Q}} E$ (in $\mathcal{M}_{F,E}$), N can in fact be realized as (the scalar extension of) an object of $\mathcal{M}_{F,E_{\mathrm{cm}}}$. □

REMARK 4.1.23. When $E = \mathbb{Q}$, Lemma 2.4.1 shows that $\mathrm{HT}_\tau(N)$ is independent of τ; this is essentially the assertion that for a smooth projective X/F, the Hodge numbers of X do not depend on the choice of embedding $F \hookrightarrow \mathbb{C}$. The next few results (culminating in Corollary 4.1.26) all have corresponding strengthenings when $E = \mathbb{Q}$.

We first record the de Rham-Betti version of the desired symmetry:

LEMMA 4.1.24. *Let N be an object of $\mathcal{M}_{F,E}$ lying in the k-component of the grading. Then for any choice of complex conjugation c in $\mathrm{Gal}(\overline{E}/\mathbb{Q})$,*

$$\mathrm{HT}_{c\circ\tau}(N) = \{k - h : h \in \mathrm{HT}_\tau(N)\}.$$

PROOF. Any such object is built by taking one of the form $M = H^k(X) \otimes E$, for X/F a smooth projective variety and applying an idempotent $\alpha \in C^0_{\mathrm{mot}}(X,X)_E$.[7] Fix an embedding $\iota \colon \overline{E} \hookrightarrow \mathbb{C}$, so that $\iota \circ \tau \colon F \hookrightarrow \mathbb{C}$. There is a functorial (Betti-de Rham) comparison isomorphism

$$M_{\mathrm{dR}} \otimes_{F,\iota\tau} \mathbb{C} \cong M_{B,\iota\tau} \otimes_{\mathbb{Q}} \mathbb{C},$$

which commutes with the action of $C^0_{\mathrm{mot}}(X,X)_E$ on M_{dR} and $M_{B,\iota\tau}$, in particular making this an isomorphism of free $E \otimes_{\mathbb{Q}} \mathbb{C}$-modules. It induces an isomorphism on the corresponding gradeds (with $q = k - p$):

$$\mathrm{gr}^p(M_{\mathrm{dR}} \otimes_{F,\iota\tau} \mathbb{C}) \cong H^{p,q}_{\iota\tau}(M) \cong H^{p,q}(X_{\iota\tau}) \otimes_{\mathbb{Q}} E,$$

again $E \otimes_{\mathbb{Q}} \mathbb{C}$-linear. Just for orientation amidst the formalism, this says that

$$\begin{aligned}\mathrm{HT}_\tau(M_{\mathrm{dR}}) &= \mathrm{HT}_{\iota\tau}(M_{\mathrm{dR}}) \\ &= \{p : H^{p,q}(X_{\iota\tau}) \neq 0, \text{counted with multiplicity } \dim_{\mathbb{C}} H^{p,q}(X_{\iota\tau})\}.\end{aligned}$$

Now, the projection $X_{\iota\tau} \times_{\mathbb{C},c} \mathbb{C} \to X_{\iota\tau}$ induces a transfer of structure isomorphism on rational Betti cohomology, and then an $E \otimes_{\mathbb{Q}} \mathbb{C}$-linear isomorphism

$$F_\infty \colon H^{p,q}_{\iota\tau}(M) \to H^{q,p}_{c\iota\tau}(M).$$

[7] And a Tate twist, but the statement of the lemma is obviously invariant under Tate twists.

Let $\alpha \in C^0_{\mathrm{mot}}(X,X)_E$ be an E-linear motivated idempotent correspondence defining an object $N = \alpha M$ of $\mathcal{M}_{F,E}$. There is a commutative diagram of $E \otimes_{\mathbb{Q}} \mathbb{C}$-linear morphisms

$$\begin{array}{ccc} H^{p,q}_{\iota\tau}(M) & \xrightarrow{F_\infty} & H^{q,p}_{c\iota\tau}(M) \\ \downarrow{\scriptstyle \alpha_{\iota\tau}} & & \downarrow{\scriptstyle \alpha_{c\iota\tau}} \\ H^{p,q}_{\iota\tau}(M) & \xrightarrow{F_\infty} & H^{q,p}_{c\iota\tau}(M). \end{array}$$

Consider the images of the two vertical maps. Since the horizontal maps are isomorphisms, and all the maps are $E \otimes_{\mathbb{Q}} \mathbb{C}$-linear, we deduce an isomorphism

$$\alpha_{\iota\tau} H^{p,q}_{\iota\tau}(M) \otimes_{E\otimes_{\mathbb{Q}}\mathbb{C},\iota} \mathbb{C} \xrightarrow{\sim} \alpha_{c\iota\tau} H^{q,p}_{c\iota\tau}(M) \otimes_{E\otimes_{\mathbb{Q}}\mathbb{C},\iota} \mathbb{C}.$$

The left-hand side is isomorphic to

$$\mathrm{gr}^p\left((\alpha M)_{\mathrm{dR}} \otimes_{F\otimes_{\mathbb{Q}}E,\iota\tau\otimes\iota} \mathbb{C}\right),$$

and letting c' be a complex conjugation in $\mathrm{Gal}(\overline{E}/\mathbb{Q})$ such that $\iota c'\tau = c\iota\tau$, the right-hand side is isomorphic to

$$\mathrm{gr}^{k-p}\left((\alpha M)_{\mathrm{dR}} \otimes_{F\otimes_{\mathbb{Q}}E,\iota c'\tau\otimes\iota} \mathbb{C}\right).$$

It follows that

$$\mathrm{HT}_{c'\tau}(\alpha M) = \{k-h : h \in \mathrm{HT}_\tau(\alpha M)\}.$$

But by the previous lemma, $\mathrm{HT}_{c'\tau}(\alpha M) = \mathrm{HT}_{c\tau}(M)$, so we are done. □

Next we observe (as in §2.4 of [**And96b**], but with a few more details) that 'p-adic' comparison isomorphisms hold unconditionally in $\mathcal{M}_{F,E}$; this is one of the main reasons for working with motivated cycles rather than absolute Hodge cycles.

LEMMA 4.1.25. *Let N be an object of $\mathcal{M}_{F,E}$. Then for all $v|\ell$, the free $E\otimes_{\mathbb{Q}}\mathbb{Q}_\ell$-module M_ℓ is a de Rham representation of Γ_{F_v}, and there is a functorial (with respect to morphisms in $\mathcal{M}_{F,E}$) isomorphism of filtered $F_v\otimes_{\mathbb{Q}}E \cong F_v\otimes_{\mathbb{Q}_\ell}(\mathbb{Q}_\ell\otimes_{\mathbb{Q}}E)$-modules*

$$F_v \otimes_F M_{\mathrm{dR}} \cong D_{\mathrm{dR}}(M_\ell|_{\Gamma_{F_v}}).$$

PROOF. It suffices to check for objects of the form $M = H(X) \otimes_{\mathbb{Q}} E$, for X/F smooth projective. Faltings' theorem (of which there are now several proofs, including the necessary corrections to the original [**Fal89**]; a 'simple' recent proof is [**Bei12**]) provides the comparison

$$H^*_{\mathrm{dR}}(X_{F_v}) \otimes_{\mathbb{Q}} E \cong D_{\mathrm{dR}}\left(H^*(X_{\overline{F_v}}, \mathbb{Q}_\ell)\right) \otimes_{\mathbb{Q}} E,$$

which is an isomorphism of filtered $F_v \otimes_{\mathbb{Q}} E$-modules. Let $\alpha \in C^0_{\mathrm{mot}}(X_1,X_2)_E$ be a motivated correspondence. Let $d_1 = \dim(X_1)$. The class α has realizations α_{dR} and α_ℓ in cohomology groups that are also compared by Faltings (note that α_ℓ is fixed by Γ_{F_v}):

$$H^{2d_1}_{\mathrm{dR}}((X_1\times X_2)_{F_v})(d_1)) \otimes_{\mathbb{Q}} E \cong D_{\mathrm{dR}}\left(H^{2d_1}((X_1\times X_2)_{\overline{F_v}}, \mathbb{Q}_\ell)(d_1)\right) \otimes_{\mathbb{Q}} E,$$

and the claim is that α_{dR} maps to α_ℓ. The comparison isomorphism is compatible with cycle class maps (see part (*b*) of the proof of [**Bei12**, Theorem 3.6]), and our cycle α is spanned by elements of the form $\mathrm{pr}^{X_1\times X_2\times Y}_{X_1\times X_2,*}(\beta \cup *\gamma)$, with Y/F another smooth projective variety and β and γ algebraic cycles on $X_1\times X_2\times Y$. Applying the comparison isomorphism also to $X_1\times X_2\times Y$, and applying compatibility with cycle classes, cup-product, (hence) Lefschetz involution, and the projection $\mathrm{pr}^{X_1\times X_2\times Y}_{X^2,*}$–this last point because of compatibility with pull-back induced by morphisms of varieties and, again, Poincaré duality–we can deduce that α_{dR} maps to α_ℓ. □

COROLLARY 4.1.26. *Let M be an object of $\mathcal{M}_{F,E}$. Fix an embedding $\iota\colon \overline{E} \hookrightarrow \overline{\mathbb{Q}}_\ell$, and use this to identify E as a subfield of $\overline{\mathbb{Q}}_\ell$. Then for all $\tau\colon F \hookrightarrow \overline{\mathbb{Q}}_\ell$, $\mathrm{HT}_\tau(M_\ell \otimes_E \overline{\mathbb{Q}}_\ell)$ depends only on $\tau_0 = \tau|_{F_{\mathrm{cm}}}$. If M is moreover pure of some weight k, then*

$$\mathrm{HT}_{\tau_0\circ c}(M_\ell \otimes_E \overline{\mathbb{Q}}_\ell) = \left\{k - h : h \in \mathrm{HT}_{\tau_0}(M_\ell \otimes_E \overline{\mathbb{Q}}_\ell)\right\}.$$

PROOF. Write $\iota^{-1}\tau$ for the embedding $F \hookrightarrow \overline{E}$ induced by τ and ι. By the previous lemma, $\mathrm{HT}_\tau(M_\ell \otimes_E \overline{\mathbb{Q}}_\ell) = \mathrm{HT}_{\iota^{-1}\tau}(M)$. The latter depends only on $\tau|_{F_{\mathrm{cm}}}$, and when M is pure satisfies the required symmetry, by Lemmas 4.1.22 and 4.1.24. □

Finally, we can show that the Galois lifting results of §3.2 apply to representations arising from objects of $\mathcal{M}_{F,E}$. The result that follows is clumsily proven and has been superseded (and greatly strengthened) by [**Pat16**, Corollary 1.1]. The reader would do better to turn there, but here is a simpler argument in the special case in which motivic Galois groups over $\overline{F}$ are connected, which is expected always to be the case; the complication in the present proof arises from the fact that the proof of Theorem 3.2.7 requires an initial reduction to the case of connected monodromy group. Fix an embedding $E \hookrightarrow \overline{\mathbb{Q}}_\ell$, inducing a place λ of E. For short-hand, we denote by $\mathcal{G}_{F,\overline{\mathbb{Q}}_\ell}$ the base-change to $\overline{\mathbb{Q}}_\ell$ of the E_λ-group $\mathcal{G}_{F,\lambda}$ defined by the λ-adic étale fiber functor on the category $\mathcal{M}_{F,E}$, and we make the analogous definition of $\mathcal{G}^M_{F,\overline{\mathbb{Q}}_\ell}$ for objects M of $\mathcal{M}_{F,E}$.

COROLLARY 4.1.27. *Let F be a totally imaginary field. Let M be an object of $\mathcal{M}_{F,E}$ and $E \hookrightarrow \overline{\mathbb{Q}}_\ell$ a fixed embedding, to which we associate the ℓ-adic representation $M_\ell \otimes_E \overline{\mathbb{Q}}_\ell$, which is a representation both of Γ_F and of the motivic group $\mathcal{G}^M_{F,\overline{\mathbb{Q}}_\ell}$. Make the following assumption*

- *$\mathcal{G}^{M_{\overline{F}}}_{\overline{F},\overline{\mathbb{Q}}_\ell}$ is connected.*

Suppose that $\widetilde{H} \to H$ is any central torus quotient of linear algebraic groups over $\overline{\mathbb{Q}}_\ell$, and that there is a factorization

$$\mathcal{G}^M_{F,\overline{\mathbb{Q}}_\ell} \hookrightarrow H \hookrightarrow \mathrm{GL}_{\overline{\mathbb{Q}}_\ell}(M_\ell \otimes_E \overline{\mathbb{Q}}_\ell),$$

where of course the composition of these inclusions is the natural representation. Then there exists a geometric lift $\tilde{\rho}$

$$\begin{array}{ccccc} & & & & \widetilde{H}(\overline{\mathbb{Q}}_\ell) \\ & & {\scriptstyle\tilde{\rho}} \nearrow & & \downarrow \\ \Gamma_F & \xrightarrow{\rho} & \mathcal{G}^M_{F,\overline{\mathbb{Q}}_\ell}(\overline{\mathbb{Q}}_\ell) & \longrightarrow & H(\overline{\mathbb{Q}}_\ell). \end{array}$$

Similarly, if F is any number field, but E is $\mathbb{Q}$, then without any assumption on the motivic group $\mathcal{G}^{M_{\overline{F}}}_{\overline{F}}$, such a $\tilde{\rho}$ exists.

REMARK 4.1.28. The assumption that $\mathcal{G}_{\overline{F},\overline{\mathbb{Q}}_\ell}^{M_{\overline{F}}}$ is connected is expected always to hold. In particular, this should not be necessary for the conclusion of the corollary to hold.

PROOF. Corollary 4.1.26 tells us that composing ρ with any irreducible algebraic representation of $\mathcal{G}_{F,\overline{\mathbb{Q}}_\ell}^{M_{\overline{F}}}$ yields an ℓ-adic representation satisfying the Hodge symmetries of Conjecture 3.2.5. But the arguments of Proposition 3.2.7 and Theorem 3.2.10 do not apply directly, because they assume these symmetries after composition with irreducible representations of the connected component $(\overline{\rho(\Gamma_F)}^{\mathrm{Zar}})^0$.[8] The simplest (but imperfect) way around this is to assume $\mathcal{G}_{\overline{F},\overline{\mathbb{Q}}_\ell}^{M_{\overline{F}}}$ is connected, and then run through the arguments of Proposition 3.2.7 and Theorem 3.2.10 with $\mathcal{G}_{F',\overline{\mathbb{Q}}_\ell}^{M_{F'}}$, for F' sufficiently large (see Lemma 4.1.17), in place of $\overline{\rho(\Gamma_{F'})}^{\mathrm{Zar}}$: reduce to the connected case (replacing F by F') as in Theorem 3.2.10, and then note that the proof of Proposition 3.2.7 only requires identifying a connected reductive group containing the image of ρ such that composition with irreducible representations of this group yields Galois representations with the desired symmetry (whether or not these Galois representations are themselves irreducible).

The second assertion (when $E = \mathbb{Q}$) follows by the same argument: by Remark 4.1.23, the τ-labeled Hodge-Tate co-character μ_τ of ρ is independent of τ. It can be interpreted as a co-character of $(\overline{\rho(\Gamma_F)}^{\mathrm{Zar}})^0$, and composing with finite-dimensional representations of this group, we obtain the necessary Hodge-Tate symmetries to apply the argument of Proposition 3.2.7. □

REMARK 4.1.29. Similarly, there is a motivic version of Corollary 3.2.8.

4.1.6. Motivic lifting: The potentially CM case. One case of the desired motivic lifting result is immediately accessible, when the corresponding Galois representations are potentially CM. We denote by $\mathcal{CM}_F$ the Tannakian category of motives (for motivated cycles) over F generated by Artin motives and potentially CM abelian varieties. As before an embedding $\sigma\colon F \hookrightarrow \mathbb{C}$ yields, through the corresponding Betti fiber functor, a Tannakian group $\mathcal{T}_F(\sigma) = \mathrm{Aut}^{\otimes}(H_\sigma|_{\mathcal{CM}_F})$. For a number field E, we can also consider, as in §4.1.4, the category $\mathcal{CM}_{F,E}$ of potentially CM motives over F with coefficients in E, with its Tannakian group $\mathcal{T}_{F,E}(\sigma)$.[9] As with $\mathcal{G}_F$, there is a sequence

$$1 \to \mathcal{T}_{\overline{F},E} \to \mathcal{T}_{F,E} \to \Gamma_F \to 1,$$

where again the projection $\mathcal{T}_{F,E} \to \Gamma_F$ has a continuous section s_λ on E_λ-points. Deligne showed in [**DMOS82**] (Chapter IV: '*Motifs et groupes de Taniyama*') that $\mathcal{T}_\mathbb{Q}$ is isomorphic to the Taniyama group constructed by Langlands as an explicit extension of $\Gamma_\mathbb{Q}$ by the connected Serre group.[10]

PROPOSITION 4.1.30. *Let F be totally imaginary. Let $\widetilde{H} \to H$ be a surjection of linear algebraic groups over a number field E with central torus kernel. For any*

[8]If this group were, contrary to conjecture, not equal to $\mathcal{G}_{\overline{F},\overline{\mathbb{Q}}_\ell}^{M_{\overline{F}}}$, then it is not obvious in general how to relate the two putative Hodge symmetries.

[9]From now on, σ will be implicit.

[10]More precisely, equipped with the data of the projection to $\Gamma_\mathbb{Q}$, a section on $\mathbf{A}_{F,f}$-points, and the co-character $\mathbb{G}_{m,\mathbb{C}} \to \mathcal{T}_\mathbb{C}$ giving the Hodge filtration on objects of $\mathcal{CM}_\mathbb{C}$, $\mathcal{T}_\mathbb{Q}$ is uniquely isomorphic to the Taniyama group, equipped with its corresponding structures.

homomorphism $\rho\colon \mathcal{T}_{F,E} \to H$*, there is a finite extension* E'/E *such that* ρ *lifts to a homomorphism*

$$\begin{array}{ccc} & & \widetilde{H}_{E'} \\ & \overset{\tilde{\rho}}{\nearrow} & \downarrow \\ \mathcal{T}_{F,E'} & \xrightarrow{\ \rho\ } & H_{E'}. \end{array}$$

PROOF. Fix a finite place λ of E and consider the λ-adic representation $\rho_\lambda = \rho \circ s_\lambda\colon \Gamma_F \to H(E_\lambda)$. Since $\mathcal{T}_{\overline{F}}$ is isomorphic to the connected Serre group, we can unconditionally apply Corollary 4.1.27 to find, for some finite extension E'_λ/E_λ a geometric lift $\tilde{\rho}_\lambda\colon \Gamma_F \to \widetilde{H}(E'_\lambda)$ of ρ_λ. Note that $\tilde{\rho}_\lambda$ is potentially abelian, since ρ_λ is, and the kernel of $\widetilde{H} \to H$ is central. Proposition IV.D.1 of [**DMOS82**] implies that $\tilde{\rho}_\lambda$ arises from a homomorphism (of groups over E'_λ) $\mathcal{T}_F \otimes_{\mathbb{Q}} E'_\lambda \cong \mathcal{T}_{F,E} \otimes_E E'_\lambda \to \widetilde{H} \otimes_E E'_\lambda$. This in turn must be definable over some finite extension E'/E (E' is thus embedded in E'_λ), i.e. there is a homomorphism $\tilde{\rho}\colon \mathcal{T}_{F,E'} \to \widetilde{H} \otimes_E E'$ whose extension to E'_λ gives rise to $\tilde{\rho}_\lambda$. We explain this technical point in Lemma 4.1.31 below. To check that $\tilde{\rho}$ is actually a lift of $\rho \otimes_E E'$, it suffices to observe:

- The restriction to $\mathcal{T}_{\overline{F},E'}$ is a lift: the restrictions of $\tilde{\rho}$ and ρ are simply the algebraic homomorphisms of the connected Serre group (tensored with E') corresponding to the labeled Hodge-Tate weights of $\tilde{\rho}_\lambda$ and ρ_λ.
- The λ'-adic lift, for the place λ' of E' induced by $E' \hookrightarrow E'_\lambda$, is a lift (by construction).
- It suffices to check that $\tilde{\rho}$ is a lift on $E'_{\lambda'}$-points. First, it suffices to check that $\tilde{\rho} \otimes_{E'} E'_{\lambda'}$ lifts $\rho \otimes_E E'_{\lambda'}$. This in turn can be checked on $E'_{\lambda'}$-points: we are free to replace $\mathcal{T}_{F,E'} \otimes_{E'} E'_{\lambda'}$ by some finite-type quotient $\mathcal{T}$ in which the $E'_{\lambda'}$-points are Zariski-dense. Then the closed subscheme $\mathcal{T} \underset{H\times H}{\times} H \hookrightarrow \mathcal{T}$ (where the two maps to $H \times H$ are the diagonal $H \to H \times H$ and the product of the two maps $\mathcal{T} \to H \times H$ given by ρ and $\tilde{\rho}$ followed by the quotient $\widetilde{H} \to H$) contains $\mathcal{T}(E'_{\lambda'})$, hence equals $\mathcal{T}$.

Then we are done, since $\mathcal{T}_{F,E'}(E'_{\lambda'}) = \mathcal{T}_{\overline{F},E'}(E'_{\lambda'}) \cdot s_{\lambda'}(\Gamma_F)$. □

Here is the promised lemma showing that $\tilde{\rho}$ may be defined over a finite extension E' of E:

LEMMA 4.1.31. *Let* K/k *be an extension of algebraically closed fields of characteristic zero. Let* T *be a (not necessarily connected) reductive group over* k*, and let* $\pi\colon \widetilde{H} \to H$ *be a surjection (defined over* k*) of reductive* k*-groups. Suppose that* $\rho\colon T \to H$ *is a* k*-morphism, and that the scalar extension* ρ_K *lifts to a* K*-morphism* $\tilde{\rho}\colon T_K \to \widetilde{H}_K$*. Then* ρ *lifts to a* k*-morphism* $T \to \widetilde{H}$*.*

PROOF. This is a simple spreading-out argument. Namely, writing K as the direct limit of its finite-type k-sub-algebras, we can descend the relation $\pi_K \circ \tilde{\rho} = \rho_K$ to some finitely-generated k-sub-algebra R of K, in particular obtaining $\tilde{\rho}_R\colon T_R \to \widetilde{H}_R$ such that $\tilde{\rho}_R \otimes_R K = \tilde{\rho}$. We then choose a non-zero k-point $\alpha\colon R \to k$ (a morphism of k-algebras), and specializing the relation $\pi_R \circ \tilde{\rho}_R = \rho_R$ via α, we obtain a lift $\tilde{\rho}_R \otimes_{R,\alpha} k$ of ρ over k.

□

REMARK 4.1.32.
- We have stated Proposition 4.1.30 only for imaginary fields for simplicity, but of course there is a variant for general number fields taking into account Corollary 3.2.8. When F is totally real, there will be such ρ that do not lift: simply take a type A Hecke character ψ of a quadratic CM extension L/F as in Example 2.5.3. Then $\mathrm{Ad}^0(\mathrm{Ind}_L^F(\psi))$ will have a potentially abelian Galois ℓ-adic realizations $\Gamma_F \to SO_3(\overline{\mathbb{Q}}_\ell)$, which arises from a representation of $\mathcal{T}_F$ by Proposition IV.D.1 of [**DMOS82**]. As we have seen, these representations do not lift geometrically to GSpin_3.
- The method of checking that $\tilde{\rho}$–once we know it exists–actually lifts ρ will recur in §4.2; this division into a geometric argument (lifting $\rho|_{\mathcal{T}_{\overline{F},E}}$ and a Galois-theoretic argument (lifting $\rho \circ s_\lambda$) seems to be the natural way to make arguments about motivic Galois groups over number fields (compare Lemma 4.1.17).
- On the automorphic side, representations of the global Weil group ought to parametrize 'potentially abelian' automorphic representations. In that case, the analogous lifting result is a theorem of Labesse ([**Lab85**]).

Before proceeding to more elaborate examples, let us clarify here that, even though we study lifting through *central* quotients, the nature of these problems is in fact highly non-abelian. That is, suppose we have a lift $\tilde{\rho}\colon \mathcal{G}_{F,E} \to \widetilde{H}_E$ of $\rho\colon \mathcal{G}_{F,E} \to H_E$ (we may assume $\widetilde{H}$ and H are reductive; let us further suppose they are connected), and let r' and r be irreducible faithful representations of $\widetilde{H}_E$ and H_E. If the derived group of H is not simply-connected, but its simply-connected cover H_{sc} injects into $\widetilde{H}$, then typically $r' \circ \tilde{\rho}$ will not lie in the Tannakian subcategory of $\mathcal{M}_{F,E}$ generated by $r \circ \rho$ and all potentially CM motives, i.e. objects of $\mathcal{CM}_{F,E}$ (of course, it does in the example of Proposition 4.1.30). We make this precise at the Galois-theoretic level:

LEMMA 4.1.33. *Let $\widetilde{H} \to H$, r', and r be as above. Let $\rho\colon \Gamma_F \to H(\overline{\mathbb{Q}}_\ell)$ be a geometric Galois representation having a geometric lift $\tilde{\rho}\colon \Gamma_F \to \widetilde{H}(\overline{\mathbb{Q}}_\ell)$. Assume that*

- *the algebraic monodromy group of ρ is H itself;*
- *the kernel of $H_{\mathrm{sc}} \cap \overline{(\tilde{\rho}(\Gamma_F))}^{\mathrm{Zar}} \to H$ is non-zero.*

Then $r' \circ \tilde{\rho}$ is not contained in the Tannakian sub-category of semi-simple geometric $\overline{\mathbb{Q}}_\ell$-representations of Γ_F generated by $r \circ \rho$ and all potentially abelian geometric representations.

PROOF. If $r' \circ \tilde{\rho}$ were contained in this category, then there would exist an irreducible potentially abelian representation τ of Γ_F and an injection $r' \circ \tilde{\rho} \hookrightarrow \tau \otimes (r_1 \circ \rho)$ for some irreducible algebraic representation r_1 of H. By Proposition 3.4.1, τ has the form $\mathrm{Ind}_L^F(\psi \cdot \omega)$ for some character ψ and irreducible Artin representation ω of Γ_L. By Frobenius reciprocity, there is a non-zero map $(r' \circ \tilde{\rho})|_{\Gamma_L} \to \psi \cdot (r_1 \circ \rho)|_{\Gamma_L} \otimes \omega$; both sides are irreducible,[11] so ω is one-dimensional, and absorbing ω into ψ we may assume $r' \circ \tilde{\rho}|_{\Gamma_L} \xrightarrow{\sim} \psi \cdot (r_1 \circ \rho)|_{\Gamma_L}$. Comparing algebraic monodromy

[11]For the right-hand side, see Proposition 3.4.1: the tensor product of a Lie-irreducible and an Artin representation is irreducible; of course, for the purposes of this lemma, we could just further restrict L.

groups, we obtain a contradiction, since by assumption $H_{\mathrm{sc}} \cap \overline{(\tilde{\rho}(\Gamma_F))}^{\mathrm{Zar}}$ cannot inject into $\mathbb{G}_m \times H$. □

In §4.2, we will see many examples of ρ with full SO monodromy group; their lifts to GSpin will satisfy the conclusions of Lemma 4.1.33.

4.2. Motivic lifting: The hyperkähler case

4.2.1. Setup. The aim of this section is to produce a lifting not merely at the level of a single ℓ-adic representation, but of actual motives, in a very special family of cases whose prototype is the primitive second cohomology of a $K3$ surface over F. So that the reader has some examples to keep in mind, we recall the following definitions:

DEFINITION 4.2.1. Let F be a subfield of $\mathbb{C}$. A *hyperkähler variety* X over F is a geometrically connected and simply-connected smooth projective variety over F of even dimension $2r$ such that $\Gamma(X, \Omega^2_X)$ is one-dimensional, generated by a differential form ω for which ω^r is non-vanishing at every point of X (i.e., as a linear functional on the top wedge power of the tangent space). A *K3 surface* over F is a hyperkähler variety X/F of dimension 2.

EXAMPLE 4.2.2. The simplest example of a K3 surface is a smooth quartic hypersurface in $\mathbb{P}^3$. Higher-dimensional examples of hyperkählers are notoriously difficult to produce. One standard family is gotten by starting with any K3 surface X, and then for any integer $r \geq 1$ considering the Hilbert scheme $X^{[r]}$ of r points on X (more precisely, the moduli of closed sub-schemes of length r); each $X^{[r]}$ is a hyperkähler variety, with of course $X^{[1]} = X$. (This construction is due to Beauville: see [**Bea83**, §6].)

More generally, we work in the axiomatized setup of [**And96a**]. By a *polarized variety* over a subfield F of $\mathbb{C}$ we mean a pair (X, η) consisting of variety X/F and an F-rational ample line bundle η on X. For a fixed $k \leq \dim X$, we will consider the motive of primitive cohomology (omitting the η-dependence from the notation),

$$\begin{aligned} &Prim^{2k}(X)(k) \\ &\quad = \ker\left(\eta^{\dim X - 2k+1} \colon H^{2k}(X)(k) \to H^{2\dim X - 2k+2}(X)(\dim X - k + 1)\right), \end{aligned}$$

as an object of $\mathcal{M}_F$. Cup-product with η lets us endow the motive $H^{2k}(X)(k)$ with the quadratic form

$$\langle x, y\rangle_\eta = (-1)^k x \cup y \cup \eta^{\dim X - 2k} \in H^{2\dim X}(X)(\dim X) \cong \mathbb{Q},$$

and we can equivalently define $Prim^{2k}(X)(k)$ as the orthogonal complement of $H^{2k-2}(X)(k-1) \cup \eta$.

The two Weil cohomologies we use are ℓ-adic (with coefficients in $\mathbb{Q}_\ell$ or some extension inside $\overline{\mathbb{Q}}_\ell$) and Betti cohomology; we will occasionally take integral Betti cohomology, where we will always work modulo torsion, so that $Prim^{2k}(X_{\mathbb{C}}, \mathbb{Z})(k)$ is a sub-module of $H^{2k}(X_{\mathbb{C}}, \mathbb{Z})(k)/(\text{torsion})$. In that case, the primitive lattice defines a polarized (by $\langle \cdot, \cdot \rangle_\eta$) integral Hodge structure of weight 0; we write $h^{p,q}$ for the Hodge numbers. André proves his theorems, which include versions of the Shafarevich and Tate conjectures, under the following axioms:

A_k: $h^{1,-1} = 1$, $h^{0,0} > 0$, and $h^{p,q} = 0$ if $|p - q| > 2$.

B_k: There exist a smooth connected F-scheme S, a point $s \in S(F)$, and a smooth projective morphism $f\colon \mathcal{X} \to S$ such that:
- $X \cong \mathcal{X}_s$;
- the Betti class $\eta_B \in H^2(X_{\mathbb{C}}^{\mathrm{an}}, \mathbb{Z})(1)/(\text{torsion})$ extends to a section of $R^2 f_{\mathbb{C}*}^{\mathrm{an}}\mathbb{Z}(1)/(\text{torsion})$;
- letting $\widetilde{S}$ denote the universal cover of $S(\mathbb{C})$, and D denote the period domain of Hodge structures on $V_{\mathbb{Z}} := Prim^{2k}(X_{\mathbb{C}}, \mathbb{Z})(k)$ polarized by $\langle \cdot, \cdot \rangle_\eta$, we require that the image of $\widetilde{S} \to D$ contain an open subset.

B_k^+: For each $t \in S(\mathbb{C})$, every Hodge class in $H^{2k}(\mathcal{X}_t, \mathbb{Q})(k)$ is an algebraic class.

A_k is essential to the method, which relies on studying the associated Kuga-Satake abelian variety, which exists for Hodge structures of this particular form. B_k is a statement about deforming X into a 'big' family. B_k^+ is of course a case of the Hodge conjecture, which is always known when $k = 1$ (the theorem of Lefschetz). But we provide these axioms merely for orientation. Of interest is the following collection of varieties for which they are known to hold:

PROPOSITION 4.2.3 (See §2-3 of [**And96a**]). *The axioms A_1, B_1, and B_1^+ are satisfied by: abelian surfaces; surfaces of general type with ample canonical bundle $\mathcal{K}_X$, $h^{2,0}(X) = 1$, and $\mathcal{K}_X \cdot \mathcal{K}_X = 1$; and polarized hyperkähler varieties with $b_2 > 3$ (in particular, $K3$ surfaces). Cubic fourfolds (polarized via $\mathcal{O}_{\mathbb{P}^5}(1)$) satisfy A_2, B_2, and B_2^+.*

This relies on the work of many people; see §2 and §3 of [**And96a**]. Finally, André observes that these axioms are independent of the choice of embedding $F \hookrightarrow \mathbb{C}$; note that for A_k this is a special case of Remark 4.1.23.

REMARK 4.2.4. The proposition does not apply to hyperkähler varieties with Betti number 3, since its argument (building on that of [**Del72**]) requires the variety X to have sufficiently robust (projective) deformation theory, whereas a hyperkähler with $b_2(X) = 3$ is rigid. In fact, it is believed that such hyperkählers do not exist. Regardless, in that case one can still say something about the motivic lifting problem, since the ℓ-adic representation $H^2(X_{\overline{F}}, \overline{\mathbb{Q}}_\ell)$ is potentially abelian. In principle, this should force the underlying motivic Galois representation to factor through $\mathcal{T}_F$, at which point we would apply Proposition 4.1.30; but this argument would require checking that two $\mathcal{G}_F$-representations with isomorphic ℓ-adic realizations are themselves isomorphic, i.e. an unknown case of the Tate conjecture. At least we have (as in Proposition 4.1.30):

LEMMA 4.2.5. *Let X/F be a smooth projective variety with $b_2(X) = 3$. If necessary, replace F with a quadratic extension trivializing* $\det H^2(X_{\overline{F}}, \mathbb{Q}_\ell)$. *Then the ℓ-adic realization $\rho_\ell\colon \mathcal{G}_F \to \mathrm{SO}(H^2(X_{\overline{F}}, \overline{\mathbb{Q}}_\ell))$ has a lift $\tilde{\rho}_\ell$ to $\mathrm{GSpin}(H^2(X_{\overline{F}}, \overline{\mathbb{Q}}_\ell))$ that arises as the ℓ-adic realization of a representation of $\mathcal{T}_{F,E}$ for suitable E.*

NOTATION 4.2.6.

- From now on we view F as a subfield of $\mathbb{C}$, with $\overline{F}$ its algebraic closure in $\mathbb{C}$. These embeddings will be used implicitly to define Betti realizations, motivic Galois groups (for Betti realizations), and étale-Betti comparisons, for varieties (or motives) over extensions of F inside $\overline{F}$.

- We write $V_{\mathbb{Q}}$ for $Prim^{2k}(X_{\mathbb{C}}, \mathbb{Q})(k)$, and for a prime ℓ, we will write V_ℓ for the ℓ-adic realization $Prim^{2k}(X_{\overline{F}}, \mathbb{Q}_\ell)(k)$. As before, $Prim^{2k}(X)(k)$ will be used to indicate the underlying motive, i.e. object of $\mathcal{M}_F$.
- For fields E containing $\mathbb{Q}$, we will sometimes denote the extension of scalars $V_{\mathbb{Q}} \otimes_{\mathbb{Q}} E$ by V_E (or similarly for $\mathbb{Q}_\ell$ and V_ℓ). This is the Betti realization of an object $Prim^{2k}(X)(k)_E$ of $\mathcal{M}_{F,E}$.
- The same subscript conventions will hold for other motives we consider, especially the direct factors of $Prim^{2k}(X)(k)$ given by the algebraic cycles (to be denoted Alg) and its orthogonal complement, the transcendental lattice (T). These motives will be discussed in §4.2.4. In the meantime, we recall that the rational Hodge structure $Prim^{2k}(X_{\mathbb{C}}, \mathbb{Q})(k)$ has a $\mathbb{Q}$-subspace spanned by its Hodge classes; the orthogonal complement $T_{\mathbb{Q}}$, the *transcendental subspace*, of this space of Hodge cycles is again a polarized $\mathbb{Q}$-Hodge structure.
- For an object M of $\mathcal{M}_F$, we will sometimes write ρ^M for the associated motivic Galois representation. For the ℓ-adic realization of this motivic Galois representation (i.e., the representation of Γ_F on $H_\ell(M)$), we write ρ^M_ℓ.
- First assume $\dim V_{\mathbb{Q}} = m$ is odd. The group $\mathrm{GSpin}(V_{\mathbb{Q}})$ may not have a rationally-defined spin representation, but it does after some extension of scalars, and over any suitably large field E, we denote by
$$r_{\mathrm{spin}} \colon \mathrm{GSpin}(V_E) \to \mathrm{GL}(W_E)$$
this algebraic representation. If m is even, we similarly denote by $W_E = W_{+,E} \oplus W_{-,E}$ the direct sum of the two half-spin representations. In contrast to $V_E = V_{\mathbb{Q}} \otimes_{\mathbb{Q}} E$, this W_E is not necessarily an extension of scalars from an underlying $\mathbb{Q}$-space.

Possibly after replacing F by a quadratic extension, we may assume that the Γ_F-representation V_ℓ is special orthogonal, i.e. $\rho_\ell \colon \Gamma_F \to \mathrm{SO}(V_\ell) \subset \mathrm{O}(V_\ell)$. Since for almost all finite places v, the frobenius fr_v acts on $H^{2k-2}(X_{\overline{F}}, \mathbb{Q}_\ell)$ and $H^{2k}(X_{\overline{F}}, \mathbb{Q}_\ell)$ with eigenvalues that are independent of ℓ (**[Del74]**), and trivially on the ℓ-adic Chern class η_ℓ, the eigenvalues of fr_v on $Prim^{2k}(X_{\overline{F}}, \mathbb{Q}_\ell)(k)$ are independent of ℓ. Consequently, if $\det \rho_\ell$ is trivial for one ℓ, then it is for all ℓ. We will from now on assume, for technical simplicity, that this determinant condition is satisfied.

4.2.2. The Kuga-Satake construction. We review the classical Kuga-Satake construction and outline André's refinement, which implies a potential version (i.e., after replacing F by a finite extension) of our motivic lifting result. See §4 and §5 of [**And96a**]. His approach is inspired by that of [**Del72**], in which Deligne used the Kuga-Satake construction (in families) to reduce the Weil conjectures for $K3$ surfaces to the (previously known) case of abelian varieties. Let $V_{\mathbb{Z}}$ be the polarized quadratic lattice $Prim^{2k}(X_{\mathbb{C}}, \mathbb{Z})(k)$ of the previous subsection. Write $m = 2n$ or $m = 2n+1$ for its rank. Basic Hodge theory implies that the pairing on $V_{\mathbb{R}}$ is negative-definite on the sub-space $(H^{1,-1} \oplus H^{-1,1})_{\mathbb{R}}$, and positive-definite on $H^{0,0}_{\mathbb{R}}$, hence has signature $(2, m-2)$. Recalling the discussion of §4.1.1, this real Hodge structure yields a homomorphism $h \colon \mathbb{S} \to \mathrm{SO}(V_{\mathbb{R}})$ that lifts uniquely to a homomorphism $\tilde{h} \colon \mathbb{S} \to \mathrm{GSpin}(V_{\mathbb{R}})$ whose composition $N_{\mathrm{spin}} \circ \tilde{h}$ with the Clifford norm N_{spin} is the usual norm $\mathbb{S} \to \mathbb{G}_{m,\mathbb{R}}$.

Let $C(V_{\mathbb{Z}})$ and $C^+(V_{\mathbb{Z}})$ denote the Clifford algebra and the even Clifford algebra associated to the quadratic space $V_{\mathbb{Z}}$. Let $L_{\mathbb{Z}}$ be a free left $C^+(V_{\mathbb{Z}})$-module of rank one. Denote by C^+ the ring $\mathrm{End}_{C^+(V_{\mathbb{Z}})}(L_{\mathbb{Z}})^{\mathrm{op}}$. Because of the 'op,' C^+ naturally acts on $L_{\mathbb{Z}}$ on the right, and we then correspondingly have $\mathrm{End}_{C^+}(L_{\mathbb{Z}}) \cong C^+(V_{\mathbb{Z}})$. The choice of a generator x_0 of $L_{\mathbb{Z}}$ induces a ring isomorphism

$$\begin{aligned} C^+(V_{\mathbb{Z}}) &\xrightarrow{\phi_{x_0}} C^+ \\ c \mapsto &(bx_0 \mapsto bcx_0). \end{aligned}$$

Via $\tilde{h}$ and the tautological representation $\mathrm{GSpin}(V_{\mathbb{R}}) \hookrightarrow C^+(V_{\mathbb{R}})^\times$, $L_{\mathbb{R}} := L_{\mathbb{Z}} \otimes_{\mathbb{Z}} \mathbb{R}$ acquires a Hodge structure of type $(1,0), (0,1)$; the Hodge type is easily read off from the fact that $\mathrm{End}_{C^+}(L_{\mathbb{Z}}) \cong C^+(V_{\mathbb{Z}})$ is also an isomorphism of Hodge-structures,[12] or by recalling the remarks of §4.1.1 and listing the weights of the spin representation. Concretely (as in the original construction of Satake [**Sat66**] and Kuga-Satake [**KS67**]), we can choose an orthogonal basis e_1, e_2 of $(H^{1,-1} \oplus H^{-1,1})_{\mathbb{R}}$, normalized so that $\langle e_i, e_i \rangle = -1$ for $i = 1, 2$. Then the automorphism of $C^+(V_{\mathbb{R}})$ given by multiplication by e_1e_2 is a complex structure, by the defining relations for the Clifford algebra. In fact, this integral Hodge structure is polarizable: this can be shown explicitly, or by a very soft argument (apply Proposition 2.11.*b.iii*$'$ of [**Del72**]). We therefore obtain a complex abelian variety, the Kuga-Satake abelian variety $KS(X) = KS(X_{\mathbb{C}}, \eta, k)$, associated to our original $V_{\mathbb{Z}}$. The right-action of C^+ on $L_{\mathbb{Z}}$ commutes with the Hodge structure, so C^+ acts as endomorphisms of $KS(X)$. Generically, this will be the full endomorphism ring; we will have to be attentive later to how much bigger the endomorphism ring can be.

One of the main technical ingredients in [**And96a**] is the following descent result, which uses rigidity properties of the Kuga-Satake construction:

LEMMA 4.2.7 (Main Lemma 1.7.1 of [**And96a**]). *Let (X, η) be a polarized variety over a subfield F of $\mathbb{C}$, satisfying properties A_k and B_k. Then there exists an abelian variety $A_{F'}$ over some finite extension F'/F such that*

- *The base-change $A_{\mathbb{C}}$ is the Kuga-Satake variety $KS(X)$;*
- *there is a subalgebra C^+ of* $\mathrm{End}(A_{F'})$ *and an isomorphism of $\mathbb{Z}_\ell[\Gamma_{F'}]$-algebras*

$$\mathrm{End}_{C^+}\left(H^1(A_{\overline{F}}, \mathbb{Z}_\ell)\right) \cong C^+(Prim^{2k}(X_{\overline{F}}, \mathbb{Z}_\ell)(k)).$$

We subsequently write L_ℓ for the ℓ-adic realization $H^1(A_{F'} \otimes \overline{F}, \mathbb{Q}_\ell)$. The main result of the 'motivated' theory of hyperkählers, (the foundation of André's results on the Tate and Shafarevich conjectures) is:

THEOREM 4.2.8 (see Theorem 6.5.2 of [**And96a**]). *For some finite extension F'/F, $Prim^{2k}(X_{F'})(k)$ is a direct factor (in $\mathcal{M}_F$) of $\underline{End}\left(H^1(A_{F'})\right)$, and both $Prim^{2k}(X_{F'})(k)$ and $H^1(A_{F'})$ have connected motivic Galois group.*

Write $\rho^A \colon \mathcal{G}_{F'} \to \mathrm{GL}(L_{\mathbb{Q}})$ and $\rho^V : \mathcal{G}_F \to \mathrm{SO}(V_{\mathbb{Q}})$ for the motivic Galois (for the Betti realization) representations associated to $H^1(A_{F'})$ and $Prim^{2k}(X)(k)$. We will somewhat sloppily use the same notation ρ^A and ρ^V for various restrictions of these representations (eg, to $\mathcal{G}_{\overline{F}}$). Note that by Corollary 4.1.20, the image of ρ^A is the Mumford-Tate group of $KS(X)$, which is easily seen to be contained in $\mathrm{GSpin}(V_{\mathbb{Q}}) = \{x \in C^+(V_{\mathbb{Q}})^\times : xV_{\mathbb{Q}}x^{-1} = V_{\mathbb{Q}}\}$. More precisely:

[12] Giving $C^+(V_{\mathbb{Z}})$ the Hodge structure induced from that on the tensor powers of $V_{\mathbb{Z}}$.

COROLLARY 4.2.9. *For some finite extension F'/F, ρ^A factors through* $\mathrm{GSpin}(V_{\mathbb{Q}})$ *and lifts $\rho^V|_{\mathcal{G}_{F'}}$. A fortiori, ρ^A_ℓ lifts $\rho^V_\ell|_{\Gamma_{F'}}$.*

PROOF. Over $\mathbb{C}$ (or $\overline{F}$), the image of ρ^A is $MT(A_{\mathbb{C}})$, hence contained in $\mathrm{GSpin}(V_{\mathbb{Q}})$. We can therefore compare the two maps

$$\mathcal{G}_{\mathbb{C}} \xrightarrow{(\rho^V, \pi\circ\rho^A)} \mathrm{SO}(V_{\mathbb{Q}}) \times \mathrm{SO}(V_{\mathbb{Q}}) \to \mathrm{GL}\left(C^+(V_{\mathbb{Q}})\right) \times \mathrm{GL}\left(C^+(V_{\mathbb{Q}})\right),$$

where π denotes the projection $\mathrm{GSpin}(V_{\mathbb{Q}}) \to \mathrm{SO}(V_{\mathbb{Q}})$. Under the motivated isomorphism

$$C^+(Prim^{2k}(X_{\mathbb{C}})(k)) \cong \underline{End}\left(H^1(A)\right)$$

the adjoint action of ρ^A on $\mathrm{End}(L_{\mathbb{Q}})$ agrees with the action of $\pi \circ \rho^A$ on $C^+(V_{\mathbb{Q}})$, so the two compositions $\mathcal{G}_{\mathbb{C}} \to \mathrm{GL}\left(C^+(V_{\mathbb{Q}})\right)$ above coincide. Now, if m is odd, $\mathrm{SO}(V_{\mathbb{Q}}) \to \mathrm{GL}(C^+(V_{\mathbb{Q}}))$ is injective, so $\rho^V = \pi \circ \rho^A$; that is, at least over $\mathbb{C}$, ρ^A lifts ρ^V. If m is even, we deduce that the compositions

$$\mathcal{G}_{\mathbb{C}} \xrightarrow{(\rho^V, \pi\circ\rho^A)} \mathrm{SO}(V_{\mathbb{Q}}) \times \mathrm{SO}(V_{\mathbb{Q}}) \to \mathrm{GL}(\wedge^2 V_{\mathbb{Q}}) \times \mathrm{GL}(\wedge^2 V_{\mathbb{Q}})$$

agree.[13] The kernel of $\wedge^2\colon \mathrm{SO}(V_{\mathbb{Q}}) \to \mathrm{GL}(\wedge^2 V_{\mathbb{Q}})$ is $\{\pm 1\}$ (and central), so we see that ρ^V and $\pi\circ\rho^A$ agree up to twisting by some character $\chi\colon \mathcal{G}_{\mathbb{C}} \to \{\pm 1\} \subset \mathrm{SO}(V_{\mathbb{Q}})$. But clearly this character factors through $\mathcal{G}^M_{\mathbb{C}}$, where $M = Prim^{2k}(X_{\mathbb{C}})(k) \oplus H^1(A)$, and by Theorem 4.2.8, $\mathcal{G}^M_{\mathbb{C}}$ is connected. Therefore χ is trivial, and ρ^A lifts ρ^V as $\mathcal{G}_{\mathbb{C}}$-representations. By Lemma 4.1.21, the same holds for the corresponding $\mathcal{G}_{\overline{F}}$-representations, and then as in Lemma 4.1.17 the same holds over some finite extension F'/F. □

Corollary 4.2.9 does not always hold with $F' = F$, and it is our task in the coming sections to achieve a motivic descent over F itself. The Kuga-Satake variety is highly redundant, and it is technically convenient to work with a smaller 'spin' abelian variety, many copies of which constitute the Kuga-Satake variety. Since the rational Clifford algebra $C^+_{\mathbb{Q}}$ may not be split, this requires a finite extension of scalars $E/\mathbb{Q}$, after which we can work in the isogeny category $AV^0_{F',E}$ of abelian varieties over F' with E-coefficients. We take F' as in the Corollary, and now to ease the notation, we write simply A for $A_{F'}$.

LEMMA 4.2.10. *There exists a number field E and an abelian variety B/F' with endomorphisms by E such that there is a decomposition in $AV^0_{F',E}$:*

$$A \otimes_{\mathbb{Q}} E \cong B^{2^n} \quad \textit{if } m = 2n+1;$$
$$A \otimes_{\mathbb{Q}} E \cong B^{2^{n-1}} \quad \textit{if } m = 2n.$$

The ℓ-adic realization $H^1(B_{\overline{F}}, \mathbb{Q}_\ell)$ is isomorphic to the composite $r_{\mathrm{spin}} \circ \rho^A_\ell$ as $(E \otimes \mathbb{Q}_\ell)[\Gamma_{F'}]$-modules, where as before ρ^A_ℓ denotes the representation $\Gamma_{F'} \to \mathrm{GSpin}(V_\ell \otimes E)$ obtained from $L_\ell \otimes E$, and r_{spin} denotes either the spin (m odd) or sum of half-spin (m even) representations of $\mathrm{GSpin}(V_\ell \otimes E)$. When $m = 2n$, B decomposes in $AV^0_{F',E}$ as $B_+ \times B_-$, corresponding to the two half-spin representations.

---

[13]The filtration on $C^+(V_{\mathbb{Q}})$ given by the image of $V^{\otimes \leq 2i}$ is motivated, since $\mathrm{O}(V)$-stable, so ρ^V and $\pi \circ \rho^A$ coincide on the $i = 1$ graded piece, which is just $\wedge^2(V_{\mathbb{Q}})$.

PROOF. Choose a number field E splitting $C^+(V_{\mathbb{Q}})$. Then, letting W_E denote either the spin (m odd) representation or the direct sum $W_{+,E} \oplus W_{-,E}$ of the two irreducible half-spin representations (m even), $C^+(V_E)$ is isomorphic as $\mathrm{GSpin}(V_E)$-representations to $W_E^{2^n}$, or to $W_{E,+}^{2^{n-1}} \oplus W_{E,-}^{2^{n-1}}$. As E-algebra it is then isomorphic either to $\mathrm{End}(W_E) \cong M_{2^n}(E)$ or $\mathrm{End}(W_{+,E}) \oplus \mathrm{End}(W_{-,E}) \cong M_{2^{n-1}}(E) \oplus M_{2^{n-1}}(E)$.[14] Using the orthogonal idempotents in $C_E^+ \cong C^+(V_E)$, we decompose the object $A_{F'} \otimes_{\mathbb{Q}} E$ of $AV^0_{F',E}$ into 2^n (when m is odd) or 2^{n-1} (when m is even) copies of an abelian variety B/F' with complex multiplication by E. □

From this lemma and Corollary 4.2.9, we deduce a partial resolution of Question 1.1.9 in this setting; this question clearly warrants further attention, but we will settle in the remainder of this book for a somewhat weaker result (see, eg, the statement of Theorem 4.2.13).

COROLLARY 4.2.11. *Assume* $\dim V_{\mathbb{Q}}$ *is odd-dimensional. Then for some finite extension* L/F'*, the* ℓ*-adic realizations* $\rho_\ell^A \colon \Gamma_L \to \mathrm{GSpin}(V_\ell)$ *of* ρ^A *form a weakly-compatible system of* GSpin*-valued representations, in the sense of Definition* 1.2.3.

PROOF. Let $\dim V_{\mathbb{Q}} = 2n+1$. We must show that for $\ell \neq \ell'$, the semi-simple parts (under Jordan decomposition) of $\rho_\ell^A(\mathrm{fr}_v)$ and $\rho_{\ell'}^A(\mathrm{fr}_v)$ belong to the same GSpin_{2n+1}-conjugacy class (by [**Del74**], these conjugacy classes are defined over $\overline{\mathbb{Q}}$, so we may compare them as classes in $\mathrm{GSpin}_{2n+1}(\overline{\mathbb{Q}})$ under some fixed embeddings of $\overline{\mathbb{Q}}$ into, respectively, $\overline{\mathbb{Q}}_\ell$ and $\overline{\mathbb{Q}}_{\ell'}$). First we observe that a semi-simple element $x \in \mathrm{GSpin}_{2n+1}$ is determined up to conjugacy by its conjugacy classes under the spin representation, standard representation, and Clifford norm: a semi-simple conjugacy class in Spin_{2n+1} is determined by its conjugacy classes under the spin and standard representations (which together imply conjugacy in all the fundamental representations of the simply-connected group Spin_{2n+1}), and further knowing $N(x)$ allows us (twisting by a choice of $N(x)^{-1/2}$) to reduce to the Spin case. From the main theorem of [**Del74**], we know weak compatibility (as $\Gamma_{F'}$-representations) of the representations V_ℓ and $H^1(B_{\overline{F}}, \mathbb{Q}_\ell)$. It remains to show that for some finite, independent-of-ℓ extension L/F', $\{N \circ \rho_\ell^A|_{\Gamma_L}\}_\ell$ is a weakly compatible system of characters.

To see this, note that by Lemma 4.1.25, each (Tate-twisted) character $\chi_\ell = (N \circ \rho_\ell^A)(1)$ is de Rham with Hodge-Tate weights zero, hence has finite-order. It suffices to show that a common finite extension L/F' trivializes χ_ℓ for all ℓ, and for this it suffices to bound the conductors of the finite-order characters χ_ℓ. Recall that the $H^1(B_{\overline{F}}, \mathbb{Q}_\ell)$ are in fact known to form a strongly compatible system, in the sense that for each finite place v of F', there is a Weil-Deligne representation (independent of ℓ) of $W_{F'_v}$ corresponding via the ℓ-adic monodromy theorem to (the Frobenius semi-simplification of) $H^1(B_{\overline{F'_v}}, \mathbb{Q}_\ell)$ (see [**Fon94**]). In particular, we obtain an independent-of-ℓ bound on the conductor of χ_ℓ at places away from ℓ, since as GSpin_{2n+1}-representation the Clifford norm N is a direct summand of $r_{\mathrm{spin}}^{\otimes 2}$ (see page 143 for the full decomposition of $r_{\mathrm{spin}}^{\otimes 2}$). Moreover, χ_ℓ is crystalline for almost all ℓ (wherever A has good reduction, for instance); it is therefore unramified (since the Hodge-Tate weights are zero) for almost all ℓ, so we obtain an independent-of-ℓ bound on the entire conductor of every χ_ℓ. Consequently, (after identification with

[14]Note that the field E can be made explicit if we know the structure of the quadratic lattice $V_{\mathbb{Z}}$. See Example 4.2.14 for a case where $E = \mathbb{Q}$.

complex characters) there are only finitely many possibilities for the characters χ_ℓ, and in particular they all become trivial upon restriction to Γ_L for some finite extension L/F'. $\square$

REMARK 4.2.12.
- Under the Standard Conjectures, the motivic Galois representation $N \circ \rho^A$ gives rise to a weakly compatible system of ℓ-adic representations, so we expect that we can take $L = F'$ in the conclusion of Corollary 4.2.11.
- For $n \geq 3$, it does not suffice to know the conjugacy classes of a semi-simple element $x \in \mathrm{GSpin}_{2n+1}$ under the spin and standard representations–further knowledge of the Clifford norm is essential. For example, let ζ be a primitive 8th root of unity, and, in the coordinates of §2.8, let $x = (\lambda_1 - \lambda_2)(\zeta^6)(\lambda_2 - \lambda_3)(\zeta^2)(2\lambda_3)(\zeta)$. Then x and $x \cdot \lambda_0(\zeta^2)$ have the same eigenvalues in the spin (i.e., highest-weight $-\chi_0 + \frac{\chi_1+\chi_2+\chi_3}{2}$) and standard representations, but they do not have the same Clifford norm, and in particular are not GSpin_7-conjugate.
- More can be said about this question, but in what follows we will not descend this weak compatibility result to a statement for Γ_F-representations, even in situations where we achieve a motivic lift $\tilde{\rho}$ of ρ^V over F itself. In short, we will not know compatibility of the Clifford norms $N \circ \tilde{\rho}_\ell$.

We state a simple case of the main result of this section; for the proof, see §4.2.5. More general versions will be proven in stages, depending on the complexity of the transcendental lattice of $V_{\mathbb{Q}}$.

THEOREM 4.2.13. *Let (X, η) be a polarized variety over a number field $F \subset \mathbb{C}$ for which $Prim^{2k}(X_{\mathbb{C}}, \mathbb{Z})(k)$ satisfies axioms A_k, B_k, and B_k^+. Possibly enlarging F by a quadratic extension, we may assume as above that for all ℓ, $\det V_\ell = 1$, so that the ℓ-adic representation ρ_ℓ^V maps Γ_F to $\mathrm{SO}(V_\ell)$. We make the following hypothesis on the monodromy:*

- *The transcendental lattice $T_{\mathbb{Q}}$ has $\mathrm{End}_{\mathbb{Q}\text{-HS}}(T_{\mathbb{Q}}) = \mathbb{Q}$.*
- $\det(T_\ell) = 1$.[15]

Then there exists a lifting of motivic Galois representations

$$\begin{array}{ccc} & & \mathrm{GSpin}(V_E) \\ & \overset{\tilde{\rho}}{\nearrow} & \downarrow \\ \mathcal{G}_{F,E} & \xrightarrow[\rho]{} & \mathrm{SO}(V_E). \end{array}$$

Moreover there exist:

- *a finite extension $E/\mathbb{Q}$;*
- *an object B of $AV^0_{F,E}$;*
- *an Artin motive M in $\mathcal{M}_{F,E}$;*

[15]Unlike the first hypothesis, this determinant condition is a technicality; it can again be arranged after an (independent of ℓ) quadratic extension. Of course, in the generic case in which $V_{\mathbb{Q}} = T_{\mathbb{Q}}$, it is no additional hypothesis.

such that the composite $r_{\mathrm{spin}} \circ \tilde{\rho}$ is isomorphic in $\mathcal{M}_{F,E}$ to some (explicit) number of copies of $M \otimes_E H^1(B)$. In particular, for each prime λ of E, there are lifts $\tilde{\rho}_\lambda$

$$\begin{array}{ccc} & & \mathrm{GSpin}(V_\ell \otimes E_\lambda) \xrightarrow{\;r_{\mathrm{spin}}\;} \mathrm{GL}_{E_\lambda}(W_{E_\lambda}) \\ & \overset{\tilde{\rho}_\lambda}{\nearrow} & \downarrow \\ \Gamma_F & \longrightarrow & \mathrm{SO}(V_\ell) \subset \mathrm{SO}(V_\ell \otimes E_\lambda) \end{array}$$

such that the composites $\{r_{\mathrm{spin}} \circ \tilde{\rho}_\lambda\}_\lambda$ form a weakly (or even strongly: see the proof of Corollary 4.2.11*) compatible system.*

This theorem is an optimal (up to the $\mathrm{O}(V_\ell) \supset \mathrm{SO}(V_\ell)$ distinction) arithmetic refinement of the (*a priori* highly transcendental) Kuga-Satake construction, showing the precise sense in which it descends to the initial field of definition F.

4.2.3. A simple case. To achieve the refined descent of Theorem 4.2.13, the basic idea is to apply our Galois-theoretic lifting results (which apply over F) to deduce Γ_F-invariance of the abelian variety B that we know to exist over F'; Faltings's isogeny theorem ([**Fal83**]) implies that this invariance is realized by actual isogenies; then we apply a generalization of a technique used by Ribet ([**Rib92**]; our generalization is Proposition 4.2.29) to study elliptic curves over $\overline{\mathbb{Q}}$ that are isogenous to all of their $\Gamma_{\mathbb{Q}}$-conjugates (so-called "$\mathbb{Q}$-curves"). Ribet's technique applies to elliptic curves without complex multiplication, and we will have to keep track of monodromy groups enough to reduce the descent problem to one for an absolutely simple abelian variety. In some cases, a somewhat 'softer' argument than Ribet's works–we give an example in Lemma 4.2.24–but in addition to being satisfyingly explicit, the Ribet method seems to be more robust.

But first we prove Theorem 4.2.13 in the simplest case, when the Hodge structure $V_{\mathbb{Q}}$ is 'generic,' $\dim V_{\mathbb{Q}} = 2n+1$ is odd, and the even Clifford algebra $C^+(V_{\mathbb{Q}})$ is split over $\mathbb{Q}$. Our working definition of 'generic' will be that $V_{\mathbb{Q}}$ contains no trivial $\mathbb{Q}$-Hodge sub-structures (i.e., it is equal to its transcendental lattice), and that $\mathrm{End}_{\mathbb{Q}\text{-HS}}(V_{\mathbb{Q}}) = \mathbb{Q}$.

EXAMPLE 4.2.14. If X/F is a sufficiently general $K3$ surface, then these hypotheses are satisfied. The $K3$-lattice $H^2(X_{\mathbb{C}}, \mathbb{Z})$ is an even unimodular lattice of rank 22 whose signature over $\mathbb{R}$ is (by Hodge theory) $(3, 19)$. The classification of even unimodular lattices implies it is isomorphic (over $\mathbb{Z}$) to $(-E_8)^{\oplus 2} \oplus U^{\oplus 3}$, where E_8 and U are the E_8-lattice and the hyperbolic plane lattice. Over $\mathbb{Q}$, the orthogonal complement of the ample class η is isomorphic to $(-E_8)^{\oplus 2} \oplus U^{\oplus 2} \oplus \langle -q(\eta) \rangle$, where q is the quadratic form and $\langle \alpha \rangle$ denotes the one-dimensional quadratic space with a generator whose square is α. Since $Prim^2(X_{\mathbb{C}}, \mathbb{Q})$ is odd-dimensional, the basic structure theory of Clifford algebras (see Chapter 9, especially Theorem 2.10, of [**Sch85**]) implies that $C^+(Prim^2(X_{\mathbb{C}}, \mathbb{Q})$ is a central simple algebra over $\mathbb{Q}$, whose Brauer class is simply twice the Brauer class of E_8 plus twice the (trivial) Brauer class of U. This is obviously the trivial class, so $C^+(V_{\mathbb{Q}})$ is in this case isomorphic to $M_{2^{10}}(\mathbb{Q})$. More generally, this argument applies to $V_{\mathbb{Q}} = Prim^2(X_{\mathbb{C}}, \mathbb{Q})(1)$ if X is a hyperkähler satisfying:

- $b_2(X) > 3$;
- the dimension of $H^2(X_{\mathbb{C}}, \mathbb{Z})$ is even;
- the number of copies of the E_8-lattice is even.

Now, our 'generic' hypothesis implies that the Hodge group of $V_{\mathbb{Q}}$ is the full $\mathrm{SO}(V_{\mathbb{Q}})$ (more generally, see Zarhin's result, quoted as Proposition 4.2.21, below); it follows without difficulty that the Mumford-Tate group $MT(A_{\mathbb{C}})$ is the full $\mathrm{GSpin}(V_{\mathbb{Q}})$, and $\mathrm{End}^0(A_{\mathbb{C}}) = C^+_{\mathbb{Q}}$. By our second simplifying hypothesis, this Clifford algebra is isomorphic to a matrix algebra $M_{2^n}(\mathbb{Q})$, and, writing $W_{\mathbb{Q}}$ for the spin representation of $\mathrm{GSpin}(V_{\mathbb{Q}})$, we have two isomorphisms of $\mathrm{GSpin}(V_{\mathbb{Q}})$-representations:

$$C^+(V_{\mathbb{Q}}) \cong W_{\mathbb{Q}}^{2^n};$$
$$C^+(V_{\mathbb{Q}})_{\mathrm{ad}} \cong \mathrm{End}(W_{\mathbb{Q}}),$$

where $\mathrm{GSpin}(V_{\mathbb{Q}})$ acts on the first $C^+(V_{\mathbb{Q}})$ by left-multiplication, and on $C^+(V_{\mathbb{Q}})_{\mathrm{ad}}$ by conjugation (i.e., through the natural $\mathrm{SO}(V_{\mathbb{Q}})$-action).

We may certainly enlarge F' to a finite extension over which all endomorphisms of $A_{\mathbb{C}}$ are defined, and the complex multiplication by $C^+_{\mathbb{Q}}$ then gives an isogeny decomposition $A \sim B^{2^n}$, where B/F' is an abelian variety with $\mathrm{End}^0(B) = \mathbb{Q}$. We can take the spin representation $W_{\mathbb{Q}}$ to be equal to $H^1(B_{\mathbb{C}}, \mathbb{Q})$ (and, extending scalars and invoking comparison isomorphisms, we get identifications with other cohomological realizations of B–in particular, the ℓ-adic realization $W_\lambda = W_{\mathbb{Q}} \otimes_{\mathbb{Q}} \mathbb{Q}_\ell$).

Proof of Theorem 4.2.13 when $V_{\mathbb{Q}}$ is generic. We sketch the argument, with some details postponed until later sections–the goal here is to review Ribet's method and outline the argument in a simple case. We will now apply the technique of [**Rib92**] in combination with our abstract Galois-theoretic lifting results. By Corollary 4.1.27, there exists a lift

$$\tilde{\rho}_\ell \colon \Gamma_F \to \mathrm{GSpin}(V_{\overline{\mathbb{Q}}_\ell})$$

of ρ_ℓ^V, and we can normalize this lift so that in the spin representation its labeled Hodge-Tate weights $\mathrm{HT}_\tau(r_{\mathrm{spin}} \circ \tilde{\rho}_\ell)$ are of 'abelian variety'-type, i.e. 2^{n-1} zeroes and 2^{n-1} ones, for each $\tau \colon F \hookrightarrow \overline{\mathbb{Q}}_\ell$ (for details, see Lemma 4.2.23). This normalization determines $\tilde{\rho}_\ell$ up to finite-order twist, and it implies that $\tilde{\rho}_\ell|_{\Gamma_{F'}}$ is a finite-order twist of ρ_ℓ^A, since they both lift $\rho_\ell^V|_{\Gamma_{F'}}$ with the same Hodge-Tate data. We may therefore replace F' by a finite extension and assume

$$\rho_\ell^A = \tilde{\rho}_\ell|_{\Gamma_{F'}}.$$

Since $\tilde{\rho}_\ell$ begins life over F, we see that ρ_ℓ^A is Γ_F-conjugation invariant. The composition of ρ_ℓ^A with the Clifford representation is 2^n copies of the ℓ-adic representation ρ_ℓ^B associated to B, so ρ_ℓ^B is also Γ_F-invariant. By Faltings's theorem, for each $\sigma \in \Gamma_F$, there exists an isogeny $\mu_\sigma \colon {}^\sigma B \to B$; we can and do arrange that μ_σ is defined first for a (finite) system of representatives σ_i in Γ_F for $\mathrm{Gal}(F'/F)$, and then defined in general by $\mu_{\sigma_i h} = \mu_{\sigma_i}$ for all $h \in \Gamma_{F'}$. The collection of μ_σ yields an obstruction class

$$[c_B] \in H^2(\Gamma_F, \mathrm{End}^0(B)^\times) = H^2(\Gamma_F, \mathbb{Q}^\times),[16]$$

given by

$$c_B(\sigma, \tau) = \mu_\sigma \circ {}^\sigma\mu_\tau \circ \mu_{\sigma\tau}^{-1}.$$

[16] The local constancy of the isogenies μ_σ implies this is a continuous cohomology class, with $\mathbb{Q}^\times$ equipped with the discrete topology.

That is, c_B measures the failure of the diagram

$$\begin{array}{ccc} {}^{\sigma\tau}B & \xrightarrow{\mu_{\sigma\tau}} & B \\ {\scriptstyle {}^{\sigma}\mu_{\tau}}\downarrow & \nearrow{\scriptstyle \mu_{\sigma}} & \\ {}^{\sigma}B & & \end{array}$$

to commute. Now, the class c_B may be non-trivial, and the abelian variety B may not descend (up to isogeny) to F: in fact, that triviality of this class is equivalent to isogeny-descent is Theorem 8.2 of [**Rib92**]. Nevertheless, Tate's vanishing result $H^2(\Gamma_F, \overline{\mathbb{Q}}^\times) = 0$ tells us that there is a continous 1-cochain $\alpha\colon \Gamma_F \to \overline{\mathbb{Q}}^\times$ whose coboundary equals c_B, i.e. $c_B(\sigma,\tau) = \alpha(\sigma\tau)\alpha(\tau)^{-1}\alpha(\sigma)^{-1}$. By continuity, α is locally constant with respect to $\Gamma_{F''}$ for some finite F''/F', and it takes values in some finite extension $\mathbb{Q}(\alpha)$ of $\mathbb{Q}$. We now consider the restriction of scalars abelian variety $C := \mathrm{Res}_{F''/F}(B)$ (here B is really the base-change to F'', but we omit this not to clutter the notation). The endomorphism algebra $\mathcal{R} = \mathrm{End}^0(C)$ of the abelian variety C is isomorphic as $\mathbb{Q}$-vector space to

$$\mathrm{Hom}_{F''}\left(\prod_{\sigma\in\mathrm{Gal}(F''/F)} {}^{\sigma}B, B\right) = \bigoplus_{\sigma} \mathbb{Q}\mu_\sigma,$$

where again we use the fact that $\mathrm{End}^0(B_{\overline{F}}) = \mathbb{Q}$. Write λ_σ for the element of $\mathcal{R}$ corresponding to μ_σ under this isomorphism.

LEMMA 4.2.15 (Lemma 6.4 of [**Rib92**]). *The algebra structure of $\mathcal{R}$ is given by*

$$\lambda_\sigma\lambda_\tau = c_B(\sigma,\tau)\lambda_{\sigma\tau},$$

so there is a $\mathbb{Q}$-algebra homomorphism $\alpha\colon \mathcal{R} \to \mathbb{Q}(\alpha)$ given by the $\mathbb{Q}$-linear extension of $\lambda_\sigma \mapsto \alpha(\sigma)$.

Since the isogeny category of abelian varieties over F is a semi-simple abelian category, we can form the object

$$M = \mathrm{Res}_{F''/F}(B) \otimes_{\mathcal{R},\alpha} \mathbb{Q}(\alpha).$$

We regard M as an object of $AV^0_{F,\mathbb{Q}(\alpha)}$, with $\mathbb{Q}(\alpha)$-rank (in the obvious sense) equal to the $\mathbb{Q}$-rank of B. Moreover, for any place $\lambda|\ell$ of $\mathbb{Q}(\alpha)$, the λ-adic realization

$$\rho_\lambda^M : \Gamma_F \to \mathrm{GL}_{\mathbb{Q}(\alpha)_\lambda}(M_\lambda)$$

has projectivization isomorphic to the canonical projective descent to Γ_F of the Γ_F-invariant, irreducible representation of $\Gamma_{F''}$ on $H^1(B_{\overline{F}}, \mathbb{Q}(\alpha)_\lambda)$.[17] So, $\rho_\lambda^M|_{\Gamma_{F''}}$ and $\rho_\ell^B|_{\Gamma_{F''}} \cong r_{\mathrm{spin}} \circ \tilde{\rho}_\ell|_{\Gamma_{F''}}$ are isomorphic up to twist, hence $r_{\mathrm{spin}} \circ \tilde{\rho}_\ell$ and ρ_λ^M are twist-equivalent as Γ_F-representations. The representation $r_{\mathrm{spin}}\colon \mathrm{GSpin}(V_{\mathbb{Q}}) \to \mathrm{GL}(W_{\mathbb{Q}})$ is the identity on the center, so after identifying ρ_λ^M to a representation on $W_\lambda \otimes \mathbb{Q}(\alpha)_\lambda$, we see that it factors through $\mathrm{GSpin}(V_\ell \otimes \mathbb{Q}(\alpha)_\lambda)$ as a lift of $\rho_\ell^V \otimes \mathbb{Q}(\alpha)_\lambda$. The required motivic lift is the representation of $\mathcal{G}_{F,\mathbb{Q}(\alpha)}$ corresponding to $H^1(M)$ (for more details on how to check this carefully, see Corollary 4.2.20); and the various ρ_λ^M form a compatible system because they are formed from Tate modules of abelian varieties. □

[17]Under the homomorphism $\mathcal{R} \to \mathrm{End}_{\Gamma_{F''}}(\oplus^\sigma H^1(B, \mathbb{Q}_\ell)$, the λ_σ permute the factors in the direct sum; the projection via $\mathcal{R} \xrightarrow{\alpha} \mathbb{Q}(\alpha)$ collapses all the factors to a single copy with scalars extended to $\mathbb{Q}(\alpha)$, i.e. to $H^1(B, \mathbb{Q}_\ell) \otimes_{\mathbb{Q}} \mathbb{Q}(\alpha)$. This implies the claim about projective descents.

4.2.4. Arithmetic descent: Preliminary reduction. Now we proceed to a more general argument, making first some preliminary reductions to the analogous lifting problem for the transcendental lattice. We must invoke André's work on the Tate conjecture for X.

THEOREM 4.2.16 (see Theorem 1.6.1 of [**And96a**]). *Let (X, η) be a polarized variety over a number field*[18] *F satisfying A_k, B_k, and B_k^+. Then:*

- *$Prim^{2k}(X_{\overline{F}}, \mathbb{Q}_\ell)(k)$ is a semi-simple Γ_F-representation*
- *the Galois invariants $Prim^{2k}(X_{\overline{F}}, \mathbb{Q}_\ell)(k)^{\Gamma_F}$ are all $\mathbb{Q}_\ell$-linear combinations of algebraic cycles;*

The Tate conjecture for $H^{2k}(X)$ then implies:

LEMMA 4.2.17. *There is an Artin motive Alg over F whose Betti realization is the subspace of $V_{\mathbb{Q}}$ spanned by algebraic cycle classes, and whose ℓ-adic representation is the (Artin) Γ_F-representation on $V_\ell^{\Gamma_{F'}}$ for any F'/F large enough (and Galois) that all of these cycle classes are defined over F'. The transcendental lattice T likewise descends to an object of $\mathcal{M}_F$. In particular, there is an orthogonal decomposition*

$$V_\ell = Alg_\ell \oplus T_\ell$$

of Γ_F-representations (not merely $\Gamma_{F'}$-representations).

PROOF. Giving an Artin motive over F is equivalent to giving a representation of Γ_F on a (finite-dimensional) $\mathbb{Q}$-vector space. In our case, the space ($\mathbb{Q}$-span) of cycles for homological equivalence

$$Z^k_{\mathrm{hom}}(X_{F'}) \hookrightarrow H^{2k}(X_{\overline{F}}, \mathbb{Q}_\ell)(k),$$

or rather its intersection with V_ℓ, does the trick. Since the object $Prim^{2k}(X)(k)$ of $\mathcal{M}_F$ is polarized, we can define T in $\mathcal{M}_F$ as the orthogonal complement of Alg. □

We will enlarge F' as in the Lemma, so that all algebraic classes in $Prim^{2k}(X_{\mathbb{C}}, \mathbb{Q})(k)$ are already defined over F', and so that the motivic group $\mathcal{G}^T_{F'}$ is connected (see Lemma 4.1.17). Note that $T_{\mathbb{Q}}$ is an orthogonal Hodge structure of type $(1, -1), (0, 0), (-1, 1)$, with $h^{1,-1} = 1$, so the Kuga-Satake construction applies to it as well (see Variant 4.1.5 of [**And96a**]). Since the Kuga-Satake variety associated to $V_{\mathbb{Q}}$ is simply an isogeny power of that associated to $T_{\mathbb{Q}}$, the latter, to be denoted $A(T)$, also descends to some finite extension F'/F.

We introduce a little more notation. After extending scalars to a sufficiently large field E (omitted from the notation unless we want to emphasize it) we denote by $(r_{\mathrm{spin},V}, W_V)$, $(r_{\mathrm{spin,Alg}}, W_{\mathrm{Alg}})$, and $(r_{\mathrm{spin},T}, W_T)$ the spin representations of these three spin groups; as before, by 'the' spin representation in the D_n case we will mean the direct sum of the two half-spin representations.

LEMMA 4.2.18. *Suppose that ρ_ℓ^V factors through $\mathrm{SO}(Alg_\ell) \times \mathrm{SO}(T_\ell) \hookrightarrow \mathrm{SO}(V_\ell)$ (at worst, ensuring this requires making a quadratic extension of F). If we have found lifts $\tilde{\rho}_\ell^{\mathrm{Alg}}$ and $\tilde{\rho}_\ell^T$ to $\mathrm{GSpin}(Alg_\ell \otimes \overline{\mathbb{Q}}_\ell)$ and $\mathrm{GSpin}(T_\ell \otimes \overline{\mathbb{Q}}_\ell)$ such that $r_{\mathrm{spin,Alg}} \circ \tilde{\rho}_\ell^{\mathrm{Alg}}$ and $r_{\mathrm{spin},T} \circ \tilde{\rho}_\ell^T$ are motivic, then ρ_ℓ^V has a lift $\tilde{\rho}_\ell \colon \Gamma_F \to \mathrm{GSpin}(V_\ell \otimes \overline{\mathbb{Q}}_\ell)$ such that $r_{\mathrm{spin},V} \circ \tilde{\rho}_\ell$ is also motivic. If the individual lifts $r_{\mathrm{spin,Alg}} \circ \tilde{\rho}_\ell^{\mathrm{Alg}}$ and $r_{\mathrm{spin},T} \circ \tilde{\rho}_\ell^T$*

[18] Faltings's theorem works for finitely-generated extensions of $\mathbb{Q}$, so this does as well.

belong to compatible systems of ℓ-adic representations, then the same holds for $r_{\mathrm{spin},V} \circ \tilde{\rho}_\ell$.[19]

PROOF. The isomorphism of *graded* algebras (for the graded tensor product $\hat{\otimes}$)

$$C(Alg_{\mathbb{Q}})\hat{\otimes}C(T_{\mathbb{Q}}) \xrightarrow{\sim} C(V_{\mathbb{Q}})$$

induces an inclusion $C^+(Alg_{\mathbb{Q}})\otimes C^+(T_{\mathbb{Q}}) \hookrightarrow C^+(V_{\mathbb{Q}})$, and then a map (not injective)

$$C^+(Alg_{\mathbb{Q}})^\times \times C^+(T_{\mathbb{Q}})^\times \to C^+(V_{\mathbb{Q}})^\times,$$

which in turn induces a commutative diagram

$$\begin{array}{ccc} \mathrm{GSpin}(Alg_{\mathbb{Q}}) \times \mathrm{GSpin}(T_{\mathbb{Q}}) & \longrightarrow & \mathrm{GSpin}(V_{\mathbb{Q}}) \\ \downarrow & & \downarrow \\ \mathrm{SO}(Alg_{\mathbb{Q}}) \times \mathrm{SO}(T_{\mathbb{Q}}) & \longrightarrow & \mathrm{SO}(V_{\mathbb{Q}}). \end{array}$$

As long as $\rho_\ell^V \colon \Gamma_F \to \mathrm{SO}(V_\ell)$ factors through $\mathrm{SO}(Alg_\ell)\times\mathrm{SO}(T_\ell)$, this shows that we can lift Alg_ℓ and T_ℓ in order to lift V_ℓ. We next want to understand the restriction to $\mathrm{GSpin}(Alg_E)\times\mathrm{GSpin}(T_E)$ of the spin representation of $\mathrm{GSpin}(V_E)$; Just for this argument, we will ignore the similitude factor, i.e. work with weights of Spin rather than GSpin. We can write bases of the character lattices of $\mathrm{SO}(Alg_E)$, $\mathrm{SO}(T_E)$, and $\mathrm{SO}(V_E)$ as, respectively, $\chi_1, \dots, \chi_a$, $\chi_{a+1}, \dots, \chi_{a+t}$, and

$$\chi_1, \dots, \chi_{a+t} \quad \text{if either } \dim(T) \text{ or } \dim(Alg) \text{ is even;}$$

$$\chi_1, \dots, \chi_{a+t+1} \quad \text{if both } \dim(T) \text{ and } \dim(Alg) \text{ are odd.}$$

The set of weights of the spin representation of $\mathfrak{so}(Alg_E)$ (and similarly for the other cases) is then all 2^a weights of the form

$$\sum_{i=1}^{a} \frac{\pm\chi_i}{2}.$$

In the case where at least one of $\dim(T)$ and $\dim(Alg)$ is even, we see that the weights of $W_{\mathrm{Alg}}\boxtimes W_T$ are precisely those of $W_V|_{\mathrm{GSpin}(Alg_E)\times\mathrm{GSpin}(T_E)}$, and therefore $W_V \cong W_{\mathrm{Alg}}\boxtimes W_T$ as $\mathrm{GSpin}(Alg_E)\times\mathrm{GSpin}(T_E)$-representations. When both $\dim(T)$ and $\dim(Alg)$ are odd, weight-counting gives $W_V \cong (W_{\mathrm{Alg}} \boxtimes W_T)^{\oplus 2}$.

Thus, if we have found lifts $\tilde{\rho}_\ell^{\mathrm{Alg}}$ and $\tilde{\rho}_\ell^T$ (to $\mathrm{GSpin}(Alg_\ell \otimes \overline{\mathbb{Q}}_\ell)$ and $\mathrm{GSpin}(T_\ell \otimes \overline{\mathbb{Q}}_\ell)$) such that $r_{\mathrm{spin,Alg}} \circ \tilde{\rho}_\ell^{\mathrm{Alg}}$ and $r_{\mathrm{spin},T} \circ \tilde{\rho}_\ell^T$ are motivic (respectively, belong to compatible systems of ℓ-adic representations), then the resulting lift to $\mathrm{GSpin}(V_\ell \otimes \overline{\mathbb{Q}}_\ell)$, in its spin representation, is a direct sum of tensor products of motivic Galois representations (respectively, Galois representations belonging to compatible systems), hence is motivic. □

COROLLARY 4.2.19. *If* $\det T_\ell = 1$ *as* Γ_F*-representation, and if we can find a lift* $\tilde{\rho}_\ell^T$ *of* ρ_ℓ^T *such that* $r_{\mathrm{spin},T} \circ \tilde{\rho}_\ell^T$ *is motivic (respectively, belongs to a compatible system), then we can do the same for* ρ_ℓ^V.

PROOF. Since Alg_ℓ is an Artin representation, Tate's vanishing result allows us to lift to an Artin representation $\tilde{\rho}_\ell^{\mathrm{Alg}} \colon \Gamma_F \to \mathrm{GSpin}(Alg_\ell \otimes \overline{\mathbb{Q}}_\ell)$. The corollary follows. □

[19] Recall that objects of $\mathcal{M}_F$ are not in general known to give rise to compatible systems.

Corollary 4.2.20. *As in Corollary 4.2.19, assume that* $\det T_\ell = 1$, *and that for some number field* E, *and place* $\lambda|\ell$ *of* E, *we can find a lift* $\tilde{\rho}_\lambda^T \colon \Gamma_F \to \mathrm{GSpin}(T_\ell \otimes_{\mathbb{Q}_\ell} E_\lambda)$ *such that* $r_{\mathrm{spin},T} \circ \tilde{\rho}_\lambda^T$ *is the* λ*-adic realization of an object* M *of* $\mathcal{M}_{F,E}$ *whose base-change to some* F'/F *is one of the spin direct factors of the Kuga-Satake motive (with scalars extended to* E*) associated to* $T_\mathbb{Q}$ *(see Lemma 4.2.10). Then possibly enlarging* E *to a finite extension* E', *there is a lifting of representations of the motivic Galois group* $\mathcal{G}_{F,E'}$:

$$\begin{array}{ccc} & & \mathrm{GSpin}(V_\mathbb{Q} \otimes E') \\ & \nearrow & \downarrow \\ \mathcal{G}_{F,E'} & \xrightarrow[\rho^V]{} & \mathrm{SO}(V_\mathbb{Q} \otimes E'). \end{array}$$

Proof. The Artin representation ρ_ℓ^{Alg} is definable over $\mathbb{Q}$, and lifts to $\mathrm{GSpin}(Alg_{E_1})$ after making some finite extension $E_1/\mathbb{Q}$. The conclusion of the corollary will hold with E' equal to the composite E_1E. We check it using the same principle as in the proof of Proposition 4.1.30: that is, we check 'geometrically' (for the restriction to $\mathcal{G}_{\overline{F},E'}$) and for the λ-adic realization. By Corollary 4.2.9, the motivic representation $\rho^{A(T)}$ of $\mathcal{G}_{\overline{F}}$, and by extension of $\mathcal{G}_{\overline{F},E'}$, factors through $\mathrm{GSpin}(T_{E'})$ and lifts $\rho^T \colon \mathcal{G}_{\overline{F},E'} \to \mathrm{SO}(T_{E'})$. Since $H^1(A(T)) \otimes_\mathbb{Q} E'$ is just some number of copies[20] of M, the same is true of ρ^M. By assumption, the λ-adic realization ρ_λ^M is just $r_{\mathrm{spin}} \circ \tilde{\rho}_\lambda^T$, so this also factors through $\mathrm{GSpin}(T_\lambda)$, lifting ρ_λ^T. Using the section $s_\lambda \colon \Gamma_F \to \mathcal{G}_{F,E'}(E'_\lambda)$, so that $\mathcal{G}_{F,E'}(E'_\lambda) = s_\lambda(\Gamma_F) \cdot \mathcal{G}_{\overline{F},E'}(E'_\lambda)$, we conclude as in Proposition 4.1.30 that ρ^M lifts $\rho^T \otimes E'$.

Combining ρ^M with the lift of ρ_ℓ^{Alg}, as in Lemma 4.2.18, we similarly find a lift to $\mathrm{GSpin}(V_\mathbb{Q} \otimes_\mathbb{Q} E')$ of our given $\rho^V \colon \mathcal{G}_{F,E'} \to \mathrm{SO}(V_\mathbb{Q} \otimes_\mathbb{Q} E')$. ☐

4.2.5. Arithmetic descent: The generic case. To summarize, we have reduced the problem of finding motivic lifts of the motive (over F) $Prim^{2k}(X)(k)$ to the corresponding problem for the transcendental lattice T. In this section we treat the 'generic' case in which the Hodge structure $T_\mathbb{Q}$ has trivial endomorphism algebra, using a variant of Ribet's method (which will return in §4.2.6) We isolate this case both to demonstrate a slightly different argument, and because the non-generic cases will require even deeper input, André's proof of the Mumford-Tate conjecture in this context (see Theorem 4.2.27). The starting-point of the analysis of the motive T is Zarhin's calculation in [**Zar83**] of the Mumford-Tate group:

Proposition 4.2.21 (Zarhin). *Let* $T_\mathbb{Q}$ *be a* $\mathbb{Q}$*-Hodge structure with orthogonal polarization and Hodge numbers* $h^{1,-1} = 1$, $h^{0.0} > 0$, *and* $h^{p,q} = 0$ *if* $|p-q| > 2$. *Moreover assume that* $T_\mathbb{Q}$ *contains no copies of the trivial* $\mathbb{Q}$*-Hodge structure. Then* $E_T = \mathrm{End}_{\mathbb{Q}\text{-HS}}(T_\mathbb{Q})$ *is a totally real or CM field, and* $T_\mathbb{Q}$ *is a simple* $MT(T_\mathbb{Q})$*-module. There is a non-degenerate* E_T*-hermitian*[21] *pairing*

$$\langle \cdot, \cdot \rangle \colon T_\mathbb{Q} \times T_\mathbb{Q} \to E_T$$

such that

$$MT(T_\mathbb{Q}) = \mathrm{Aut}(T_\mathbb{Q}, \langle \cdot, \cdot \rangle_E) \subset \mathrm{SO}(T_\mathbb{Q}).$$

[20] 2^t for $\dim T$ odd; 2^{t-1} for $\dim T$ even.

[21] In the case of totally real E_T, this means symmetric.

For the rest of this section, we assume that $\mathrm{End}_{\mathbb{Q}\text{-HS}}(T_{\mathbb{Q}}) = \mathbb{Q}$; the Mumford-Tate group $MT(T_{\mathbb{Q}})$ is then the full $\mathrm{SO}(T_{\mathbb{Q}})$.

LEMMA 4.2.22. *Assume* $\mathrm{End}_{\mathbb{Q}\text{-HS}}(T_{\mathbb{Q}}) = \mathbb{Q}$. *Then the Mumford-Tate group, and therefore the motivic Galois group, of the Kuga-Satake variety* $A(T)$ *is equal to* $\mathrm{GSpin}(T_{\mathbb{Q}})$. *Consequently,* $\mathrm{End}^0(A(T)) = C^+_{\mathbb{Q}} \cong C^+(T_{\mathbb{Q}})$.

PROOF. Easy. □

We now choose a number field $E/\mathbb{Q}$ splitting $C^+(T_{\mathbb{Q}})$, and consider the decomposition in $AV^0_{F',E}$

$$A(T) \sim B(T)^{2^t} \quad \text{if } \dim(T_{\mathbb{Q}}) = 2t+1;$$

$$A(T) \sim B(T)^{2^{t-1}} \sim (B_+(T) \times B_-(T))^{2^{t-1}} \quad \text{if } \dim(T_{\mathbb{Q}}) = 2t,$$

as in Lemma 4.2.10. We saw in Corollary 4.2.9 that $\rho^{A(T)}$ factors through $\mathrm{GSpin}(T_{\mathbb{Q}})$ and lifts ρ^T; viewing $\rho^{A(T)}$ in $\mathrm{GSpin}(T_{\mathbb{Q}})$, we then have the relation $r_{\mathrm{spin}} \circ (\rho^{A(T)} \otimes E) = \rho^{B(T)}$, taking the Betti realization of $B(T)$ to be our model for the spin representation. We let B_0 equal $B(T)$ in the odd case and $B_+(T)$ in the even case. Similarly, we let r_0 denote r_{spin} or one of the half-spin representations (which we may assume corresponds to $B_+(T)$). We also for convenience fix an embedding $E \hookrightarrow \overline{\mathbb{Q}}_\ell$.

LEMMA 4.2.23. *(Without any assumption on* $\mathrm{End}_{\mathbb{Q}\text{-HS}}(T_{\mathbb{Q}})$*) There exist a lift*

$$\tilde{\rho}^T_\ell \colon \Gamma_F \to \mathrm{GSpin}(T_\ell \otimes \overline{\mathbb{Q}}_\ell)$$

of ρ^T_ℓ *and a finite extension* F''/F' *such that*

$$r_{\mathrm{spin}} \circ \tilde{\rho}^T_\ell|_{\Gamma_{F''}} \cong H^1(B(T)_{\overline{F}}, \mathbb{Q}_\ell) \otimes_E \overline{\mathbb{Q}}_\ell|_{\Gamma_{F''}}.$$

PROOF. As in the arguments of §3.2, we first choose a lift $\Gamma_F \to \mathrm{GSpin}(T_\ell \otimes \overline{\mathbb{Q}}_\ell)$ with finite-order Clifford norm. In the root datum notation of §2.8, the Hodge-Tate cocharacters μ_τ of ρ^T_ℓ, for all $\tau \colon F \hookrightarrow \overline{\mathbb{Q}}_\ell$, are (conjugate to) λ_1, and the finite-order Clifford norm lifts have Sen operators (up to conjugacy) corresponding to the elements $\tilde{\mu}_\tau = \lambda_1$ of the co-character lattice tensored with $\mathbb{Q}$. We can modify this initial lift by twisting by $\lambda_0 \circ \omega'$, where $\omega' \colon \Gamma_F \to \overline{\mathbb{Q}}_\ell^\times$ is a Galois character whose square differs from the inverse ω^{-1} of the cyclotomic character by a finite-order twist (such ω' exists for any F). This gives a new lift $\tilde{\rho}^T_\ell$ whose composition with the Clifford norm (which, recall, is $2\chi_0$ in our notation) is ω^{-1}, up to a finite-order twist, and whose τ-labeled Hodge-Tate weights in the spin representation are 1 and 0 with equal multiplicity (compare the argument of §4.1.1). Then $r_{\mathrm{spin}} \circ \tilde{\rho}^T_\ell|_{\Gamma_{F'}}$ differs from $H^1(B(T)_{\overline{F}}, \mathbb{Q}_\ell) \otimes_E \overline{\mathbb{Q}}_\ell$ by a finite-order twist, hence they are isomorphic after some additional finite base-change F''/F'. □

LEMMA 4.2.24. *There is a factor* M *of* $\mathrm{Res}_{F''/F}(B_0)$ *(viewed as an object of* $AV^0_{F,E}$*), having endomorphisms by a finite extension* E'/E*, and an embedding* $E' \hookrightarrow \overline{\mathbb{Q}}_\ell$ *extending our fixed* $E \hookrightarrow \overline{\mathbb{Q}}_\ell$ *such that the associated* ℓ*-adic realization* $M_\ell := H^1(M_{\overline{F}}, \mathbb{Q}_\ell) \otimes_{E'} \overline{\mathbb{Q}}_\ell$ *is isomorphic as* $\overline{\mathbb{Q}}_\ell[\Gamma_F]$*-representation to* $r_0 \circ \tilde{\rho}^T_\ell$.

PROOF. By the (E-linear) Tate conjecture,

$$\mathrm{End}_{AV^0_{F,E}}\left(\mathrm{Res}_{F''/F}(B_0)\right) \otimes_E \overline{\mathbb{Q}}_\ell \cong \mathrm{End}_{\overline{\mathbb{Q}}_\ell[\Gamma_F]}\left(\mathrm{Ind}^F_{F''}(H^1(B_{0,\overline{F}}, \mathbb{Q}_\ell) \otimes_E \overline{\mathbb{Q}}_\ell)\right).$$

The Galois representation being induced is $r_0 \circ \tilde{\rho}_\ell^T$, and inside this endomorphism ring we can consider

$$\mathrm{Hom}_{\overline{\mathbb{Q}}_\ell[\Gamma_F]}\left(r_0 \circ \tilde{\rho}_\ell^T, \mathrm{Ind}_{F''}^F(r_0 \circ \tilde{\rho}_\ell^T)\right),$$

which by Frobenius reciprocity is just

$$\mathrm{End}_{\overline{\mathbb{Q}}_\ell[\Gamma_{F''}]}\left(r_0 \circ \tilde{\rho}_\ell^T\right) = \overline{\mathbb{Q}}_\ell,$$

since $\mathrm{End}_{AV^0_{F'',E}}(B_0) = E$. In other words, there is a unique $\overline{\mathbb{Q}}_\ell$-line in $\mathrm{End}_{AV^0_{F,E}}\left(\mathrm{Res}_{F''/F}(B_0)\right) \otimes_E \overline{\mathbb{Q}}_\ell$ consisting of projectors onto the $r_0 \circ \tilde{\rho}_\ell^T$-isotypic piece of the ℓ-adic representation. Decomposing the semi-simple E-algebra $\mathrm{End}_{AV^0_{F,E}}\left(\mathrm{Res}_{F''/F}(B_0)\right)$ into simple factors, we see that this line lives in a unique simple component (tensored with $\overline{\mathbb{Q}}_\ell$), which itself must be just a finite field extension E' of E (else the $r_0 \circ \tilde{\rho}_\ell^T$-isotypic piece would have multiplicity greater than 1); it then corresponds to exactly one of the simple factors of $E' \otimes_E \overline{\mathbb{Q}}_\ell$, i.e. a particular embedding $E' \hookrightarrow \overline{\mathbb{Q}}_\ell$. We can therefore take M to be the abelian variety corresponding to this factor E'; M has complex multiplication by E', and via this specified embedding $E' \hookrightarrow \overline{\mathbb{Q}}_\ell$, the ℓ-adic realization M_ℓ is isomorphic to $r_0 \circ \tilde{\rho}_\ell^T$. □

REMARK 4.2.25. This is at its core the same proof as given in §4.2.3; the latter proof is probably more transparent, but this one is somewhat 'softer.' I don't think it translates as well to the more general context of §4.2.6, however.

COROLLARY 4.2.26. *Theorem* 4.2.13 *holds for the motive* $Prim^{2k}(X)(k)$ *over* F.

PROOF. When $\dim(T_{\mathbb{Q}})$ is odd, we are done, by the previous lemma and Corollaries 4.2.19 and 4.2.20. When $\dim(T_{\mathbb{Q}})$ is even, we take the output M in $AV^0_{F,E'}$ of the previous lemma, view it in $\mathcal{M}_{F,E'}$, and form the twisted dual $M^\vee(-1)$. Here there are two cases: if $\dim T_{\mathbb{Q}}$ is not divisible by four, this object corresponds to the composition of the other half-spin representation with $\tilde{\rho}_\ell^T$: the two half-spin representations of GSpin_{2n} have highest weights $-\chi_0 + \frac{1}{2}(\sum_{i=1}^{n-1} \chi_i + \chi_n)$ (for r_0) and $-\chi_0 + \frac{1}{2}(\sum_{i=1}^{n-1} \chi_i - \chi_n)$ (for the other half-spin representation), so the lowest weight of $r_0^\vee \otimes (-2\chi_0)$ is $-\chi_0 - \frac{1}{2}(\sum_{i=1}^{n} \chi_i)$, which, when n is even, is visibly the lowest weight of the other half-spin representation. In that case, we can apply the earlier corollaries to $M \oplus M^\vee(-1)$. If $\dim T_{\mathbb{Q}}$ is divisible by four (so the half-spin representations are self-dual), apply Lemma 4.2.24 to the composition of $\tilde{\rho}_\ell^T$ with the *other* half-spin representation as well; so instead of a single motive M, we now have two motives M_+ and M_- (enlarge the coefficient fields of M_+ and M_-, viewed as subfields of $\overline{\mathbb{Q}}_\ell$ via the respective embeddings produced by Lemma 4.2.24, to some common over-field) such that the ℓ-adic realization $(M_+ \oplus M_-)_\ell$ is isomorphic to $r_{\mathrm{spin}} \circ \tilde{\rho}_\ell^T$. Then as before, we may apply Corollary 4.2.20. □

4.2.6. Non-generic cases: $\dim(T_{\mathbb{Q}})$ **odd.** To study the non-generic case in which $\mathrm{End}_{\mathbb{Q}\text{-HS}}(T_{\mathbb{Q}})$ is larger than $\mathbb{Q}$, we have to understand the ℓ-adic algebraic monodromy groups of the representations ρ_ℓ^T, i.e. the ℓ-adic analogue of Zarhin's result. André has proven ([**And96a**, Theorem 1.6.1]) the Mumford-Tate conjecture in this context; a gap in the argument of [**And96a**, §7.4] is completed in a preprint

([**Moo15**]) of Ben Moonen, in the course of generalizing some of André's work. Here is the precise result:

THEOREM 4.2.27 (André). *Let (X,η) be a polarized variety over a finitely-generated extension F of $\mathbb{Q}$ satisfying A_k, B_k, and B_k^+. Then the inclusion $\overline{\rho_\ell(\Gamma_F)}^{\mathrm{Zar}} \hookrightarrow \mathcal{G}_F^V \otimes_{\mathbb{Q}} \mathbb{Q}_\ell$ of the algebraic monodromy group into the ℓ-adic motivic Galois group of $V = Prim^{2k}(X)(k)$ is an isomorphism on connected components of the identity.*

Recall that over the field F', we may assume the groups $\overline{\rho_\ell(\Gamma_{F'})}^{\mathrm{Zar}}$ and $\mathcal{G}_{F'}^V$ are connected, and therefore isomorphic. Combining Theorem 6.5.1 of [**And96a**] with André's result that Hodge cycles on abelian varieties are motivated, and with Zarhin's description ([**Zar83**]) of the Mumford-Tate group of the transcendental lattice $T^{2k}(X_{\mathbb{C}},\mathbb{Q})(k)$, we obtain (see Corollary 1.5.2 of [**And96a**]):

COROLLARY 4.2.28. *The semisimple $\mathbb{Q}$-algebra $E_T := \mathrm{End}_{\mathcal{G}_{F'}^T}(T)$ is a totally real or CM field, and there is a natural E_T-hermitian pairing $\langle\cdot,\cdot\rangle_{E_T} : T\times T \to E_T$. The motivic group (which equals the Mumford-Tate group, and equals, after $\otimes_{\mathbb{Q}}\mathbb{Q}_\ell$, the ℓ-adic algebraic monodromy group) $\mathcal{G}_{F'}^T$ is then isomorphic to the full orthogonal (E_T totally real) or unitary (E_T CM) group*

$$\mathrm{Aut}(T,\langle\cdot,\cdot\rangle_{E_T}) \hookrightarrow \mathrm{SO}(T_{\mathbb{Q}}).$$

Before continuing, we formulate a variant of Ribet's method with coefficients:

PROPOSITION 4.2.29. *Let F'/F be an extension of number fields, and let $E/\mathbb{Q}$ be a finite extension. Suppose we are given an object B of $AV^0_{F',E}$ such that for some embedding $E \hookrightarrow \overline{\mathbb{Q}}_\ell$, the associated ℓ-adic realization $B_\ell = H^1(B_{\overline{F}},\mathbb{Q}_\ell)\otimes_E \overline{\mathbb{Q}}_\ell$ satisfies the invariance condition ${}^\sigma B_\ell \cong B_\ell$ for all $\sigma \in \Gamma_F$. Further assume that $\mathrm{End}_{AV^0_{\overline{F},E}}(B_{\overline{F}}) = E$ (in particular, $\mathrm{End}_{AV^0_{F',E}}(B) = E$). Then there exist a finite extension E'/E, an extension $E' \hookrightarrow \overline{\mathbb{Q}}_\ell$ of the embedding $E \hookrightarrow \overline{\mathbb{Q}}_\ell$, a number field F''/F', and an object M of $AV^0_{F,E'}$ such that*

$$\left(H^1(M_{\overline{F}},\mathbb{Q}_\ell)\otimes_{E'}\overline{\mathbb{Q}}_\ell\right)|_{\Gamma_{F''}} \cong B_\ell|_{\Gamma_{F''}}.$$

That is, B, up to twist, has a motivic descent to F.

PROOF. This is proven as on page 130, using the E-linear variant of Faltings' theorem:

$$\mathrm{Hom}_{AV^0_{F',E}}({}^\sigma B, B)\otimes_E \overline{\mathbb{Q}}_\ell \xrightarrow{\sim} \mathrm{Hom}_{\overline{\mathbb{Q}}_\ell[\Gamma_{F'}]}\left(H^1({}^\sigma B_{\overline{F}},\mathbb{Q}_\ell)\otimes_E\overline{\mathbb{Q}}_\ell, H^1(B_{\overline{F}},\mathbb{Q}_\ell)\otimes_E\overline{\mathbb{Q}}_\ell\right).$$

In the notation of the earlier proof, $M = \mathrm{Res}_{F''/F}(B)\otimes_{\mathcal{R},\alpha}\mathbb{Q}(\alpha)$ is the required motivic descent, to an isogeny abelian variety over F with $\mathbb{Q}(\alpha)$-multiplication (so in the statement of the proposition, $E' \hookrightarrow \overline{\mathbb{Q}}_\ell$ is $\mathbb{Q}(\alpha) \subset \overline{\mathbb{Q}} \hookrightarrow \overline{\mathbb{Q}}_\ell$, extending the initial $E \hookrightarrow \overline{\mathbb{Q}}_\ell$). □

We now assume that $\dim_{\mathbb{Q}} T_{\mathbb{Q}}$ is *odd*, say of the form cd where $d = 2d_0+1 = [E_T : \mathbb{Q}]$. In particular, E_T is totally real. Let $\tilde{\rho}_\ell^T$ and $B(T)$ be as in Lemma 4.2.23, and replace F' by a large enough extension to satisfy the conclusion of that lemma, and such that $\overline{(\rho_\ell^T(\Gamma_{F'}))}^{\mathrm{Zar}}$ is connected. Recall that $B(T)$ is an object of $AV^0_{F',E}$ for some finite extension $E/\mathbb{Q}$ large enough to split $C^+(T_{\mathbb{Q}})$. Corollary 4.2.28 implies that

$$\overline{(\rho_\ell^T \otimes \overline{\mathbb{Q}}_\ell)(\Gamma_{F'})}^{\mathrm{Zar}} \cong \prod_1^d \mathrm{SO}_c(\overline{\mathbb{Q}}_\ell) \subset \mathrm{SO}_{cd}(\overline{\mathbb{Q}}_\ell).$$

Restricting the spin representation W_{cd} of $\mathfrak{so}_{cd}(\overline{\mathbb{Q}}_\ell)$ to $\prod_1^d \mathfrak{so}_c(\overline{\mathbb{Q}}_\ell)$, we obtain (via a weight calculation as in Lemma 4.2.18, and writing W_c for the spin representation of $\mathfrak{so}_c$)

$$W_{cd}|_{\prod \mathfrak{so}_c} \cong (\boxtimes_1^d W_c)^{2^{d_0}}.$$

The Lie algebra of the lift $\tilde{\rho}_\ell^T$ is one copy of the additive group $\mathfrak{g}_a$ times the Lie algebra of $\rho_\ell^T \otimes \overline{\mathbb{Q}}_\ell$, and this $\mathfrak{g}_a$ acts by scalars in the spin representation, so $r_{\mathrm{spin}} \circ \tilde{\rho}_\ell^T|_{\Gamma_{F'}}$ has an analogous decomposition as 2^{d_0} copies of some (absolutely, Lie) irreducible representation W' of $\Gamma_{F'}$. For a suitable enlargement E' of E (and extension $E' \hookrightarrow \overline{\mathbb{Q}}_\ell$), and F'' of F', we can realize W' as $H^1(B(T)', \mathbb{Q}_\ell) \otimes_{E'} \overline{\mathbb{Q}}_\ell$ for an object $B(T)'$ of $AV^0_{F'',E'}$. More precisely, we need the decomposition of $W_{cd}|_{\prod \mathfrak{so}_c}$ to be defined over E'; the first claim then follows from Faltings. The extension F''/F' is needed to decompose $B(T)_{\overline{F}} \otimes_E E'$ over some finite extension. Then $\mathrm{End}_{AV^0_{F'',E'}}(B(T)') = E'$, and the isomorphism $r_{\mathrm{spin}} \circ \tilde{\rho}_\ell^T \cong (W')^{2^{d_0}}$ implies Γ_F-conjugation invariance of W'. Thus we can apply Ribet's method to deduce:

LEMMA 4.2.30. *There exist a finite extension E''/E', an embedding $E'' \hookrightarrow \overline{\mathbb{Q}}_\ell$ extending the given $E' \hookrightarrow \overline{\mathbb{Q}}_\ell$, and an object M of $AV^0_{F,E''}$ such that*

$$\left(H^1(M_{\overline{F}}, \mathbb{Q}_\ell) \otimes_{E''} \overline{\mathbb{Q}}_\ell|_{\Gamma_{F'''}}\right) \cong W'|_{\Gamma_{F'''}}$$

for some still further finite extension F'''/F''.

Let M_ℓ denote the associated ℓ-adic realization (via $E'' \hookrightarrow \overline{\mathbb{Q}}_\ell$). Since $r_{\mathrm{spin}} \circ \tilde{\rho}_\ell^T$ is Lie-isotypic, and M_ℓ is a descent to Γ_F of its unique (after finite restriction) Lie-irreducible constituent, Corollary 3.4.2 shows that there is an Artin representation ω of Γ_F such that

$$r_{\mathrm{spin}} \circ \tilde{\rho}_\ell^T \cong M_\ell \otimes \omega.$$

Possibly enlarging the field of coefficients yet again, we deduce:

THEOREM 4.2.31. *Suppose $\dim T_{\mathbb{Q}}$ is odd, and that for some (hence for all) ℓ, $\det T_\ell = 1$ as Γ_F-representation. Then there exist*

- *a number field $\widetilde{E}$ and an embedding $\widetilde{E} \hookrightarrow \overline{\mathbb{Q}}_\ell$;*
- *an object $\widetilde{M}$ of $\mathcal{M}_{F,\widetilde{E}}$ that is a tensor product of an Artin motive and (the image of) an object of $AV^0_{F,\widetilde{E}}$;*
- *and a lifting $\tilde{\rho}^T \colon \mathcal{G}_{F,\widetilde{E}} \to \mathrm{GSpin}(T_{\widetilde{E}})$ of ρ^T;*

such that $r_{\mathrm{spin}} \circ \tilde{\rho}^T$ is isomorphic to $\widetilde{M}$. On $\tilde{\lambda}$-adic realizations (for places $\tilde{\lambda}$ of $\widetilde{E}$), $\tilde{\rho}^T$ gives rise to lifts $\tilde{\rho}_{\tilde{\lambda}}^T$ of $\rho_\ell^T \otimes_{\mathbb{Q}_\ell} \widetilde{E}_{\tilde{\lambda}}$ such that $\{r_{\mathrm{spin}} \circ \tilde{\rho}_{\tilde{\lambda}}^T\}_{\tilde{\lambda}}$ is a weakly-compatible system. Moreover, the same conclusions hold with the motive $V = Prim^{2k}(X)(k)$ in place of its transcendental lattice T (and, again, a possible enlargement of $\widetilde{E}$).

PROOF. The number field $\widetilde{E}$ is the composite (inside the ambient $\overline{\mathbb{Q}}_\ell$) of the field E'' and the field needed to define the Artin representation ω. To conclude the proof of the theorem, we make three observations:

- $\mathcal{M}_{F,\widetilde{E}}$ is Tannakian (note that we already know that $M \otimes_{E''} \widetilde{E}$ and ω are motivic);
- M and ω both give rise to compatible systems of ℓ-adic representations;
- Corollary 4.2.20 applies to lift the representations of motivic Galois groups.

□

4.2.7. Non-generic cases: $\dim T_{\mathbb{Q}}$ even. We do not fully treat the case of even-rank transcendental lattice, but here give a couple examples, describing the 'shape' of the Galois representations in light of Proposition 3.4.1.

First, continue to assume E_T is totally real. Let $\dim T_{\mathbb{Q}} = 2n = cd$, with $d = [E_T : \mathbb{Q}]$. For any N, denote by $W_{2N,\pm}$ (for each choice of $\pm$) the two half-spin representations of $\mathfrak{so}_{2N}$, and continue to write W_{2N+1} for the spin representation of $\mathfrak{so}_{2N+1}$.

LEMMA 4.2.32. *When c is even, the restriction $W_{cd,+}|_{\prod \mathfrak{so}_c}$ is given by*

$$W_{cd,+}|_{\prod_1^d \mathfrak{so}_c} \cong \bigoplus_{\substack{\epsilon=(\epsilon_i)\in\{\pm\}^d \\ \prod \epsilon_i = 1}} \boxtimes_{i=1}^d W_{c,\epsilon_i},$$

where the indexing set ranges over all choices of signs with "$-$" occurring an even number of times. This is a direct sum of distinct Lie-irreducible representations.

When c is odd, so $d = 2d_0$ is even,

$$W_{cd,+}|_{\prod_1^d \mathfrak{so}_c} \cong 2^{d_0-1} \boxtimes_1^d W_c,$$

a single Lie-irreducible representation occurring with multiplicity 2^{d_0-1}.

Now, recall (Lemma 4.2.23) that after a sufficient base-change F'/F, we can find an abelian variety $B_+(T)$ (with coefficients in a number field E, embedded in $\overline{\mathbb{Q}}_\ell$; we may by extending scalars assume E is large enough that the above decomposition of spin representations is defined over E), and a lift $\tilde{\rho}_\ell^T$ such that $r_+ \circ \tilde{\rho}_\ell^T|_{\Gamma_{F'}}$ is isomorphic to $H^1(B_+(T), \mathbb{Q}_\ell) \otimes_E \overline{\mathbb{Q}}_\ell$. We assume F' sufficiently large that this Galois representation is a sum of Lie-irreducible representations.

PROPOSITION 4.2.33. *Suppose c is even. Then motivic lifting holds for ρ^T.*

PROOF. Since c is even, the previous lemma shows that $r_+ \circ \tilde{\rho}_\ell^T$ is Lie-multiplicity-free, hence is a direct sum of inductions of non-conjugate, Lie-irreducible Galois representations. If $\pi_0(\rho_\ell^T)$ is trivial, in which case $\tilde{\rho}_\ell^T$ also has connected monodromy group,[22] then no inductions occur in this decomposition, so each factor $\boxtimes_1^d W_{c,\epsilon_i}$ in Lemma 4.2.32 corresponds to a Lie-irreducible factor of $r_+ \circ \tilde{\rho}_\ell^T$, as Γ_F-representation. Each of these factors is, over F', of the form

$$H^1(B_\epsilon, \mathbb{Q}_\ell) \otimes_E \overline{\mathbb{Q}}_\ell,$$

where B_ϵ is an object of $AV_{F',E}^0$ with endomorphism algebra just E itself (the usual application of Faltings' theorem, using the Lie-multiplicity-free property, and the fact that the spin representation decomposition holds over E). By Γ_F-invariance of each factor, we can apply Proposition 4.2.29 to produce an object M_ϵ of AV_{F,E_ϵ}^0, for some finite extension E_ϵ/E inside $\overline{\mathbb{Q}}_\ell$, with $M_{\epsilon,\ell}|_{\Gamma_{F'}}$ isomorphic to a finite-order twist of $H^1(B_\epsilon, \mathbb{Q}_\ell) \otimes_{E_\epsilon} \overline{\mathbb{Q}}_\ell$. By Lie-irreducibility, some finite-order twist (a

[22]Since it contains the center of GSpin; the sort of example this avoids is $\rho_{f,\ell} \otimes \omega^{\frac{1-k}{2}}$, where f is a classical modular form of odd weight k.

character of $\mathrm{Gal}(F'/F)$, in fact, and, again, we may have to enlarge E_ϵ) M'_ϵ of M_ϵ has ℓ-adic realization isomorphic to the corresponding factor of $r_+ \circ \tilde{\rho}_\ell^T$. Inside the ambient $\overline{\mathbb{Q}}_\ell$, we take the compositum E' of the various E_ϵ, extending scalars on each M'_ϵ. Then the object $\oplus_\epsilon M'_\epsilon$ of $\mathcal{M}_{F,E'}$ satisfies the hypotheses of Corollary 4.2.20, so we deduce the existence of the desired motivic lift.

We sketch the case of non-connected monodromy. Take the (motivic) Lie-irreducible factors of $r_+ \circ \tilde{\rho}_\ell^T|_{\Gamma_{F'}}$, and partition them into Γ_F-orbits. Fix a representative W_i of each orbit, and consider the stabilizer Γ_{F_i} in Γ_F of W_i. Arguing as in Proposition 4.2.33, we can apply Ribet's method to descend each W_i to an object M_i of $\mathcal{M}_{F_i,E_i}$ for some extension E_i of E inside $\overline{\mathbb{Q}}_\ell$. Moreover, twisting yields an M'_i whose ℓ-adic realization is isomorphic to a factor of $r_+ \circ \tilde{\rho}_\ell^T$: the isomorphisms $\mathrm{Hom}_{\Gamma_{F'}}(M_1, V) = \mathrm{Hom}_{\Gamma_{F_i}}(\mathrm{Ind}_{F'}^{F_i}(M_i), V) = \mathrm{Hom}_{\Gamma_{F_i}}(M_i \otimes \overline{\mathbb{Q}}_\ell[\mathrm{Gal}(F'/F_i)], V)$, with $V = r_+ \circ \tilde{\rho}_\ell^T$, imply this claim, using the fact that M_i is Lie-irreducible and after finite restriction occurs with multiplicity one in $r_+ \circ \tilde{\rho}_\ell^T$. Then we can induce (the representation of motivic Galois groups) from F_i to F to obtain our motives over F. This completes the case of even c (and E_T totally real). □

Having demonstrated the available techniques in a couple of quite different situations (namely, where the lifts range between the extremes of being Lie-multiplicity-free and Lie-isotypic), we stop here, remarking only that when E_T is CM, the analysis must begin not from Lemma 4.2.32 but from the restriction $W_{cd,+}|_{\prod \mathfrak{gl}_c}$, the product ranging over pairs of complex-conjugate embeddings $E_T \hookrightarrow \overline{\mathbb{Q}}_\ell$. This restriction is given by:

LEMMA 4.2.34. *Suppose $d = 2d_0$ is even, with notation otherwise as above. We denote irreducible representations of $\mathfrak{gl}_c$ by $W(r)$, where W is an irreducible representation of $\mathfrak{sl}_c$, and (r) indicates that the restriction to the center $\mathfrak{g}_a \subset \mathfrak{gl}_c$ is multiplication by r. Then, letting V_i denote the standard representation of the ith copy of $\mathfrak{sl}_c$*

$$W_{cd,+}|_{\prod_1^{d_0} \mathfrak{gl}_c} \cong \bigoplus_{\substack{i_1,\dots,i_{d_0}: \\ \sum i_j \in 2\mathbb{Z}}} \wedge^{i_1} V_1^*(\frac{c}{2} - i_1) \boxtimes \dots \boxtimes \wedge^{i_{d_0}} V_{d_0}^*(\frac{c}{2} - i_{d_0}).$$

REMARK 4.2.35.
- Note that the representations occurring here do not necessarily extend to representations of GL_c, since c may be odd; they do extend on the (connected) double-cover of GL_c.
- In particular, we see that when E_T is CM, $r_+ \circ \tilde{\rho}_\ell^T$ is Lie-multiplicity-free. This suggests proceeding as in Proposition 4.2.33, although we will stop here.

4.3. Towards a generalized Kuga-Satake theory

4.3.1. A conjecture. It is fair to assume that one could establish a motivic lifting result for the remaining hyperkähler cases. More important, these lifting results clamor for generalization. Motivated by the Fontaine-Mazur conjecture, Theorem 3.2.10, Proposition 4.1.30, and Theorem 4.2.31, we are led to the following much more ambitious conjecture:

CONJECTURE 4.3.1. *Let F and E be number fields, and let $\widetilde{H} \twoheadrightarrow H$ be a surjection, with central torus kernel, of linear algebraic groups over E. Suppose we are given a homomorphism $\rho\colon \mathcal{G}_{F,E} \to H$. Then if F is imaginary, there is a* *finite extension E'/E and a homomorphism $\tilde{\rho}\colon \mathcal{G}_{F,E'} \to \widetilde{H}_{E'}$ lifting $\rho \otimes_E E'$. If F is totally real, then such a lift exists if and only if the Hodge number parity obstruction of Corollary 3.2.8 vanishes.*

To indicate the scope of this conjecture, let X/F be any smooth projective variety, and consider for any $k \leq \dim X$ the motive $H^{2k}(X)(k)$ (or $Prim^{2k}(X)(k)$, having chosen an ample line bundle). This gives rise to an orthogonal representation of $\mathcal{G}_F$, and the conjecture in this case (for GSpin $\to$ SO, or the variant with the full orthogonal group in place of SO) amounts to a generalization of the Kuga-Satake construction to *arbitrary* orthogonally-polarized motivic (over F) Hodge structures. For other choices of $\widetilde{H}$ and H (for instance, $\mathrm{GL}_n \to \mathrm{PGL}_n$, where necessarily we will have coefficient field larger than $\mathbb{Q}$), the conjectured generalization is even more mysterious (compare Lemma 4.1.33).

Note the role of Lemma 4.1.31 in building our confidence in this conjecture. Let $\widetilde{H} \to H$ be a morphism of groups, say over $\overline{\mathbb{Q}}$, as in the conjecture. One way of formulating the Fontaine-Mazur conjecture is that $\mathcal{G}_F \otimes_{\mathbb{Q}} \overline{\mathbb{Q}}_\ell$ should be isomorphic to the Tannakian group for the Tannakian category $\mathrm{Rep}^{g,ss}_{\overline{\mathbb{Q}}_\ell}(\Gamma_F)$ of semi-simple geometric $\overline{\mathbb{Q}}_\ell$-representations of Γ_F. In particular, a geometric lift $\tilde{\rho}_\ell\colon \Gamma_F \to \widetilde{H}(\overline{\mathbb{Q}}_\ell)$ of $\rho_\ell\colon \Gamma_F \to H(\overline{\mathbb{Q}}_\ell)$ should arise from an algebraic homomorphism of $\overline{\mathbb{Q}}_\ell$-groups $\mathcal{G}_F \otimes \overline{\mathbb{Q}}_\ell \to \widetilde{H}_{\overline{\mathbb{Q}}_\ell}$. Lemma 4.1.31 then tells us that if ρ arises from some $\mathcal{G}_F \otimes \overline{\mathbb{Q}} \to H$, then we can find a lift $\tilde{\rho}\colon \mathcal{G}_F \otimes \overline{\mathbb{Q}} \to \widetilde{H}$ of homomorphisms of $\overline{\mathbb{Q}}$-groups.

Now that we know to look for such a thing, we conclude by giving one more example in which it is easy to construct a 'generalized Kuga-Satake motive.'

4.3.2. Motivic lifting: Abelian varieties. If $A/\mathbb{C}$ is an abelian surface, then $H^2(A, \mathbb{Q}) \cong \wedge^2 H^1(A, \mathbb{Q})$ has Hodge numbers $h^{2,0} = 1$ and $h^{1,-1} = 4$. The classical Kuga-Satake construction then associates to A another abelian variety, $KS(A)$, with rational cohomology $C^+(H^2(A, \mathbb{Q}))$ (or the analogue with $Prim^2(A, \mathbb{Q})$, if we have fixed a polarization), and a theorem of Morrison ([**Mor85**]) asserts that $KS(A)$ is isogenous to A^8 (or A^4 if we work with $Prim^2(A, \mathbb{Q})$). We now show that this construction can be generalized to abelian varieties of any dimension.

Let F be any subfield of $\mathbb{C}$, and let A/F be an abelian variety of dimension g with a fixed polarization. Consider the algebraic representation

$$\rho\colon \mathcal{G}_F \to \mathrm{GSp}(H^1(A_{\mathbb{C}}, \mathbb{Q})) \cong \mathrm{GSp}_{2g},$$

or, for the theory over $\mathbb{C}$, its restriction to $\mathcal{G}_{\mathbb{C}}$. We will compose ρ with a homomorphism $\mathrm{GSp}_{2g} \to \mathrm{GSpin}_N$ for suitable N to produce the generalized Kuga-Satake motive. Let $W = H^1(A_{\mathbb{C}}, \mathbb{Q})$, and let $r_{e_1+e_2}\colon \mathrm{Sp}(V) \to \mathrm{GL}(V_{e_1+e_2})$, in the weight notation of §2.8, denote the irreducible representation of $\mathrm{Sp}(V)$ obtained as the complement of the trivial representation in $\wedge^2(W)$. This absolutely irreducible representation, defined over $\mathbb{Q}$, has image in $\mathrm{SO}(V_{e_1+e_2})$, where the quadratic form is induced from the pairing canonically induced on $\wedge^2(W)$, and which coincides with the induced polarization on $Prim^2(A_{\mathbb{C}}, \mathbb{Q})$. Since $\mathrm{Sp}(W)$ is simply-connected, there is a lift to an algebraic homomorphism $\tilde{r}\colon \mathrm{Sp}(W) \to \mathrm{Spin}(V_{e_1+e_2})$. Over $\mathbb{C}$ this follows from standard Lie theory, and such a homomorphism descends to $\overline{\mathbb{Q}}$ (see Lemma 4.1.31); since the map of Lie algebras $\mathfrak{sp}(W) \to \mathfrak{so}(V_{e_1+e_2})$ is defined over $\mathbb{Q}$, we see that the map over $\overline{\mathbb{Q}}$ is $\Gamma_{\mathbb{Q}}$-invariant, hence descends to a morphism over $\mathbb{Q}$. The dimension of $V_{e_1+e_2}$ is $\binom{2g}{2} - 1$, which is odd ($= 2n+1$) if g is even, and even ($= 2n$) if g is odd. In either case, we have the representation r_{ϖ_n} of $\mathrm{Spin}(V_{e_1+e_2})_{\overline{\mathbb{Q}}}$ having highest weight $\varpi_n = \frac{\sum_{i=1}^n \chi_i}{2}$.[23]

LEMMA 4.3.2. *Let c denote the non-trivial (central) element of the kernel of $\mathrm{Spin}(V_{e_1+e_2}) \to \mathrm{SO}(V_{e_1+e_2})$. If $g \equiv 2, 3 \pmod 4$, then $\tilde{r}(-1) = c$, and if $g \equiv 0, 1 \pmod 4$, then $\tilde{r}(-1) = 1$.*

PROOF. In all cases, $r_{e_1+e_2}(-1) = 1$, so $\tilde{r}(-1)$ equals either 1 or c. Since c acts as -1 in any of the spin representations (c *is* the element -1 of the Clifford algebra), it suffices to compute $r_{\varpi_n} \circ \tilde{r}(-1)$. The weights of $V_{e_1+e_2}$ are

$$\{\pm(e_i + e_j)\}_{1 \leq i < j \leq g} \cup \{e_i - e_j\}_{i \neq j},$$

except with one copy of the weight zero deleted (so that zero has multiplicity $g - 1$ rather than g). It follows that $r_{\varpi_n} \circ \tilde{r}$ has a weight equal to

$$\frac{1}{2}\left(\sum_{1 \leq i < j \leq g} (e_i + e_j) + \sum_{1 \leq i < j \leq g} (e_i - e_j)\right) = \sum_{i=1}^{g} (g - i) e_i.$$

In particular, $r_{\varpi_n} \circ \tilde{r}(-1)$ is multiplication by $(-1)^{g(g-1)/2}$, and the lemma follows. $\square$

COROLLARY 4.3.3. *$\tilde{r}\colon \mathrm{Sp}(W) \to \mathrm{Spin}(V_{e_1+e_2})$ extends to an algebraic homomorphism $\mathrm{GSp}(W) \to \mathrm{GSpin}(V_{e_1+e_2})$. If $g \equiv 2, 3 \pmod 4$, then this map can be chosen so the Clifford norm coincides with the symplectic multiplier; if $g \equiv 0, 1 \pmod 4$, then this map can be chosen to factor through $\mathrm{Spin}(V_{e_1+e_2})$. The composition*

$$\mathcal{G}_F \xrightarrow{\rho} \mathrm{GSp}(W) \xrightarrow{\tilde{r}} \mathrm{GSpin}(V_{e_1+e_2}) \hookrightarrow \mathrm{GL}(C^+(V_{e_1+e_2}))$$

defines the generalized Kuga-Satake lift of A.

PROOF. This follows immediately from Lemma 4.3.2 and the identifications:

$$\frac{\mathbb{G}_m \times \mathrm{Sp}(W)}{\langle(-1, -1)\rangle} \xrightarrow{\sim} \mathrm{GSp}(W)$$

$$\frac{\mathbb{G}_m \times \mathrm{Spin}(V_{e_1+e_2})}{\langle(-1, c)\rangle} \xrightarrow{\sim} \mathrm{GSpin}(V_{e_1+e_2}).$$

When $g \equiv 2, 3 \pmod 4$, we take the map $\mathbb{G}_m \to \mathbb{G}_m$ to be the identity, and when $g \equiv 0, 1 \pmod 4$, we take it to be trivial. $\square$

REMARK 4.3.4.
- Repeating the above arguments with $\wedge^2(W)$ in place of $V_{e_1+e_2}$, we can similarly construct lifts $\mathcal{G}_F \to \mathrm{GSpin}(\wedge^2(W))$.
- When $g = 2$, this recovers the classical construction. In that case, the composition $r_{\varpi_n} \circ \tilde{r}$ is the identity, and, decomposing $C^+(V_{e_1+e_2})$ as 4 copies of r_{ϖ_n} (as $\mathrm{GSpin}(V_{e_1+e_2})$ representation), the identification (up to isogeny) $KS(A) \sim A^4$ is nearly a tautology.

The motivic formalism now tells us that $r_{\varpi_n} \circ \tilde{r} \circ \rho$ is (the Betti realization of) an object of $\mathcal{M}_F$. Since the ℓ-adic realizations of ρ form a weakly (in fact, strictly) compatible system, the same is true for the ℓ-adic realizations of this Kuga-Satake motive. In this case, however, we can say more, and will realize this explicitly

[23]We use the common fundamental weight notation here. This is the spin representation in the odd case; one of the half-spin representations in the even case.

(and unconditionally) as a Grothendieck motive. The first step is to compute the plethysm $r_{\varpi_n} \circ \tilde{r}$; we will do this, but first mention an equivalent, structurally appealing plethysm. First we treat the D_n case, that is when $d = 2g = \dim H^1(X)$ satisfies $2n = \binom{d}{2}$; this amounts to $\dim X$ being even. Let $V = H^2(X) = \wedge^2 W$. We use the following (common) notation for fundamental weights of D_n:

- $\varpi_i = \chi_1 + \ldots + \chi_i$, and $r_{\varpi_i} = \wedge^i V$, for $i = 1, \ldots, n-2$;
- $\varpi_{n-1} = \frac{\chi_1 + \ldots + \chi_{n-1} - \chi_n}{2}$, $\varpi_n = \frac{\sum_1^n \chi_i}{2}$, and $r_{\varpi_{n-1}}$, r_{ϖ_n} are the two half-spin representations.

As representations of $\mathfrak{so}_{2n}$, we have the following identity:

$$\bigoplus_{i=0}^{2n} \wedge^i(V) = \left(r_{\varpi_{n-1}} \oplus r_{\varpi_n}\right)^{\otimes 2}.$$

The plethysm problem that we expect to solve, then, is to describe a representation $r_{KS(X)}$ of $\mathrm{Sp}(W)$ such that

$$\bigoplus_{i=0}^{2n} \wedge^i(\wedge^2 W) = (r_{KS(X)})^{\otimes 2}.$$

(More ambitiously, we could attempt this with $\wedge^{2k}(W)$ instead.) Similarly, in the B_n case, we have fundamental weights $\varpi_i = \sum_1^i \chi_i$ for $i = 1, \ldots, n-1$, and $\varpi_n = \frac{\sum \chi_i}{2}$, corresponding, respectively to the wedge powers $\wedge^i V$ of the standard representation (r_{ϖ_1}) and the spin representation. As representations of $\mathfrak{so}_{2n+1}$, we have the identity

$$\bigoplus_{i=0}^{n} \wedge^i(V) = r_{\varpi_n}^{\otimes 2},$$

so in this case we want a representation r_{KS} of $\mathrm{Sp}(W)$ satisfying

$$\bigoplus_{i=0}^{n} \wedge^i(\wedge^2(W)) = r_{KS}^{\otimes 2}.$$

PROPOSITION 4.3.5. *Writing ω_i $(i = 1, \ldots, g)$ for the usual fundamental weights*[24] *of C_g, then in all cases (g odd or even) we have*

$$\wedge^\bullet(\wedge^2(W)) \cong 2^g (V_{\omega_1 + \ldots + \omega_{g-1}})^{\otimes 2},$$

as $\mathrm{Sp}(W)$-representations. Here $\wedge^\bullet$ denotes the full exterior algebra. This is deduced from the above discussion and the two calculations:

- *(even) As $\mathrm{Sp}(W)$-representations,*
$$(r_{\varpi_{n-1}} \oplus r_{\varpi_n}) \circ \tilde{r} \cong 2^{g/2} V_{\omega_1 + \ldots + \omega_{g-1}}.$$
- *(odd) As $\mathrm{Sp}(W)$-representations,*
$$r_{\varpi_n} \circ \tilde{r} \cong 2^{(g-1)/2} V_{\omega_1 + \ldots + \omega_{g-1}}.$$

PROOF. We treat the even case, the odd case being essentially identical. There are $g^2 - g$ non-zero weights in $\wedge^2(W)$, and the weight zero occurs with multiplicity g. The weights of $(r_{\varpi_{n-1}} \oplus r_{\varpi_n}) \circ \tilde{r}$ are sums of plus or minus any $n = g^2 - \frac{g}{2}$ weights of $\wedge^2(W)$, then the total divided by two. It follows that the highest weight of $(r_{\varpi_{n-1}} \oplus r_{\varpi_n}) \circ \tilde{r}$ is, as previously computed, $\sum_{i=1}^{g-1} \omega_i$, but moreover that it

[24] $\omega_i = e_1 + \ldots + e_i$ in the standard coordinate system

occurs with multiplicity $2^{g/2}$ (here $\frac{g}{2} = n - (g^2 - g)$; we can choose the weight $+0$ or -0 this many times). It follows that $(r_{\varpi_{n-1}} \oplus r_{\varpi_n}) \circ \tilde{r}$ contains $2^{g/2} V_{\omega_1 + \ldots + \omega_{g-1}}$; by dimension count (see the following Lemma 4.3.6), they are isomorphic. □

LEMMA 4.3.6. *With the above notation, the dimension of the irreducible representation $V_{\omega_1+\ldots+\omega_{g-1}}$ of $\mathrm{Sp}(W)$ is $2^{g(g-1)}$.*

PROOF. Simplifying the Weyl dimension formula, we find

$$\dim V_{\omega_1+\ldots+\omega_{g-1}} = 2^{|\Phi^+|} \prod_{1 \le i \le j \le g} \frac{2g+1-(i+j)}{2g+2-(i+j)},$$

where the number $|\Phi^+|$ of positive roots is 2^{g^2}. The product telescopes: fix an i, and the corresponding product over j is equal to $\frac{1}{2}$. The lemma follows. □

COROLLARY 4.3.7. *Let F be any subfield of $\mathbb{C}$, and let X/F be an abelian variety of any dimension g, giving rise to a representation*

$$\rho_{H^1(X)} \colon \mathcal{G}_F \to \mathrm{GSp}(H^1(X, \mathbb{Q})).$$

Then the Kuga-Satake lift (Corollary 4.3.3 and remark following) of the representation

$$\mathcal{G}_F \to \mathrm{SO}(H^2(X, \mathbb{Q})(1))$$

can be explicitly realized in the spin (or sum of half-spin) representation as $2^{\lfloor \frac{g}{2} \rfloor}$ copies of the composition $r_{\omega_1+\ldots+\omega_{g-1}} \circ \rho_{H^1(X)}$ with the highest weight $\omega_1 + \ldots + \omega_{g-1}$ representation of $\mathrm{Sp}(H^1(X, \mathbb{Q}))$.[25]

Let us call this object of $\mathcal{M}_F$ the Kuga-Satake motive $KS(X)$. Then $KS(X)$ is in fact a Grothendieck motive, for either numerical or homological equivalence.

PROOF. It remains to check that $KS(X)$ can be cut out by algebraic cycles. We start with the explicit description (due to Weyl; see §17.3 of [**FH91**]) of the representation $V_{\omega_1+\cdots+\omega_{g-1}}$. From now on, abbreviate $\lambda = \sum_1^{g-1} \omega_i$, and $r = \sum_1^{g-1} i$ (the length of λ); in fact, what follows applies to an arbitrary partition λ. Then

$$V_\lambda = W^{\langle r \rangle} \cap \mathbb{S}_\lambda(W)$$

as $\mathrm{Sp}(W)$-representation. We have to explain this notation: $\mathbb{S}_\lambda(W)$ denotes the Schur functor associated to the partition λ, which explicitly is equal to the image of the Young symmetrizer c_λ acting on $W^{\otimes r}$; and $W^{\langle r \rangle}$ is the subspace of $W^{\otimes r}$ given by intersecting the kernels of all the contractions $(1 \le p < q \le r)$

$$c_{p,q} \colon W^{\otimes r} \to W^{\otimes(r-2)}$$
$$v_1 \otimes \ldots \otimes v_r \mapsto \langle v_p, v_q \rangle \cdot v_1 \otimes \ldots \otimes \hat{v}_p \otimes \ldots \otimes \hat{v}_q \otimes \ldots \otimes v_r,$$

where $\langle \cdot, \cdot \rangle$ represents the symplectic form on W. The result will now follow from some basic facts about algebraic cycles and, crucially, the Lefschetz and Künneth Standard Conjectures for abelian varieties (due to Lieberman; for a proof, see [**Kle68**]). We will write π_X^i for the algebraic cycle on $X \times X$ inducing $H(X) \twoheadrightarrow H^i(X) \hookrightarrow H(X)$. Fix a polarization of X, giving rise to a Lefschetz operator L_X. Recall Jannsen's fundamental result ([**Jan92**]) that numerical motives form a semi-simple abelian category. This and the Künneth Standard Conjecture for

[25] Extended to GSp on the center by the prescription of Corollary 4.3.3. Throughout this argument we will ignore these extra scalars to simplify the notation.

abelian varieties imply that the category of numerical motives generated by abelian varieties over F is a semi-simple Tannakian category. We deduce that the following are (numerical) sub-motives of $H(X^r)$:

- $M = (X, \pi_X^1, 0)$, i.e. the object corresponding to $H^1(X)$;
- The kernel of each contraction $H^1(X)^{\otimes r} \to H^1(X)^{\otimes(r-2)}$. For notational simplicity, take $p = 1, q = 2$. If we were working in cohomology, we would compute the kernel of $c_{1,2}$ as

$$\ker\left(H^1(X) \otimes H^1(X) \xrightarrow{\langle\cdot,\cdot\rangle} \mathbb{Q}(-1)\right) \otimes H^1(X)^{\otimes r-2},$$

  where recall that the polarization $\langle\cdot,\cdot\rangle$ is defined by $\langle w_1, w_2\rangle = L_X^{d-1}(w_1 \cup w_2)$. Now, $W^{\otimes 2} = \mathrm{Sym}^2(W) \oplus \wedge^2(W)$, and cup-product kills $\mathrm{Sym}^2(W)$, while mapping $\wedge^2(W)$ isomorphically to $H^2(X)$. The kernel of $c_{1,2}$ is therefore isomorphic to

$$\left(\mathrm{Sym}^2(H^1(X)) \oplus Prim^2(X)\right) \otimes H^1(X)^{\otimes r-2}.$$

  To realize the analogous numerical motive, we can define $\mathrm{Sym}^2(M)$ as

$$(X \times X, \frac{1}{2}(1 + (12)) \cdot (\pi_X^1 \times \pi_X^1), 0).$$

  For the projector, we have taken the product of two commuting idempotents, the first one being the Young symmetrizer associated to the Schur functor Sym^2. There is also a projector p^2 onto primitive cohomology (see 1.4.4, 2.3, and $2A11$ of [**Kle68**]), so $(X, p^2, 0)$ realizes $Prim^2(X)$ as a numerical motive. Finally, since direct sums and tensor products exist in our category, we can therefore describe the kernel of any $c_{p,q}$ as a numerical motive $M_{p,q}$.
- In general, the Young symmetrizer c_λ is not quite an idempotent. Write $c_\lambda^2 = x^{-1}c_\lambda$ (x depends on λ, but we have fixed a λ), making xc_λ an idempotent algebraic correspondence on X^r. Now, c_λ commutes with the r-fold Künneth projector $(\pi_X^1)^r$ (this is easily checked at the level of cohomological correspondences, so *a fortiori* holds for numerical equivalence), so

$$(X^r, xc_\lambda \cdot (\pi_X^1)^r, 0)$$

  defines a numerical motive, denoted $\mathbb{S}_\lambda(H^1(X))$.
- We would like to conclude the proof by intersecting the objects $M_{p,q}$ ($p < q$) and $\mathbb{S}_\lambda(H^1(X))$. This is possible for numerical motives since the category is abelian. Finally, since all of the cycles considered in the proof are cycles on (disjoint unions of) abelian varieties in characteristic zero, where numerical and homological equivalence coincide,[26] we also deduce the existence of a homological motive corresponding to $KS(X)$.

□

REMARK 4.3.8. A weight calculation yields the Hodge numbers of $KS(X)$.

---

[26] In the presence of the Standard Conjecture of Hodge type, the Lefschetz conjecture implies 'num = hom': see Proposition 5.1 of [**Kle94**].

4.3.3. Coda. Identifying among all rational Hodge structures those that are motivic is one of the fundamental problems of complex algebraic geometry, and this generalized Kuga-Satake theory would systematically construct new *motivic* Hodge structures from old, in a way not achievable by simply playing the Tannakian game. I hope that investigation of these phenomena will provide a stimulating testing-ground for thinking about Hodge theory in non-classical weights.

It is also tempting to ask what should be true if we replace F by other fields, especially finitely-generated subfields of $\mathbb{C}$. Our motivic descent in the hyperkähler case works as written, except for the critical absence of Tate's basic vanishing result; in other contexts, it might be hoped that a similar descent works, conditional on the relevant cases of the Tate conjecture. I suspect only a qualitative *potential* lifting result will hold in this generality–that would suffice to imply the analogous lifting conjecture for $\mathbb{C}$ itself. But another clearly important question to ask is: what, if any, is the analogue of Tate's vanishing theorem when the number field F is replaced by any field finitely-generated over $\mathbb{Q}$?

Bibliography

[AC89] J. Arthur and L. Clozel, *Simple algebras, base change, and the advanced theory of the trace formula*, Annals of Mathematics Studies, vol. 120, Princeton University Press, Princeton, NJ, 1989. MR1007299

[And96a] Y. André, *On the Shafarevich and Tate conjectures for hyper-Kähler varieties*, Math. Ann. **305** (1996), no. 2, 205–248. MR1391213

[And96b] Y. André, *Pour une théorie inconditionnelle des motifs* (French), Inst. Hautes Études Sci. Publ. Math. **83** (1996), 5–49. MR1423019

[And04] Y. André, *Une introduction aux motifs (motifs purs, motifs mixtes, périodes)* (French, with English and French summaries), Panoramas et Synthèses [Panoramas and Syntheses], vol. 17, Société Mathématique de France, Paris, 2004. MR2115000

[AP06] J. D. Adler and D. Prasad, *On certain multiplicity one theorems*, Israel J. Math. **153** (2006), 221–245. MR2254643

[APP16] J. Adler, S. Patrikis, and D. Prasad, *Restrictions of representations*, work in progress (2016).

[Art02] J. Arthur, *A note on the automorphic Langlands group*, Canad. Math. Bull. **45** (2002), no. 4, 466–482. Dedicated to Robert V. Moody. MR1941222

[Art13] J. Arthur, *The endoscopic classification of representations*, American Mathematical Society Colloquium Publications, vol. 61, American Mathematical Society, Providence, RI, 2013. Orthogonal and symplectic groups. MR3135650

[BC] O. Brinon and B. Conrad, *Cmi summer school notes on p-adic hodge theory: preliminary version.*

[Bea83] A. Beauville, *Variétés Kähleriennes dont la première classe de Chern est nulle* (French), J. Differential Geom. **18** (1983), no. 4, 755–782 (1984). MR730926

[Bei12] A. Beilinson, *p-adic periods and derived de Rham cohomology*, J. Amer. Math. Soc. **25** (2012), no. 3, 715–738. MR2904571

[Ber04] L. Berger, *An introduction to the theory of p-adic representations* (English, with English and French summaries), Geometric aspects of Dwork theory. Vol. I, II, Walter de Gruyter, Berlin, 2004, pp. 255–292. MR2023292

[BG11] K. Buzzard and T. Gee, *The conjectural connections between automorphic representations and Galois representations*, Automorphic forms and Galois representations. Vol. 1, London Math. Soc. Lecture Note Ser., vol. 414, Cambridge Univ. Press, Cambridge, 2014, pp. 135–187. MR3444225

[Bha12] B. Bhatt, *p-adic derived de rham cohomology*, preprint available as arXiv:1204.6560 (2012).

[BLGGT14] T. Barnet-Lamb, T. Gee, D. Geraghty, and R. Taylor, *Potential automorphy and change of weight*, Ann. of Math. (2) **179** (2014), no. 2, 501–609. MR3152941

[Bor79] A. Borel, *Automorphic L-functions*, Automorphic forms, representations and L-functions (Proc. Sympos. Pure Math., Oregon State Univ., Corvallis, Ore., 1977), Proc. Sympos. Pure Math., XXXIII, Amer. Math. Soc., Providence, R.I., 1979, pp. 27–61. MR546608

[BR93] D. Blasius and J. D. Rogawski, *Motives for Hilbert modular forms*, Invent. Math. **114** (1993), no. 1, 55–87. MR1235020

[Car79] P. Cartier, *Representations of p-adic groups: a survey*, Automorphic forms, representations and L-functions (Proc. Sympos. Pure Math., Oregon State Univ., Corvallis, Ore., 1977), Proc. Sympos. Pure Math., XXXIII, Amer. Math. Soc., Providence, R.I., 1979, pp. 111–155. MR546593

[Cas] W. Casselman, *Introduction to admissible representations of p-adic groups*, preprint, available at http://www.math.ubc.ca/ cass/research/publications.html.

[CCO14] C.-L. Chai, B. Conrad, and F. Oort, *Complex multiplication and lifting problems*, Mathematical Surveys and Monographs, vol. 195, American Mathematical Society, Providence, RI, 2014. MR3137398

[CHT08] L. Clozel, M. Harris, and R. Taylor, *Automorphy for some l-adic lifts of automorphic mod l Galois representations*, Publ. Math. Inst. Hautes Études Sci. **108** (2008), 1–181. With Appendix A, summarizing unpublished work of Russ Mann, and Appendix B by Marie-France Vignéras. MR2470687

[CKPSS04] J. W. Cogdell, H. H. Kim, I. I. Piatetski-Shapiro, and F. Shahidi, *Functoriality for the classical groups*, Publ. Math. Inst. Hautes Études Sci. **99** (2004), 163–233. MR2075885

[Clo86] L. Clozel, *On limit multiplicities of discrete series representations in spaces of automorphic forms*, Invent. Math. **83** (1986), no. 2, 265–284. MR818353

[Clo90] L. Clozel, *Motifs et formes automorphes: applications du principe de fonctorialité* (French), Automorphic forms, Shimura varieties, and L-functions, Vol. I (Ann Arbor, MI, 1988), Perspect. Math., vol. 10, Academic Press, Boston, MA, 1990, pp. 77–159. MR1044819

[Con11] Brian Conrad, *Lifting global representations with local properties*, Preprint available at http://math.stanford.edu/ conrad/ (2011).

[Cp86] J. W. S. Cassels and A. Fröhlich (eds.), *Algebraic number theory*, Academic Press, Inc. [Harcourt Brace Jovanovich, Publishers], London, 1986. Reprint of the 1967 original. MR911121

[Del71] P. Deligne, *Formes modulaires et représentations l-adiques* (French), Séminaire Bourbaki. Vol. 1968/69: Exposés 347–363, Lecture Notes in Math., vol. 175, Springer, Berlin, 1971, pp. Exp. No. 355, 139–172. MR3077124

[Del72] P. Deligne, *La conjecture de Weil pour les surfaces $K3$* (French), Invent. Math. **15** (1972), 206–226. MR0296076

[Del74] P. Deligne, *La conjecture de Weil. I* (French), Inst. Hautes Études Sci. Publ. Math. **43** (1974), 273–307. MR0340258

[Del79] P. Deligne, *Valeurs de fonctions L et périodes d'intégrales* (French), Automorphic forms, representations and L-functions (Proc. Sympos. Pure Math., Oregon State Univ., Corvallis, Ore., 1977), Proc. Sympos. Pure Math., XXXIII, Amer. Math. Soc., Providence, R.I., 1979, pp. 313–346. With an appendix by N. Koblitz and A. Ogus. MR546622

[Del90] P. Deligne, *Catégories tannakiennes* (French), The Grothendieck Festschrift, Vol. II, Progr. Math., vol. 87, Birkhäuser Boston, Boston, MA, 1990, pp. 111–195. MR1106898

[DM11] P. Deligne and James S. Milne, *Tannakian categories (corrected texed version)*, http://www.jmilne.org/math/xnotes/index.html (2011).

[DMOS82] P. Deligne, J. S. Milne, A. Ogus, and K.-y. Shih, *Hodge cycles, motives, and Shimura varieties*, Lecture Notes in Mathematics, vol. 900, Springer-Verlag, Berlin-New York, 1982. MR654325

[Fal83] G. Faltings, *Endlichkeitssätze für abelsche Varietäten über Zahlkörpern* (German), Invent. Math. **73** (1983), no. 3, 349–366. MR718935

[Fal89] G. Faltings, *Crystalline cohomology and p-adic Galois-representations*, Algebraic analysis, geometry, and number theory (Baltimore, MD, 1988), Johns Hopkins Univ. Press, Baltimore, MD, 1989, pp. 25–80. MR1463696

[FF12] L. Fargues and J.-M. Fontaine, *Vector bundles and p-adic Galois representations*, Fifth International Congress of Chinese Mathematicians. Part 1, 2, AMS/IP Stud. Adv. Math., 51, pt. 1, vol. 2, Amer. Math. Soc., Providence, RI, 2012, pp. 77–113. MR2908062

[FH91] W. Fulton and J. Harris, *Representation theory*, Graduate Texts in Mathematics, vol. 129, Springer-Verlag, New York, 1991. A first course; Readings in Mathematics. MR1153249

[Fli06] Y. Z. Flicker, *Automorphic representations of low rank groups*, World Scientific Publishing Co. Pte. Ltd., Hackensack, NJ, 2006. MR2251430

[Fon94] J.-M. Fontaine, *Représentations l-adiques potentiellement semi-stables* (French), Astérisque **223** (1994), 321–347. Périodes p-adiques (Bures-sur-Yvette, 1988). MR1293977

[Gal00] F. Galluzzi, *Abelian fourfold of Mumford-type and Kuga-Satake varieties*, Indag. Math. (N.S.) **11** (2000), no. 4, 547–560. MR1909819

[GJ78] S. Gelbart and H. Jacquet, *A relation between automorphic representations of* GL(2) *and* GL(3), Ann. Sci. École Norm. Sup. (4) **11** (1978), no. 4, 471–542. MR533066

[GK82] S. S. Gelbart and A. W. Knapp, *L-indistinguishability and R groups for the special linear group*, Adv. in Math. **43** (1982), no. 2, 101–121. MR644669

[Gro98] B. H. Gross, *On the Satake isomorphism*, Galois representations in arithmetic algebraic geometry (Durham, 1996), London Math. Soc. Lecture Note Ser., vol. 254, Cambridge Univ. Press, Cambridge, 1998, pp. 223–237. MR1696481

[GT10] W. T. Gan and S. Takeda, *The local Langlands conjecture for Sp(4)*, Int. Math. Res. Not. IMRN **15** (2010), 2987–3038. MR2673717

[HC50] Harish-Chandra, *Lie algebras and the Tannaka duality theorem*, Ann. of Math. (2) **51** (1950), 299–330. MR0033811

[Jan92] U. Jannsen, *Motives, numerical equivalence, and semi-simplicity*, Invent. Math. **107** (1992), no. 3, 447–452. MR1150598

[Kat87] N. M. Katz, *On the monodromy groups attached to certain families of exponential sums*, Duke Math. J. **54** (1987), no. 1, 41–56. MR885774

[Kim03] H. H. Kim, *Functoriality for the exterior square of* GL_4 *and the symmetric fourth of* GL_2, J. Amer. Math. Soc. **16** (2003), no. 1, 139–183. With appendix 1 by Dinakar Ramakrishnan and appendix 2 by Kim and Peter Sarnak. MR1937203

[Kle68] S. L. Kleiman, *Algebraic cycles and the Weil conjectures*, Dix exposés sur la cohomologie des schémas, Adv. Stud. Pure Math., vol. 3, North-Holland, Amsterdam, 1968, pp. 359–386. MR292838

[Kle94] S. L. Kleiman, *The standard conjectures*, Motives (Seattle, WA, 1991), Proc. Sympos. Pure Math., vol. 55, Amer. Math. Soc., Providence, RI, 1994, pp. 3–20. MR1265519

[Kna02] A. W. Knapp, *Lie groups beyond an introduction*, 2nd ed., Progress in Mathematics, vol. 140, Birkhäuser Boston, Inc., Boston, MA, 2002. MR1920389

[KS67] M. Kuga and I. Satake, *Abelian varieties attached to polarized* K_3*-surfaces*, Math. Ann. **169** (1967), 239–242. MR0210717

[Lab85] J.-P. Labesse, *Cohomologie, L-groupes et fonctorialité* (French), Compositio Math. **55** (1985), no. 2, 163–184. MR795713

[Lan89] R. P. Langlands, *On the classification of irreducible representations of real algebraic groups*, Representation theory and harmonic analysis on semisimple Lie groups, Math. Surveys Monogr., vol. 31, Amer. Math. Soc., Providence, RI, 1989, pp. 101–170. MR1011897

[Lap99] E. M. Lapid, *Some results on multiplicities for* SL(n), Israel J. Math. **112** (1999), 157–186. MR1714998

[Lar94] M. Larsen, *On the conjugacy of element-conjugate homomorphisms*, Israel J. Math. **88** (1994), no. 1-3, 253–277. MR1303498

[LL79] J.-P. Labesse and R. P. Langlands, *L-indistinguishability for* SL(2), Canad. J. Math. **31** (1979), no. 4, 726–785. MR540902

[LR98] E. Lapid and J. Rogawski, *On twists of cuspidal representations of* GL(2), Forum Math. **10** (1998), no. 2, 175–197. MR1611951

[LY] T. Liu and J.-K. Yu, *On automorphy of certain Galois representations of* GO_4*-type*, J. Number Theory **161** (2016), 49–74. With an appendix by Liang Xiao. MR3435718

[Moo15] B. Moonen, *On the Tate and Mumford-Tate conjectures in codimension 1 for varieties with* $h^{2,0} = 1$, Duke Math. J. **166** (2017), no. 4, 739–799. MR3619305

[Mor85] D. R. Morrison, *The Kuga-Satake variety of an abelian surface*, J. Algebra **92** (1985), no. 2, 454–476. MR778462

[MS14] R. A. Moy and J. Specter, *There exist non-CM Hilbert modular forms of partial weight 1*, Int. Math. Res. Not. IMRN **24** (2015), 13047–13061, DOI 10.1093/imrn/rnv089. MR3436137

[Mum70] D. Mumford, *Abelian varieties*, Tata Institute of Fundamental Research Studies in Mathematics, No. 5, Published for the Tata Institute of Fundamental Research, Bombay; Oxford University Press, London, 1970. MR0282985

[Pat] S. Patrikis, *Generalized Kuga-Satake theory and rigid local systems I: the middle convolution*, Recent advances in Hodge theory, London Math. Soc. Lecture Note Ser., vol. 427, Cambridge Univ. Press, Cambridge, 2016, pp. 368–392. MR3409882

[Pat14a] S. Patrikis, *Generalized Kuga-Satake theory and rigid local systems, II: rigid Hecke eigensheaves*, Algebra Number Theory **10** (2016), no. 7, 1477–1526, DOI 10.2140/ant.2016.10.1477. MR3554239

[Pat14b] S. Patrikis, *On the sign of regular algebraic polarizable automorphic representations*, Math. Ann. **362** (2015), no. 1-2, 147–171. MR3343873

[Pat16] S. Patrikis, *Generalized Kuga-Satake theory and good reduction properties of Galois representations*, Algebra Number Theory **11** (2017), no. 10, 2397–2423, DOI 10.2140/ant.2017.11.2397. MR3744361

[PT15] S. Patrikis and R. Taylor, *Automorphy and irreducibility of some l-adic representations*, Compos. Math. **151** (2015), no. 2, 207–229. MR3314824

[Raj98] C. S. Rajan, *On strong multiplicity one for l-adic representations*, Internat. Math. Res. Notices **3** (1998), 161–172. MR1606395

[Raj02] C. S. Rajan, *On the image and fibres of solvable base change*, Math. Res. Lett. **9** (2002), no. 4, 499–508. MR1928869

[Ram00] D. Ramakrishnan, *Modularity of the Rankin-Selberg L-series, and multiplicity one for* SL(2), Ann. of Math. (2) **152** (2000), no. 1, 45–111. MR1792292

[Rib92] K. A. Ribet, *Abelian varieties over* **Q** *and modular forms*, Algebra and topology 1992 (Taejŏn), Korea Adv. Inst. Sci. Tech., Taejŏn, 1992, pp. 53–79. MR1212980

[Sat66] I. Satake, *Clifford algebras and families of abelian varieties*, Nagoya Math. J. **27** (1966), 435–446. MR0210716

[Sch85] W. Scharlau, *Quadratic and Hermitian forms*, Grundlehren der Mathematischen Wissenschaften [Fundamental Principles of Mathematical Sciences], vol. 270, Springer-Verlag, Berlin, 1985. MR770063

[Sch94] A. J. Scholl, *Classical motives*, Motives (Seattle, WA, 1991), Proc. Sympos. Pure Math., vol. 55, Amer. Math. Soc., Providence, RI, 1994, pp. 163–187. MR1265529

[Sch13] P. Scholze, *p-adic Hodge theory for rigid-analytic varieties*, Forum Math. Pi **1** (2013), e1, 77. MR3090230

[Sen81] S. Sen, *Continuous cohomology and p-adic Galois representations*, Invent. Math. **62** (1980/81), no. 1, 89–116. MR595584

[Ser77] J.-P. Serre, *Modular forms of weight one and Galois representations*, Algebraic number fields: L-functions and Galois properties (Proc. Sympos., Univ. Durham, Durham, 1975), Academic Press, London, 1977, pp. 193–268. MR0450201

[Ser81] J.-P. Serre, *Quelques applications du théorème de densité de Chebotarev* (French), Inst. Hautes Études Sci. Publ. Math. **54** (1981), 323–401. MR644559

[Ser94] J.-P. Serre, *Propriétés conjecturales des groupes de Galois motiviques et des représentations l-adiques* (French), Motives (Seattle, WA, 1991), Proc. Sympos. Pure Math., vol. 55, Amer. Math. Soc., Providence, RI, 1994, pp. 377–400. MR1265537

[Ser98] J.-P. Serre, *Abelian l-adic representations and elliptic curves*, McGill University lecture notes written with the collaboration of Willem Kuyk and John Labute, W. A. Benjamin, Inc., New York-Amsterdam, 1968. MR0263823

[sga73] *Théorie des topos et cohomologie étale des schémas. Tome 3* (French), Lecture Notes in Mathematics, Vol. 305, Springer-Verlag, Berlin-New York, 1973. Séminaire de Géométrie Algébrique du Bois-Marie 1963–1964 (SGA 4); Dirigé par M. Artin, A. Grothendieck et J. L. Verdier. Avec la collaboration de P. Deligne et B. Saint-Donat. MR0354654

[Shi11] S. W. Shin, *Galois representations arising from some compact Shimura varieties*, Ann. of Math. (2) **173** (2011), no. 3, 1645–1741. MR2800722

[Shi12] S. W. Shin, *Automorphic Plancherel density theorem*, Israel J. Math. **192** (2012), no. 1, 83–120. MR3004076

[Spr79] T. A. Springer, *Reductive groups*, Automorphic forms, representations and L-functions (Proc. Sympos. Pure Math., Oregon State Univ., Corvallis, Ore., 1977), Proc. Sympos. Pure Math., XXXIII, Amer. Math. Soc., Providence, R.I., 1979, pp. 3–27. MR546587

[Spr09] T. A. Springer, *Linear algebraic groups*, 2nd ed., Modern Birkhäuser Classics, Birkhäuser Boston, Inc., Boston, MA, 2009. MR2458469

[SR72] N. Saavedra Rivano, *Catégories Tannakiennes* (French), Lecture Notes in Mathematics, Vol. 265, Springer-Verlag, Berlin-New York, 1972. MR0338002

[Ste74] R. Steinberg, *Conjugacy classes in algebraic groups*, Lecture Notes in Mathematics, Vol. 366, Springer-Verlag, Berlin-New York, 1974. Notes by Vinay V. Deodhar. MR0352279

[Tat79] J. Tate, *Number theoretic background*, Automorphic forms, representations and L-functions (Proc. Sympos. Pure Math., Oregon State Univ., Corvallis, Ore., 1977), Proc. Sympos. Pure Math., XXXIII, Amer. Math. Soc., Providence, R.I., 1979, pp. 3–26. MR546607

[Tay12] R. Taylor, *The image of complex conjugation in l-adic representations associated to automorphic forms*, Algebra Number Theory **6** (2012), no. 3, 405–435. MR2966704

[Voi05] C. Voisin, *A generalization of the Kuga-Satake construction*, Pure Appl. Math. Q. **1** (2005), no. 3, Special Issue: In memory of Armand Borel., 415–439. MR2201323

[Wal82] M. Waldschmidt, *Sur certains caractères du groupe des classes d'idèles d'un corps de nombres* (French), Seminar on Number Theory, Paris 1980-81 (Paris, 1980/1981), Progr. Math., vol. 22, Birkhäuser Boston, Boston, MA, 1982, pp. 323–335. MR693328

[Wei56] A. Weil, *On a certain type of characters of the idèle-class group of an algebraic number-field*, Proceedings of the international symposium on algebraic number theory, Tokyo & Nikko, 1955, Science Council of Japan, Tokyo, 1956, pp. 1–7. MR0083523

[Win95] J.-P. Wintenberger, *Relèvement selon une isogénie de systèmes l-adiques de représentations galoisiennes associés aux motifs* (French, with English and French summaries), Invent. Math. **120** (1995), no. 2, 215–240. MR1329040

[Zar83] Yu. G. Zarhin, *Hodge groups of $K3$ surfaces*, J. Reine Angew. Math. **341** (1983), 193–220. MR697317

Index of symbols

Index of terms and concepts

Editorial Information

To be published in the *Memoirs*, a paper must be correct, new, nontrivial, and significant. Further, it must be well written and of interest to a substantial number of mathematicians. Piecemeal results, such as an inconclusive step toward an unproved major theorem or a minor variation on a known result, are in general not acceptable for publication.

Papers appearing in *Memoirs* are generally at least 80 and not more than 200 published pages in length. Papers less than 80 or more than 200 published pages require the approval of the Managing Editor of the Transactions/Memoirs Editorial Board. Published pages are the same size as those generated in the style files provided for $\mathcal{A}_{\mathcal{M}}\mathcal{S}$-LaTeX or $\mathcal{A}_{\mathcal{M}}\mathcal{S}$-TeX.

Information on the backlog for this journal can be found on the AMS website starting from `http://www.ams.org/memo`.

A Consent to Publish is required before we can begin processing your paper. After a paper is accepted for publication, the Providence office will send a Consent to Publish and Copyright Agreement to all authors of the paper. By submitting a paper to the *Memoirs*, authors certify that the results have not been submitted to nor are they under consideration for publication by another journal, conference proceedings, or similar publication.

Information for Authors

Memoirs is an author-prepared publication. Once formatted for print and on-line publication, articles will be published as is with the addition of AMS-prepared frontmatter and backmatter. Articles are not copyedited; however, confirmation copy will be sent to the authors.

Initial submission. The AMS uses Centralized Manuscript Processing for initial submissions. Authors should submit a PDF file using the Initial Manuscript Submission form found at `www.ams.org/submission/memo`, or send one copy of the manuscript to the following address: Centralized Manuscript Processing, MEMOIRS OF THE AMS, 201 Charles Street, Providence, RI 02904-2294 USA. If a paper copy is being forwarded to the AMS, indicate that it is for *Memoirs* and include the name of the corresponding author, contact information such as email address or mailing address, and the name of an appropriate Editor to review the paper (see the list of Editors below).

The paper must contain a *descriptive title* and an *abstract* that summarizes the article in language suitable for workers in the general field (algebra, analysis, etc.). The *descriptive title* should be short, but informative; useless or vague phrases such as "some remarks about" or "concerning" should be avoided. The *abstract* should be at least one complete sentence, and at most 300 words. Included with the footnotes to the paper should be the 2010 *Mathematics Subject Classification* representing the primary and secondary subjects of the article. The classifications are accessible from `www.ams.org/msc/`. The Mathematics Subject Classification footnote may be followed by a list of *key words and phrases* describing the subject matter of the article and taken from it. Journal abbreviations used in bibliographies are listed in the latest *Mathematical Reviews* annual index. The series abbreviations are also accessible from `www.ams.org/msnhtml/serials.pdf`. To help in preparing and verifying references, the AMS offers MR Lookup, a Reference Tool for Linking, at `www.ams.org/mrlookup/`.

Electronically prepared manuscripts. The AMS encourages electronically prepared manuscripts, with a strong preference for $\mathcal{A}_{\mathcal{M}}\mathcal{S}$-LaTeX. To this end, the Society has prepared $\mathcal{A}_{\mathcal{M}}\mathcal{S}$-LaTeX author packages for each AMS publication. Author packages include instructions for preparing electronic manuscripts, samples, and a style file that generates the particular design specifications of that publication series. Though $\mathcal{A}_{\mathcal{M}}\mathcal{S}$-LaTeX is the highly preferred format of TeX, author packages are also available in $\mathcal{A}_{\mathcal{M}}\mathcal{S}$-TeX.

Authors may retrieve an author package for *Memoirs of the AMS* from `www.ams.org/journals/memo/memoauthorpac.html`. The *AMS Author Handbook* is available in PDF format from the author package link. The author package can also be obtained free of charge by sending email to tech-support@ams.org or from the Publication Division, American Mathematical Society, 201 Charles St., Providence, RI 02904-2294, USA. When requesting an author package, please specify $\mathcal{AMS}$-LaTeX or $\mathcal{AMS}$-TeX and the publication in which your paper will appear. Please be sure to include your complete mailing address.

After acceptance. The source files for the final version of the electronic manuscript should be sent to the Providence office immediately after the paper has been accepted for publication. The author should also submit a PDF of the final version of the paper to the editor, who will forward a copy to the Providence office.

Accepted electronically prepared files can be submitted via the web at www.ams.org/submit-book-journal/, sent via FTP, or sent on CD to the Electronic Prepress Department, American Mathematical Society, 201 Charles Street, Providence, RI 02904-2294 USA. TeX source files and graphic files can be transferred over the Internet by FTP to the Internet node ftp.ams.org (130.44.1.100). When sending a manuscript electronically via CD, please be sure to include a message indicating that the paper is for the *Memoirs*.

Electronic graphics. Comprehensive instructions on preparing graphics are available at www.ams.org/authors/journals.html. A few of the major requirements are given here.

Submit files for graphics as EPS (Encapsulated PostScript) files. This includes graphics originated via a graphics application as well as scanned photographs or other computer-generated images. If this is not possible, TIFF files are acceptable as long as they can be opened in Adobe Photoshop or Illustrator.

Authors using graphics packages for the creation of electronic art should also avoid the use of any lines thinner than 0.5 points in width. Many graphics packages allow the user to specify a "hairline" for a very thin line. Hairlines often look acceptable when proofed on a typical laser printer. However, when produced on a high-resolution laser imagesetter, hairlines become nearly invisible and will be lost entirely in the final printing process.

Screens should be set to values between 15% and 85%. Screens which fall outside of this range are too light or too dark to print correctly. Variations of screens within a graphic should be no less than 10%.

Any graphics created in color will be rendered in grayscale for the printed version unless color printing is authorized by the Managing Editor and the Publisher. In general, color graphics will appear in color in the online version.

Inquiries. Any inquiries concerning a paper that has been accepted for publication should be sent to memo-query@ams.org or directly to the Electronic Prepress Department, American Mathematical Society, 201 Charles St., Providence, RI 02904-2294 USA.

SELECTED PUBLISHED TITLES IN THIS SERIES

1230 **Alexander Nagel, Fulvio Ricci, Elias M. Stein, and Stephen Wainger,** Algebras of Singular Integral Operators with Kernels Controlled by Multiple Norms, 2018

1229 **T. Alazard, N. Burq, and C. Zuily,** Strichartz Estimates and the Cauchy Problem for the Gravity Water Waves Equations, 2018

1228 **Nawaf Bou-Rabee and Eric Vanden-Eijnden,** Continuous-Time Random Walks for the Numerical Solution of Stochastic Differential Equations, 2018

1227 **Alexandru D. Ionescu and Fabio Pusateri,** Global Regularity for 2D Water Waves with Surface Tension, 2018

1226 **Paul Feehan and Thomas G. Leness,** An SO(3)-Monopole Cobordism Formula Relating Donaldson and Seiberg-Witten Invariants, 2018

1225 **A. Agrachev, D. Barilari, and L. Rizzi,** Curvature: A Variational Approach, 2018

1224 **Werner Hoffmann and Satoshi Wakatsuki,** On the Geometric Side of the Arthur Trace Formula for the Symplectic Group of Rank 2, 2018

1223 **Sergey Fomin and Professor Dylan Thurston,** Cluster Algebras and Triangulated Surfaces Part II: Lambda Lengths, 2018

1222 **Maurice Duits and Kurt Johansson,** On Mesoscopic Equilibrium for Linear Statistics in Dyson's Brownian Motion, 2018

1221 **Francesco Lin,** A Morse-Bott Approach to Monopole Floer Homology and the Triangulation Conjecture, 2018

1220 **Paata Ivanisvili, Dmitriy M. Stolyarov, Vasily I. Vasyunin, and Pavel B. Zatitskiy,** Bellman Function for Extremal Problems in BMO II: Evolution, 2018

1219 **Olivier Frécon,** Algebraic $\overline{\mathbb{Q}}$-Groups as Abstract Groups, 2018

1218 **Gabriella Pinzari,** Perihelia Reduction and Global Kolmogorov Tori in the Planetary Problem, 2018

1217 **Chin-Yu Hsiao,** Szegő Kernel Asymptotics for High Power of CR Line Bundles and Kodaira Embedding Theorems on CR Manifolds, 2018

1216 **Robert Lipshitz, Peter Ozsváth, and Dylan P. Thurston,** Bordered Heegaard Floer Homology, 2018

1215 **Lior Fishman, David Simmons, and Mariusz Urbański,** Diophantine Approximation and the Geometry of Limit Sets in Gromov Hyperbolic Metric Spaces, 2018

1214 **David Chataur, Martintxo Saralegi-Aranguren, and Daniel Tanré,** Intersection Cohomology, Simplicial Blow-Up and Rational Homotopy, 2018

1213 **Sebastiano Carpi, Yasuyuki Kawahigashi, Roberto Longo, and Mihály Weiner,** From Vertex Operator Algebras to Conformal Nets and Back, 2018

1212 **Roelof Bruggeman, Youngju Choie, and Nikolaos Diamantis,** Holomorphic Automorphic Forms and Cohomology, 2018

1211 **Shouhei Honda,** Elliptic PDEs on Compact Ricci Limit Spaces and Applications, 2018

1210 **Zhou Gang, Dan Knopf, and Israel Michael Sigal,** Neckpinch Dynamics for Asymmetric Surfaces Evolving by Mean Curvature Flow, 2018

1209 **Cristian Anghel, Iustin Coandă, and Nicolae Manolache,** Globally Generated Vector Bundles with Small c_1 on Projective Spaces, 2018

1208 **Matthew Harrison-Trainor,** Degree Spectra of Relations on a Cone, 2018

1207 **Alastair J. Litterick,** On Non-Generic Finite Subgroups of Exceptional Algebraic Groups, 2018

1206 **Anne-Laure Dalibard and Laure Saint-Raymond,** Mathematical Study of Degenerate Boundary Layers: A Large Scale Ocean Circulation Problem, 2018

For a complete list of titles in this series, visit the
AMS Bookstore at **www.ams.org/bookstore/memoseries/**.